IECC

INTERNATIONAL
ENERGY CONSERVATION
CODE®

A Member of the International Code Family®

2021

2021 International Energy Conservation Code®

Date of First Publication: January 29, 2021

First Printing: January 2021
Second Printing: September 2021

ISBN: 978-1-60983-961-1 (soft-cover edition)
ISBN: 978-1-955052-51-1 (PDF download)

COPYRIGHT © 2021
by
INTERNATIONAL CODE COUNCIL, INC.

ALL RIGHTS RESERVED. This 2021 *International Energy Conservation Code®* is a copyrighted work owned by the International Code Council, Inc. ("ICC"). Without advance written permission from the ICC, no part of this book may be reproduced, distributed or transmitted in any form or by any means, including, without limitation, electronic, optical or mechanical means (by way of example, and not limitation, photocopying or recording by or in an information storage retrieval system). For information on use rights and permissions, please contact: ICC Publications, 4051 Flossmoor Road, Country Club Hills, IL 60478. Phone 1-888-ICC-SAFE (422-7233).

Trademarks: "International Code Council," the "International Code Council" logo, "ICC," the "ICC" logo, "International Energy Conservation Code," "IECC" and other names and trademarks appearing in this book are registered trademarks of the International Code Council, Inc., and/or its licensors (as applicable), and may not be used without permission.

PRINTED IN THE USA

PREFACE

Introduction

The *International Energy Conservation Code®* (IECC®) establishes minimum requirements for energy-efficient buildings using prescriptive and performance-related provisions. It is founded on broad-based principles that make possible the use of new materials and new energy-efficient designs. This 2021 edition is fully compatible with all of the *International Codes®* (I-Codes®) published by the International Code Council® (ICC®), including the *International Building Code®* (IBC®), *International Existing Building Code®* (IEBC®), *International Fire Code®* (IFC®), *International Fuel Gas Code®* (IFGC®), *International Green Construction Code®* (IgCC®), *International Mechanical Code®* (IMC®), *International Plumbing Code®* (IPC®), *International Private Sewage Disposal Code®* (IPSDC®), *International Property Maintenance Code®* (IPMC®), *International Residential Code®* (IRC®), *International Swimming Pool and Spa Code®* (ISPSC®), *International Wildland-Urban Interface Code®* (IWUIC®), *International Zoning Code®* (IZC®) and *International Code Council Performance Code®* (ICCPC®).

This code contains separate provisions for commercial buildings and for low-rise residential buildings (3 stories or less in height above grade). Each set of provisions, IECC—Commercial Provisions and IECC—Residential Provisions, is separately applied to buildings within its respective scope. Each set of provisions is to be treated separately. Each contains a Scope and Administration chapter, a Definitions chapter, a General Requirements chapter, a chapter containing energy efficiency requirements and an Existing Buildings chapter containing provisions applicable to buildings within its scope.

The I-Codes, including the IECC, are used in a variety of ways in both the public and private sectors. Most industry professionals are familiar with the I-Codes as the basis of laws and regulations in communities across the US and in other countries. However, the impact of the codes extends well beyond the regulatory arena, as they are used in a variety of nonregulatory settings, including:

- Voluntary compliance programs such as those promoting sustainability, energy efficiency and disaster resistance.
- The insurance industry, to estimate and manage risk, and as a tool in underwriting and rate decisions.
- Certification and credentialing of individuals involved in the fields of building design, construction and safety.
- Certification of building and construction-related products.
- US federal agencies, to guide construction in an array of government-owned properties.
- Facilities management.
- "Best practices" benchmarks for designers and builders, including those who are engaged in projects in jurisdictions that do not have a formal regulatory system or a governmental enforcement mechanism.
- College, university and professional school textbooks and curricula.
- Reference works related to building design and construction.

In addition to the codes themselves, the code development process brings together building professionals on a regular basis. It provides an international forum for discussion and deliberation about building design, construction methods, safety, performance requirements, technological advances and innovative products.

Development

This 2021 edition presents the code as originally issued, with changes reflected in the 2000 through 2018 editions and further changes approved through the ICC Code Development Process through 2019. A new edition such as this is promulgated every 3 years.

This code is founded on principles intended to establish provisions consistent with the scope of an energy conservation code that adequately conserves energy; provisions that do not unnecessarily increase construction costs; provisions that do not restrict the use of new materials, products or methods of construction; and provisions that do not give preferential treatment to particular types or classes of materials, products or methods of construction.

Maintenance

The IECC is kept up to date through the review of proposed changes submitted by code enforcement officials, industry representatives, design professionals and other interested parties. Proposed changes are carefully considered through an open code development process in which all interested and affected parties may participate.

The ICC Code Development Process reflects principles of openness, transparency, balance, due process and consensus, the principles embodied in OMB Circular A-119, which governs the federal government's use of private-sector standards. The ICC process is open to anyone; there is no cost to participate, and people can participate without travel cost through the ICC's cloud-based app, cdpACCESS®. A broad cross-section of interests are represented in the ICC Code Development Process. The codes, which are updated regularly, include safeguards that allow for emergency action when required for health and safety reasons.

In order to ensure that organizations with a direct and material interest in the codes have a voice in the process, the ICC has developed partnerships with key industry segments that support the ICC's important public safety mission. Some code development committee members were nominated by the following industry partners and approved by the ICC Board:

- National Association of Home Builders (NAHB)
- National Multifamily Housing Council (NMHC)

The code development committees evaluate and make recommendations regarding proposed changes to the codes. Their recommendations are then subject to public comment and council-wide votes. The ICC's governmental members—public safety officials who have no financial or business interest in the outcome—cast the final votes on proposed changes.

The contents of this work are subject to change through the code development cycles and by any governmental entity that enacts the code into law. For more information regarding the code development process, contact the Codes and Standards Development Department of the ICC.

While the I-Code development procedure is thorough and comprehensive, the ICC, its members and those participating in the development of the codes disclaim any liability resulting from the publication or use of the I-Codes, or from compliance or noncompliance with their provisions. The ICC does not have the power or authority to police or enforce compliance with the contents of this code.

Code Development Committee Responsibilities (Letter Designations in Front of Section Numbers)

In each code development cycle, proposed changes to this code are considered at the Committee Action Hearings by the applicable International Code Development Committee. The IECC—Commercial Provisions (sections designated with a "C" prior to the section number) are primarily maintained by the Commercial Energy Code Development Committee. The IECC—Residential Provisions (sections designated with an "R" prior to the section number) are maintained by the Residential Energy Code Development Committee. This is designated in the chapter headings by a [CE] and [RE], respectively.

Maintenance responsibilities for the IECC are designated as follows:

[CE] = International Commercial Energy Conservation Code Development Committee

[RE] = International Residential Energy Conservation Code Development Committee

For the development of the 2024 edition of the I-Codes, there will be two groups of code development committees and they will meet in separate years, as shown in the following Code Development Hearings Table.

Code change proposals submitted for code sections that have a letter designation in front of them will be heard by the respective committee responsible for such code sections. Because different committees hold Committee Action Hearings in different years, proposals for several I-Codes will be heard by committees in both the 2021 (Group A) and the 2022 (Group B) code development cycles.

All code change proposals for the IECC are considered in the Group B hearings.

It is very important that anyone submitting code change proposals understands which code development committee is responsible for the section of the code that is the subject of the code change proposal. For further information on the Code Development Committee responsibilities, please visit the ICC website at www.iccsafe.org/current-code-development-cycle.

CODE DEVELOPMENT HEARINGS

Group A Codes (Heard in 2021, Code Change Proposals Deadline: January 11, 2021)	Group B Codes (Heard in 2022, Code Change Proposals Deadline: January 10, 2022)
International Building Code — Egress (Chapters 10, 11, Appendix E) — Fire Safety (Chapters 7, 8, 9, 14, 26) — General (Chapters 2–6, 12, 27–33, Appendices A, B, C, D, K, N)	Administrative Provisions (Chapter 1 of all codes except IECC, IRC and IgCC; IBC Appendix O; the appendices titled "Board of Appeals" for all codes except IECC, IRC, IgCC, ICCPC and IZC; administrative updates to currently referenced standards; and designated definitions)
International Fire Code	**International Building Code** — Structural (Chapters 15–25, Appendices F, G, H, I, J, L, M)
International Fuel Gas Code	**International Existing Building Code**
International Mechanical Code	**International Energy Conservation Code—Commercial**
International Plumbing Code	**International Energy Conservation Code—Residential** — IECC—Residential — IRC—Energy (Chapter 11)
International Property Maintenance Code	**International Green Construction Code** (Chapter 1)
International Private Sewage Disposal Code	**International Residential Code** — IRC—Building (Chapters 1–10, Appendices AE, AF, AH, AJ, AK, AL, AM, AO, AQ, AR, AS, AT, AU, AV, AW)
International Residential Code — IRC—Mechanical (Chapters 12–23) — IRC—Plumbing (Chapters 25–33, Appendices AG, AI, AN, AP)	
International Swimming Pool and Spa Code	
International Wildland-Urban Interface Code	
International Zoning Code	

Note: Proposed changes to the ICCPC will be heard by the code development committee noted in brackets [ ] in the text of the ICCPC.

Marginal Markings

Solid vertical lines in the margins within the body of the code indicate a technical change from the requirements of the 2018 edition. Deletion indicators in the form of an arrow (➡) are provided in the margin where an entire section, exception or table has been deleted or an item in a list of items or row of a table has been deleted.

A single asterisk [*] placed in the margin indicates that text or a table has been relocated within the code. A double asterisk [**] placed in the margin indicates that the text or table immediately following it has been relocated there from elsewhere in the code. The following tables indicate such relocations in the 2021 edition of the IECC.

IECC COMMERCIAL RELOCATIONS

2021 LOCATION	2018 LOCATION
C402.2.1.5	C402.2.1.1
C404.6.1.1	C404.7
C405.2.3.1	C405.2.2.2
C405.2.4	C405.2.3
C501.2	C501.4
C503.2.2.1	C401.2.1

IECC RESIDENTIAL RELOCATIONS

2021 LOCATION	2018 LOCATION
Table R402.1.2	Table R402.1.4
Table R402.1.3	Table R402.1.2
R403.3.2	R403.3.7
R403.3.3	R403.3.6
R403.3.3.1	R403.3.6.1
R403.3.4	R403.3.2
R403.3.4.1	R403.3.2.1
R403.3.5	R403.3.3
R403.3.6	R403.3.4
R403.3.7	R403.3.5

Coordination of the International Codes

The coordination of technical provisions is one of the strengths of the ICC family of model codes. The codes can be used as a complete set of complementary documents, which will provide users with full integration and coordination of technical provisions. Individual codes can also be used in subsets or as stand-alone documents. To make sure that each individual code is as complete as possible, some technical provisions that are relevant to more than one subject area are duplicated in some of the model codes. This allows users maximum flexibility in their application of the I-Codes.

Italicized Terms

Terms italicized in code text, other than document titles, are defined in Chapter 2. The terms selected to be italicized have definitions that the user should read carefully to better understand the code. Where italicized, the Chapter 2 definition applies. If not italicized, common-use definitions apply.

Adoption

The ICC maintains a copyright in all of its codes and standards. Maintaining copyright allows the ICC to fund its mission through sales of books, in both print and electronic formats. The ICC welcomes adoption of its codes by jurisdictions that recognize and acknowledge the ICC's copyright in the code, and further acknowledge the substantial shared value of the public/private partnership for code development between jurisdictions and the ICC.

The ICC also recognizes the need for jurisdictions to make laws available to the public. All I-Codes and I-Standards, along with the laws of many jurisdictions, are available for free in a nondownloadable form on the ICC's website. Jurisdictions should contact the ICC at adoptions@iccsafe.org to learn how to adopt and distribute laws based on the IECC in a manner that provides necessary access, while maintaining the ICC's copyright.

To facilitate adoption, two sections of this code contain blanks for fill-in information that needs to be supplied by the adopting jurisdiction as part of the adoption legislation. For this code, please see:

Section C101.1. Insert: **[NAME OF JURISDICTION]**.

Section R101.1. Insert: **[NAME OF JURISDICTION]**.

Effective Use of the International Energy Conservation Code

The IECC is a model code that regulates minimum energy conservation requirements for new buildings. The IECC addresses energy conservation requirements for all aspects of energy use in both commercial and residential construction, including heating and ventilating, lighting, water heating, and power usage for appliances and building systems.

The IECC is a design document. For example, before one constructs a building, the designer must determine the minimum insulation R-values and fenestration U-factors for the building exterior envelope. Depending on whether the building is for residential use or for commercial use, the IECC sets forth minimum requirements for exterior envelope insulation, window and door U-factors and SHGC ratings, duct insulation, lighting and power efficiency, and water distribution insulation.

ARRANGEMENT AND FORMAT OF THE 2021 IECC

The IECC contains two separate sets of provisions—one for commercial buildings and one for residential buildings. Each set of provisions is applied separately to buildings within their scope. The IECC—Commercial Provisions apply to all buildings except for residential buildings three stories or less in height. The IECC—Residential Provisions apply to detached one- and two-family dwellings and multiple single-family dwellings as well as Group R-2, R-3 and R-4 buildings three stories or less in height. These scopes are based on the definitions of "Commercial building" and "Residential building," respectively, in Chapter 2 of each set of provisions. Note that the IECC—Commercial Provisions therefore contain provisions for residential buildings four stories or greater in height.

The following table shows how the IECC is divided. The ensuing chapter-by-chapter synopsis details the scope and intent of the provisions of the IECC.

Chapter Topics

Chapter	Subjects
1–2	Administration and definitions
3	Climate zones and general materials requirements
4	Energy efficiency requirements
5	Existing buildings
6	Referenced standards
CA	Board of appeals
CB	Solar-ready zone
CC	Net zero energy

Chapter 1 Scope and Administration

Chapters 1 [CE] and 1 [RE] contain provisions for the application, enforcement and administration of subsequent requirements of the code. In addition to establishing the scope of the code, Chapter 1 identifies which buildings and structures come under its purview. Chapter 1 is largely concerned with maintaining "due process of law" in enforcing the energy conservation criteria contained in the body of this code. Only through careful observation of the administrative provisions can the code official reasonably expect to demonstrate that "equal protection under the law" has been provided.

Chapter 2 Definitions

Terms that are defined in the code are listed alphabetically in Chapters 2 [CE] and 2 [RE]. While a defined term may be used in one chapter or another, the meaning provided in Chapter 2 is applicable throughout the code.

Additional definitions regarding climate zones are found in Tables C301.3 and R301.3. These are not listed in Chapter 2.

Where understanding of a term's definition is especially key to or necessary for understanding of a particular code provision, the term is shown in italics. This is true only for those terms that have a meaning that is unique to the code. In other words, the generally understood meaning of a term or phrase might not be sufficient or consistent with the meaning prescribed by the code; therefore, it is essential that the code-defined meaning be known.

Guidance regarding tense, gender and plurality of defined terms as well as guidance regarding terms not defined in this code is provided.

Chapter 3 General Requirements

Chapters 3 [CE] and 3 [RE] specify the climate zones that will serve to establish the exterior design conditions. In addition, Chapter 3 provides interior design conditions that are used as a basis for assumptions in heating and cooling load calculations, and provides basic material requirements for insulation materials and fenestration materials. Climate has a major impact on the energy use of most buildings. The code establishes many requirements such as wall and roof insulation R-values, window and door thermal transmittance (U-factors) and provisions that affect the mechanical systems based on the climate where the building is located. This chapter contains information that will be used to properly assign the building location into the correct climate zone and is used as the basis for establishing or eliminating requirements.

Chapter 4 Energy Efficiency

Chapter 4 [CE] contains the energy-efficiency-related requirements for the design and construction of most types of commercial buildings and residential buildings greater than three stories in height above grade. This chapter defines requirements for the portions of the building and building systems that impact energy use in new commercial construction and new residential construction greater than three stories in height, and promotes the effective use of energy. In addition to energy conservation requirements for the building envelope, this chapter contains requirements that impact energy efficiency for the HVAC systems, the electrical systems and the plumbing systems. It should be noted, however, that requirements are contained in other codes that have an impact on energy conservation. For instance, requirements for water flow rates are regulated by the *International Plumbing Code*.

Chapter 4 [RE] contains the energy-efficiency-related requirements for the design and construction of residential buildings regulated under this code. It should be noted that the definition of a residential building in this code is unique for this code. In this code, residential buildings include detached one- and two-family dwellings and multiple single-family dwellings as well as R-2, R-3 or R-4 buildings three stories or less in height. All other buildings, including residential buildings greater than three stories in height, are regulated by the energy conservation requirements in the IECC—Commercial Provisions. The applicable portions of a residential building must comply with the provisions within this chapter for energy efficiency. This chapter defines requirements for the portions of the building and building systems that impact energy use in new residential construction and promotes the effective use of energy. The provisions within the chapter promote energy efficiency in the building envelope, the heating and cooling system and the service water-heating system of the building.

Chapter 5 Existing Buildings

Chapters 5 [CE] and [RE] contain the technical energy efficiency requirements for existing buildings. Chapter 5 provisions address the maintenance of buildings in compliance with the code as well as how additions, alterations, repairs and changes of occupancy need to be addressed from the standpoint of energy efficiency. Specific provisions are provided for historic buildings.

Chapter 6 Referenced Standards

The code contains numerous references to standards that are used to regulate materials and methods of construction. Chapters 6 [CE] and 6 [RE] list all standards referenced in their respective portions of the code. The standards are part of the code to the extent of the reference to the standard. Compliance with the referenced standard is necessary for compliance with this code. By providing specifically adopted standards, the construction and installation requirements necessary for compliance with the code can be readily determined. The basis for code compliance is, therefore, established and available on an equal basis to the code official, contractor, designer and owner.

Chapter 6 is organized in a manner that makes it easy to locate specific standards. It lists all of the referenced standards, alphabetically, by acronym of the promulgating agency of the standard. Each agency's standards are then listed in either alphabetical or numeric order based on the standard identification. The list also contains the title of the standard; the edition (date) of the standard referenced; any addenda included as part of the ICC adoption; and the section or sections of this code that reference the standard.

Appendices

The appendices, while not part of the code, can become part of the code when specifically included in the adopting ordinance.

Chapter 1 requires the establishment of a board of appeals to hear appeals regarding determinations made by the code official. Appendices CA and RA provide qualification standards for members of the board as well as operational procedures of such board.

Appendices CB and RB address provisions for solar capacity in new structures.

Appendices CC and RC provide requirements intended bring about net zero annual energy consumption in their respective structures.

ABBREVIATIONS AND NOTATIONS

The following table contains a list of common abbreviations and units of measurement used in this code. Some of the abbreviations are for terms defined in Chapter 2. Others are terms used in various tables and text of the code.

Abbreviations and Notations

AFUE	Annual fuel utilization efficiency
bhp	Brake horsepower (fans)
Btu	British thermal unit
Btu/h × ft²	Btu per hour per square foot
C-factor	See Chapter 2—Definitions
CDD	Cooling degree days
cfm	Cubic feet per minute
cfm/ft²	Cubic feet per minute per square foot
ci	Continuous insulation
COP	Coefficient of performance
DCV	Demand control ventilation
°C	Degrees Celsius
°F	Degrees Fahrenheit
DWHR	Drain water heat recovery
DX	Direct expansion
E_c	Combustion efficiency
E_v	Ventilation efficiency
E_t	Thermal efficiency
EER	Energy efficiency ratio
EF	Energy factor
ERI	Energy rating index
F-factor	See Chapter 2—Definitions
FDD	Fault detection and diagnostics
FEI	Fan energy index
FL	Full load
ft²	Square foot
gpm	Gallons per minute
HDD	Heating degree days
hp	Horsepower
HSPF	Heating seasonal performance factor
HVAC	Heating, ventilating and air conditioning

(continued)

Abbreviations and Notations—continued

IEER	Integrated energy efficiency ratio
IPLV	Integrated Part Load Value
Kg/m^2	Kilograms per square meter
kW	Kilowatt
LPD	Light power density (lighting power allowance)
L/s	Liters per second
Ls	Liner system
m^2	Square meters
MERV	Minimum efficiency reporting value
NAECA	National Appliance Energy Conservation Act
NPLV	Nonstandard Part Load Value
Pa	Pascal
PF	Projection factor
pcf	Pounds per cubic foot
psf	Pounds per square foot
PTAC	Packaged terminal air conditioner
PTHP	Packaged terminal heat pump
R-value	See Chapter 2—Definitions
SCOP	Sensible coefficient of performance
SEER	Seasonal energy efficiency ratio
SHGC	Solar Heat Gain Coefficient
SPVAC	Single packaged vertical air conditioner
SPVHP	Single packaged vertical heat pump
SRI	Solar reflectance index
SWHF	Service water heat recovery factor
U-factor	See Chapter 2—Definitions
VAV	Variable air volume
VRF	Variable refrigerant flow
VT	Visible transmittance
W	Watts
w.c.	Water column
w.g.	Water gauge

TABLE OF CONTENTS

IECC—COMMERCIAL PROVISIONS .... C-i

CHAPTER 1   SCOPE AND ADMINISTRATION............C1-1

CHAPTER 2   DEFINITIONS.................C2-1

CHAPTER 3   GENERAL REQUIREMENTS ...C3-1

CHAPTER 4   COMMERCIAL ENERGY EFFICIENCY C4-1

CHAPTER 5   EXISTING BUILDINGS C5-1

CHAPTER 6   REFERENCED STANDARDS ...C6-1

APPENDIX CA   BOARD OF APPEALS—COMMERCIAL .. APPENDIX CA-1

APPENDIX CB   SOLAR-READY ZONE—COMMERCIAL .. APPENDIX CB-1

APPENDIX CC   ZERO ENERGY COMMERCIAL BUILDING PROVISIONS..... APPENDIX CC-1

INDEX.................INDEX C-1

IECC—RESIDENTIAL PROVISIONS .... R-i

CHAPTER 1   SCOPE AND ADMINISTRATION R1-1

CHAPTER 2   DEFINITIONS R2-1

CHAPTER 3   GENERAL REQUIREMENTS... R3-1

CHAPTER 4   RESIDENTIAL ENERGY EFFICIENCY R4-1

CHAPTER 5   EXISTING BUILDINGS R5-1

CHAPTER 6   REFERENCED STANDARDS ... R6-1

APPENDIX RA   BOARD OF APPEALS—RESIDENTIAL....APPENDIX RA-1

APPENDIX RB   SOLAR-READY PROVISIONS—DETACHED ONE- AND TWO-FAMILY DWELLINGS AND TOWNHOUSES ... APPENDIX RB-1

APPENDIX RC   ZERO ENERGY RESIDENTIAL BUILDING PROVISIONS .....APPENDIX RC-1

INDEX .... INDEX R-1

IECC—COMMERCIAL PROVISIONS

TABLE OF CONTENTS

CHAPTER 1 SCOPE AND ADMINISTRATION C1-1

PART 1—SCOPE AND APPLICATION C1-1
Section
C101 Scope and General Requirements C1-1
C102 Alternative Materials, Design and Methods of Construction and Equipment C1-1

PART 2—ADMINISTRATION AND ENFORCEMENT C1-1
Section
C103 Construction Documents C1-1
C104 Fees C1-3
C105 Inspections C1-3
C106 Notice of Approval C1-4
C107 Validity C1-4
C108 Referenced Standards..................... C1-4
C109 Stop Work Order C1-4
C110 Board of Appeals........................ C1-4

CHAPTER 2 DEFINITIONS C2-1
Section
C201 General.................................... C2-1
C202 General Definitions C2-1

CHAPTER 3 GENERAL REQUIREMENTS ... C3-1
Section
C301 Climate Zones C3-1
C302 Design Conditions C3-36
C303 Materials, Systems and Equipment C3-36

CHAPTER 4 COMMERCIAL ENERGY EFFICIENCY C4-1
Section
C401 General.................................... C4-1
C402 Building Envelope Requirements C4-1
C403 Building Mechanical Systems C4-14
C404 Service Water Heating.................... C4-55
C405 Electrical Power and Lighting Systems C4-59
C406 Additional Efficiency Requirements......... C4-76
C407 Total Building Performance C4-84
C408 Maintenance Information and System Commissioning.................. C4-91

CHAPTER 5 EXISTING BUILDINGS C5-1
Section
C501 General C5-1
C502 Additions................................. C5-1
C503 Alterations............................... C5-2
C504 Repairs C5-3
C505 Change of Occupancy or Use C5-3

CHAPTER 6 REFERENCED STANDARDS C6-1

APPENDIX CA BOARD OF APPEALS— COMMERCIAL ...APPENDIX CA-1
Section
CA101 General................... APPENDIX CA-1

APPENDIX CB SOLAR-READY ZONE— COMMERCIAL ...APPENDIX CB-1
Section
CB101 Scope .... APPENDIX CB-1
CB102 General Definition .... APPENDIX CB-1
CB103 Solar-ready Zone .... APPENDIX CB-1

APPENDIX CC ZERO ENERGY COMMERCIAL BUILDING PROVISIONS......APPENDIX CC-1
Section
CC101 General................... APPENDIX CC-1
CC102 Definitions .... APPENDIX CC-1
CC103 Minimum Renewable Energy .. APPENDIX CC-1

INDEX ....INDEX C-1

CHAPTER 1 [CE]
SCOPE AND ADMINISTRATION

User note:

About this chapter: Chapter 1 establishes the limits of applicability of the code and describes how the code is to be applied and enforced. Chapter 1 is in two parts: Part 1—Scope and Application and Part 2—Administration and Enforcement. Section C101 identifies what buildings, systems, appliances and equipment fall under its purview and references other I-Codes as applicable. Standards and codes are scoped to the extent referenced.

The code is intended to be adopted as a legally enforceable document and it cannot be effective without adequate provisions for its administration and enforcement. The provisions of Chapter 1 establish the authority and duties of the code official appointed by the authority having jurisdiction and also establish the rights and privileges of the design professional, contractor and property owner.

PART 1—SCOPE AND APPLICATION

SECTION C101
SCOPE AND GENERAL REQUIREMENTS

C101.1 Title. This code shall be known as the *Energy Conservation Code* of **[NAME OF JURISDICTION]**, and shall be cited as such. It is referred to herein as "this code."

C101.2 Scope. This code applies to *commercial buildings* and the buildings' sites and associated systems and equipment.

C101.3 Intent. This code shall regulate the design and construction of buildings for the effective use and conservation of energy over the useful life of each building. This code is intended to provide flexibility to permit the use of innovative approaches and techniques to achieve this objective. This code is not intended to abridge safety, health or environmental requirements contained in other applicable codes or ordinances.

C101.4 Applicability. Where, in any specific case, different sections of this code specify different materials, methods of construction or other requirements, the most restrictive shall govern. Where there is a conflict between a general requirement and a specific requirement, the specific requirement shall govern.

C101.4.1 Mixed residential and commercial buildings. Where a building includes both *residential building* and *commercial building* portions, each portion shall be separately considered and meet the applicable provisions of IECC—Commercial Provisions or IECC—Residential Provisions.

C101.5 Compliance. *Residential buildings* shall meet the provisions of IECC—Residential Provisions. *Commercial buildings* shall meet the provisions of IECC—Commercial Provisions.

C101.5.1 Compliance materials. The *code official* shall be permitted to approve specific computer software, worksheets, compliance manuals and other similar materials that meet the intent of this code.

SECTION C102
ALTERNATIVE MATERIALS, DESIGN AND METHODS OF CONSTRUCTION AND EQUIPMENT

C102.1 General. The provisions of this code are not intended to prevent the installation of any material or to prohibit any design or method of construction not specifically prescribed by this code, provided that any such alternative has been *approved*. The code official shall have the authority to approve an alternative material, design or method of construction upon the written application of the owner or the owner's authorized agent. The *code official* shall first find that the proposed design is satisfactory and complies with the intent of the provisions of this code, and that the material, method or work offered is, for the purpose intended, not less than the equivalent of that prescribed in this code in quality, strength, effectiveness, *fire resistance*, durability, energy conservation and safety. The *code official* shall respond to the applicant, in writing, stating the reasons why the alternative was approved or was not *approved*.

C102.1.1 Above code programs. The *code official* or other authority having jurisdiction shall be permitted to deem a national, state or local energy efficiency program as exceeding the energy efficiency required by this code. Buildings *approved* in writing by such an energy efficiency program shall be considered to be in compliance with this code. The requirements identified in Table C407.2 shall be met.

PART 2—ADMINISTRATION AND ENFORCEMENT

SECTION C103
CONSTRUCTION DOCUMENTS

C103.1 General. Construction documents and other supporting data shall be submitted in one or more sets, or in a digital format where allowed by the building official, with each application for a permit. The construction documents shall be prepared by a registered design professional where required by the statutes of the jurisdiction in which the project is to be constructed. Where special conditions exist, the *code official* is authorized to require necessary construction documents to be prepared by a registered design professional.

Exception: The *code official* is authorized to waive the requirements for construction documents or other supporting data if the *code official* determines they are not necessary to confirm compliance with this code.

C103.2 Information on construction documents. Construction documents shall be drawn to scale on suitable material. Electronic media documents are permitted to be submitted where *approved* by the *code official*. Construction documents shall be of sufficient clarity to indicate the location, nature and extent of the work proposed, and show in sufficient detail pertinent data and features of the building, systems and equipment as herein governed. Details shall include, but are not limited to, the following as applicable:

1. Energy compliance path.
2. Insulation materials and their *R*-values.
3. Fenestration *U*-factors and solar heat gain coefficients (SHGCs).
4. Area-weighted *U*-factor and solar heat gain coefficient (SHGC) calculations.
5. Mechanical system design criteria.
6. Mechanical and service water-heating systems and equipment types, sizes and efficiencies.
7. Economizer description.
8. Equipment and system controls.
9. Fan motor horsepower (hp) and controls.
10. Duct sealing, duct and pipe insulation and location.
11. Lighting fixture schedule with wattage and control narrative.
12. Location of *daylight* zones on floor plans.
13. Air barrier and air sealing details, including the location of the air barrier.

C103.2.1 Building thermal envelope depiction. The *building thermal envelope* shall be represented on the construction drawings.

C103.3 Examination of documents. The *code official* shall examine or cause to be examined the accompanying construction documents and shall ascertain whether the construction indicated and described is in accordance with the requirements of this code and other pertinent laws or ordinances. The *code official* is authorized to utilize a registered design professional, or other *approved* entity not affiliated with the building design or construction, in conducting the review of the plans and specifications for compliance with the code.

C103.3.1 Approval of construction documents. When the *code official* issues a permit where construction documents are required, the construction documents shall be endorsed in writing and stamped "Reviewed for Code Compliance." Such *approved* construction documents shall not be changed, modified or altered without authorization from the *code official*. Work shall be done in accordance with the *approved* construction documents.

One set of construction documents so reviewed shall be retained by the *code official*. The other set shall be returned to the applicant, kept at the site of work and shall be open to inspection by the *code official* or a duly authorized representative.

C103.3.2 Previous approvals. This code shall not require changes in the construction documents, construction or designated occupancy of a structure for which a lawful permit has been heretofore issued or otherwise lawfully authorized, and the construction of which has been pursued in good faith within 180 days after the effective date of this code and has not been abandoned.

C103.3.3 Phased approval. The *code official* shall have the authority to issue a permit for the construction of part of an energy conservation system before the construction documents for the entire system have been submitted or *approved*, provided that adequate information and detailed statements have been filed complying with all pertinent requirements of this code. The holders of such permit shall proceed at their own risk without assurance that the permit for the entire energy conservation system will be granted.

C103.4 Amended construction documents. Changes made during construction that are not in compliance with the *approved* construction documents shall be resubmitted for approval as an amended set of construction documents.

C103.5 Retention of construction documents. One set of *approved* construction documents shall be retained by the *code official* for a period of not less than 180 days from date of completion of the permitted work, or as required by state or local laws.

C103.6 Building documentation and closeout submittal requirements. The construction documents shall specify that the documents described in this section be provided to the building owner or owner's authorized agent within 90 days of the date of receipt of the certificate of occupancy.

C103.6.1 Record documents. Construction documents shall be updated to convey a record of the completed work. Such updates shall include mechanical, electrical and control drawings that indicate all changes to size, type and location of components, equipment and assemblies.

C103.6.2 Compliance documentation. Energy code compliance documentation and supporting calculations shall be delivered in one document to the building owner as part of the project record documents or manuals, or as a standalone document. This document shall include the specific energy code edition utilized for compliance determination for each system, documentation demonstrating compliance with Section C303.1.3 for each fenestration product installed, and the interior lighting power compliance path, building area or space-by-space, used to calculate the lighting power allowance.

For projects complying with Item 2 of Section C401.2, the documentation shall include:

1. The envelope insulation compliance path.

2. All compliance calculations including those required by Sections C402.1.5, C403.8.1, C405.3 and C405.5.

For projects complying with Section C407, the documentation shall include that required by Sections C407.3.1 and C407.3.2.

C103.6.3 Systems operation control. Training shall be provided to those responsible for maintaining and operating equipment included in the manuals required by Section C103.6.2.

The training shall include:

1. Review of manuals and permanent certificate.
2. Hands-on demonstration of all normal maintenance procedures, normal operating modes, and all emergency shutdown and startup procedures.
3. Training completion report.

SECTION C104
FEES

C104.1 Fees. A permit shall not be issued until the fees prescribed in Section C104.2 have been paid, nor shall an amendment to a permit be released until the additional fee, if any, has been paid.

C104.2 Schedule of permit fees. A fee for each permit shall be paid as required, in accordance with the schedule as established by the applicable governing authority.

C104.3 Work commencing before permit issuance. Any person who commences any work before obtaining the necessary permits shall be subject to an additional fee established by the *code official* that shall be in addition to the required permit fees.

C104.4 Related fees. The payment of the fee for the construction, *alteration*, removal or demolition of work done in connection to or concurrently with the work or activity authorized by a permit shall not relieve the applicant or holder of the permit from the payment of other fees that are prescribed by law.

C104.5 Refunds. The *code official* is authorized to establish a refund policy.

SECTION C105
INSPECTIONS

C105.1 General. Construction or work for which a permit is required shall be subject to inspection by the code official, his or her designated agent or an *approved agency*, and such construction or work shall remain visible and able to be accessed for inspection purposes until *approved*. Approval as a result of an inspection shall not be construed to be an approval of a violation of the provisions of this code or of other ordinances of the jurisdiction. Inspections presuming to give authority to violate or cancel the provisions of this code or of other ordinances of the jurisdiction shall not be valid. It shall be the duty of the permit applicant to cause the work to remain visible and able to be accessed for inspection purposes. Neither the code official nor the jurisdiction shall be liable for expense entailed in the removal or replacement of any material, product, system or building component required to allow inspection to validate compliance with this code.

C105.2 Required inspections. The *code official*, his or her designated agent or an *approved agency*, upon notification, shall make the inspections set forth in Sections C105.2.1 through C105.2.6.

C105.2.1 Footing and foundation insulation. Inspections shall verify the footing and foundation insulation *R*-value, location, thickness, depth of burial and protection of insulation as required by the code, *approved* plans and specifications.

C105.2.2 Thermal envelope. Inspections shall verify the correct type of insulation, *R*-values, location of insulation, fenestration, *U*-factor, SHGC and VT, and that air leakage controls are properly installed, as required by the code, *approved* plans and specifications.

C105.2.3 Plumbing system. Inspections shall verify the type of insulation, *R*-values, protection required, controls and heat traps as required by the code, *approved* plans and specifications.

C105.2.4 Mechanical system. Inspections shall verify the installed HVAC equipment for the correct type and size, controls, insulation, *R*-values, system and damper air leakage, minimum fan efficiency, energy recovery and economizer as required by the code, *approved* plans and specifications.

C105.2.5 Electrical system. Inspections shall verify lighting system controls, components and meters as required by the code, *approved* plans and specifications.

C105.2.6 Final inspection. The final inspection shall include verification of the installation and proper operation of all required building controls, and documentation verifying activities associated with required *building commissioning* have been conducted in accordance with Section C408.

C105.3 Reinspection. A building shall be reinspected where determined necessary by the *code official*.

C105.4 Approved inspection agencies. The *code official* is authorized to accept reports of third-party inspection agencies not affiliated with the building design or construction, provided that such agencies are *approved* as to qualifications and reliability relevant to the building components and systems that they are inspecting.

C105.5 Inspection requests. It shall be the duty of the holder of the permit or their duly authorized agent to notify the *code official* when work is ready for inspection. It shall be the duty of the permit holder to provide access to and means for inspections of such work that are required by this code.

C105.6 Reinspection and testing. Where any work or installation does not pass an initial test or inspection, the necessary corrections shall be made to achieve compliance with this code. The work or installation shall then be resubmitted to the *code official* for inspection and testing.

SECTION C106
NOTICE OF APPROVAL

C106.1 Approval. After the prescribed tests and inspections indicate that the work complies in all respects with this code, a notice of approval shall be issued by the code *official*.

C106.2 Revocation. The *code official* is authorized to suspend or revoke, in writing, a notice of approval issued under the provisions of this code wherever the certificate is issued in error, or on the basis of incorrect information supplied, or where it is determined that the *building* or structure, premise, or portion thereof is in violation of any ordinance or regulation or any of the provisions of this code.

SECTION C107
VALIDITY

C107.1 General. If a portion of this code is held to be illegal or void, such a decision shall not affect the validity of the remainder of this code.

SECTION C108
REFERENCED STANDARDS

C108.1 Referenced codes and standards. The codes and standards referenced in this code shall be those listed in Chapter 6, and such codes and standards shall be considered as part of the requirements of this code to the prescribed extent of each such reference and as further regulated in Sections C108.1.1 and C108.1.2.

C108.1.1 Conflicts. Where conflicts occur between provisions of this code and referenced codes and standards, the provisions of this code shall apply.

C108.1.2 Provisions in referenced codes and standards. Where the extent of the reference to a referenced code or standard includes subject matter that is within the scope of this code, the provisions of this code, as applicable, shall take precedence over the provisions in the referenced code or standard.

C108.2 Applications of references. References to chapter or section numbers, or to provisions not specifically identified by number, shall be construed to refer to such chapter, section or provision of this code.

C108.3 Other laws. The provisions of this code shall not be deemed to nullify any provisions of local, state or federal law.

SECTION C109
STOP WORK ORDER

C109.1 Authority. Where the *code official* finds any work regulated by this code being performed in a manner contrary to the provisions of this code or in a dangerous or unsafe manner, the *code official* is authorized to issue a stop work order.

C109.2 Issuance. The stop work order shall be in writing and shall be given to the owner of the property, the owner's authorized agent or the person performing the work. Upon issuance of a stop work order, the cited work shall immediately cease. The stop work order shall state the reason for the order and the conditions under which the cited work is authorized to resume.

C109.3 Emergencies. Where an emergency exists, the *code official* shall not be required to give a written notice prior to stopping the work.

C109.4 Failure to comply. Any person who shall continue any work after having been served with a stop work order, except such work as that person is directed to perform to remove a violation or unsafe condition, shall be subject to fines established by the authority having jurisdiction.

SECTION C110
BOARD OF APPEALS

C110.1 General. In order to hear and decide appeals of orders, decisions or determinations made by the *code official* relative to the application and interpretation of this code, there shall be and is hereby created a board of appeals. The *code official* shall be an ex officio member of said board but shall not have a vote on any matter before the board. The board of appeals shall be appointed by the governing body and shall hold office at its pleasure. The board shall adopt rules of procedure for conducting its business, and shall render all decisions and findings in writing to the appellant with a duplicate copy to the *code official*.

C110.2 Limitations on authority. An application for appeal shall be based on a claim that the true intent of this code or the rules legally adopted thereunder have been incorrectly interpreted, the provisions of this code do not fully apply or an equally good or better form of construction is proposed. The board shall not have authority to waive requirements of this code.

C110.3 Qualifications. The board of appeals shall consist of members who are qualified by experience and training and are not employees of the jurisdiction.

CHAPTER 2 [CE]
DEFINITIONS

User note:
About this chapter: Codes, by their very nature, are technical documents. Every word, term and punctuation mark can add to or change the meaning of a technical requirement. It is necessary to maintain a consensus on the specific meaning of each term contained in the code. Chapter 2 performs this function by stating clearly what specific terms mean for the purposes of the code.

SECTION C201
GENERAL

C201.1 Scope. Unless stated otherwise, the following words and terms in this code shall have the meanings indicated in this chapter.

C201.2 Interchangeability. Words used in the present tense include the future; words in the masculine gender include the feminine and neuter; the singular number includes the plural and the plural includes the singular.

C201.3 Terms defined in other codes. Terms that are not defined in this code but are defined in the *International Building Code*, *International Fire Code*, *International Fuel Gas Code*, *International Mechanical Code*, *International Plumbing Code* or the *International Residential Code* shall have the meanings ascribed to them in those codes.

C201.4 Terms not defined. Terms not defined by this chapter shall have ordinarily accepted meanings such as the context implies.

SECTION C202
GENERAL DEFINITIONS

ABOVE-GRADE WALL. See "*Wall, above-grade.*"

ACCESS (TO). That which enables a device, appliance or equipment to be reached by *ready access* or by a means that first requires the removal or movement of a panel or similar obstruction.

ADDITION. An extension or increase in the *conditioned space* floor area, number of stories or height of a building or structure.

AIR BARRIER. One or more materials joined together in a continuous manner to restrict or prevent the passage of air through the *building thermal envelope* and its assemblies.

AIR CURTAIN. A device, installed at the *building entrance*, that generates and discharges a laminar air stream intended to prevent the infiltration of external, unconditioned air into the conditioned spaces, or the loss of interior, conditioned air to the outside.

ALTERATION. Any construction, retrofit or renovation to an existing structure other than repair or addition. Also, a change in a building, electrical, gas, mechanical or plumbing system that involves an extension, addition or change to the arrangement, type or purpose of the original installation.

APPROVED. Acceptable to the code official.

APPROVED AGENCY. An established and recognized agency that is regularly engaged in conducting tests or furnishing inspection services, or furnishing product certification, where such agency has been approved by the *code official*.

AUTOMATIC. Self-acting, operating by its own mechanism when actuated by some impersonal influence, as, for example, a change in current strength, pressure, temperature or mechanical configuration (see "*Manual*").

BELOW-GRADE WALL. See "*Wall, below-grade.*"

BIOGAS. A mixture of hydrocarbons that is a gas at 60°F (15.5°C) and 1 atmosphere of pressure that is produced through the anaerobic digestion of organic matter.

BIOMASS. Nonfossilized and biodegradable organic material originating from plants, animals and/or microorganisms, including products, by-products, residues and waste from agriculture, forestry and related industries as well as the nonfossilized and biodegradable organic fractions of industrial and municipal wastes, including gases and liquids recovered from the decomposition of nonfossilized and biodegradable organic material.

BOILER, MODULATING. A boiler that is capable of more than a single firing rate in response to a varying temperature or heating load.

BOILER SYSTEM. One or more boilers, their piping and controls that work together to supply steam or hot water to heat output devices remote from the boiler.

BUBBLE POINT. The refrigerant liquid saturation temperature at a specified pressure.

BUILDING. Any structure used or intended for supporting or sheltering any use or occupancy, including any mechanical systems, service water-heating systems and electric power and lighting systems located on the building site and supporting the building.

BUILDING COMMISSIONING. A process that verifies and documents that the selected building systems have been designed, installed and function according to the owner's project requirements and construction documents, and to minimum code requirements.

BUILDING ENTRANCE. Any door, set of doors, doorway or other form of portal that is used to gain access to the building from the outside by the public.

BUILDING SITE. A contiguous area of land that is under the ownership or control of one entity.

DEFINITIONS

BUILDING THERMAL ENVELOPE. The basement walls, exterior walls, floors, ceilings, roofs and any other building element assemblies that enclose *conditioned space* or provide a boundary between *conditioned space* and exempt or unconditioned space.

CAPTIVE KEY OVERRIDE. A lighting control that will not release the key that activates the override when the lighting is on.

CAVITY INSULATION. Insulating material located between framing members.

***C*-FACTOR (THERMAL CONDUCTANCE).** The coefficient of heat transmission (surface to surface) through a building component or assembly, equal to the time rate of heat flow per unit area and the unit temperature difference between the warm side and cold side surfaces (Btu/h × ft^2 × °F) [W/(m^2 × K)].

CHANGE OF OCCUPANCY. A change in the use of a building or a portion of a building that results in any of the following:

1. A *change of occupancy* classification.
2. A change from one group to another group within an occupancy classification.
3. Any change in use within a group for which there is a change in the application of the requirements of this code.

CIRCULATING HOT WATER SYSTEM. A specifically designed water distribution system where one or more pumps are operated in the service hot water piping to circulate heated water from the water-heating equipment to the fixture supply and back to the water-heating equipment.

CLIMATE ZONE. A geographical region based on climatic criteria as specified in this code.

CODE OFFICIAL. The officer or other designated authority charged with the administration and enforcement of this code, or a duly authorized representative.

COEFFICIENT OF PERFORMANCE (COP) – COOLING. The ratio of the rate of heat removal to the rate of energy input, in consistent units, for a complete refrigerating system or some specific portion of that system under designated operating conditions.

COEFFICIENT OF PERFORMANCE (COP) – HEATING. The ratio of the rate of heat delivered to the rate of energy input, in consistent units, for a complete heat pump system, including the compressor and, if applicable, auxiliary heat, under designated operating conditions.

COMMERCIAL BUILDING. For this code, all buildings that are not included in the definition of "*Residential building.*"

COMPUTER ROOM. A room whose primary function is to house equipment for the processing and storage of electronic data which has a design total information technology equipment (ITE) equipment power density less than or equal to 20 watts per square foot (20 watts per 0.092 m^2) of conditioned area or a design total ITE equipment load less than or equal to 10 kW.

CONDENSING UNIT. A factory-made assembly of refrigeration components designed to compress and liquefy a specific refrigerant. The unit consists of one or more refrigerant compressors, refrigerant condensers (air-cooled, evaporatively cooled or water-cooled), condenser fans and motors (where used) and factory-supplied accessories.

CONDITIONED FLOOR AREA. The horizontal projection of the floors associated with the *conditioned space*.

CONDITIONED SPACE. An area, room or space that is enclosed within the *building thermal envelope* and is directly or indirectly heated or cooled. Spaces are indirectly heated or cooled where they communicate through openings with conditioned spaces, where they are separated from conditioned spaces by uninsulated walls, floors or ceilings, or where they contain uninsulated ducts, piping or other sources of heating or cooling.

CONTINUOUS INSULATION (ci). Insulating material that is continuous across all structural members without thermal bridges other than fasteners and service openings. It is installed on the interior or exterior or is integral to any opaque surface of the building envelope.

CRAWL SPACE WALL. The opaque portion of a wall that encloses a crawl space and is partially or totally below grade.

CURTAIN WALL. Fenestration products used to create an external nonload-bearing wall that is designed to separate the exterior and interior environments.

DATA CENTER. A room or series of rooms that share data center systems, whose primary function is to house equipment for the processing and storage of electronic data and that has a design total ITE equipment power density exceeding 20 watts per square foot (20 watts per 0.092 m^2) of conditioned area and a total design ITE equipment load greater than 10 kW.

DATA CENTER SYSTEMS. HVAC systems and equipment, or portions thereof, used to provide cooling or ventilation in a data center.

DAYLIGHT RESPONSIVE CONTROL. A device or system that provides automatic control of electric light levels based on the amount of daylight in a space.

DAYLIGHT ZONE. That portion of a building's interior floor area that is illuminated by natural light.

DEMAND CONTROL VENTILATION (DCV). A ventilation system capability that provides for the automatic reduction of outdoor air intake below design rates when the actual occupancy of spaces served by the system is less than design occupancy.

DEMAND RECIRCULATION WATER SYSTEM. A water distribution system where one or more pumps prime the service hot water piping with heated water upon a demand for hot water.

DIRECT DIGITAL CONTROL (DDC). A type of control where controlled and monitored analog or binary data, such as temperature and contact closures, are converted to digital format for manipulation and calculations by a digital computer or microprocessor, then converted back to analog or binary form to control physical devices.

DEFINITIONS

DUCT. A tube or conduit utilized for conveying air. The air passages of self-contained systems are not to be construed as air ducts.

DUCT SYSTEM. A continuous passageway for the transmission of air that, in addition to ducts, includes duct fittings, dampers, plenums, fans and accessory air-handling equipment and appliances.

DWELLING UNIT. A single unit providing complete independent living facilities for one or more persons, including permanent provisions for living, sleeping, eating, cooking and sanitation.

DYNAMIC GLAZING. Any fenestration product that has the fully reversible ability to change its performance properties, including *U*-factor, solar heat gain coefficient (SHGC) or visible transmittance (VT).

ECONOMIZER, AIR. A duct and damper arrangement and automatic control system that allows a cooling system to supply outside air to reduce or eliminate the need for mechanical cooling during mild or cold weather.

ECONOMIZER, WATER. A system where the supply air of a cooling system is cooled indirectly with water that is itself cooled by heat or mass transfer to the environment without the use of mechanical cooling.

ENCLOSED SPACE. A volume surrounded by solid surfaces such as walls, floors, roofs and openable devices, such as doors and operable windows.

ENERGY ANALYSIS. A method for estimating the annual energy use of the *proposed design* and *standard reference design* based on estimates of energy use.

ENERGY COST. The total estimated annual cost for purchased energy for the building functions regulated by this code, including applicable demand charges.

ENERGY RECOVERY VENTILATION SYSTEM. Systems that employ air-to-air heat exchangers to recover energy from exhaust air for the purpose of preheating, precooling, humidifying or dehumidifying outdoor ventilation air prior to supplying the air to a space, either directly or as part of an HVAC system.

ENERGY SIMULATION TOOL. An *approved* software program or calculation-based methodology that projects the annual energy use of a building.

ENTHALPY RECOVERY RATIO. Change in the enthalpy of the *outdoor air* supply divided by the difference between the *outdoor air* and entering exhaust air enthalpy, expressed as a percentage.

ENTRANCE DOOR. A vertical fenestration product used for occupant ingress, egress and access in nonresidential buildings, including, but not limited to, exterior entrances utilizing latching hardware and automatic closers and containing over 50 percent glazing specifically designed to withstand heavy-duty usage.

EQUIPMENT ROOM. A space that contains either electrical equipment, mechanical equipment, machinery, water pumps or hydraulic pumps that are a function of the building's services.

EXTERIOR WALL. Walls including both above-grade walls and basement walls.

FAN, EMBEDDED. A fan that is part of a manufactured assembly where the assembly includes functions other than air movement.

FAN ARRAY. Multiple fans in parallel between two plenum sections in an air distribution system.

FAN BRAKE HORSEPOWER (BHP). The horsepower delivered to the fan's shaft. Brake horsepower does not include the mechanical drive losses, such as that from belts and gears.

FAN ENERGY INDEX (FEI). The ratio of the electric input power of a reference fan to the electric input power of the actual fan as calculated in accordance with AMCA 208.

FAN NAMEPLATE ELECTRICAL INPUT POWER. The nominal electrical input power rating stamped on a fan assembly nameplate.

FAN SYSTEM BHP. The sum of the fan brake horsepower of all fans that are required to operate at fan system design conditions to supply air from the heating or cooling source to the *conditioned spaces* and return it to the source or exhaust it to the outdoors.

FAN SYSTEM DESIGN CONDITIONS. Operating conditions that can be expected to occur during normal system operation that result in the highest supply fan airflow rate to conditioned spaces served by the system, other than during air economizer operation.

FAN SYSTEM ELECTRICAL INPUT POWER. The sum of the fan electrical power of all fans that are required to operate at fan system design conditions to supply air from the heating or cooling source to the conditioned spaces and/or return it to the source or exhaust it to the outdoors.

FAN SYSTEM MOTOR NAMEPLATE HP. The sum of the motor nameplate horsepower of all fans that are required to operate at design conditions to supply air from the heating or cooling source to the *conditioned spaces* and return it to the source or exhaust it to the outdoors.

FAULT DETECTION AND DIAGNOSTICS (FDD) SYSTEM. A software platform that utilizes building analytic algorithms to convert data provided by sensors and devices to automatically identify faults in building systems and provide a prioritized list of actionable resolutions to those faults based on cost or energy avoidance, comfort and maintenance impact.

FENESTRATION. Products classified as either skylights or vertical fenestration.

 Skylights. Glass or other transparent or translucent glazing material installed at a slope of less than 60 degrees (1.05 rad) from horizontal, including unit skylights, tubular daylighting devices and glazing materials in solariums, sunrooms, roofs, greenhouses and sloped walls.

 Vertical fenestration. Windows that are fixed or operable, opaque doors, glazed doors, glazed block and combination opaque and glazed doors composed of glass or other transparent or translucent glazing materials and installed at a slope of not less than 60 degrees (1.05 rad) from horizontal.

FENESTRATION PRODUCT, FIELD-FABRICATED. A fenestration product whose frame is made at the construction site of standard dimensional lumber or other materials that were not previously cut or otherwise formed with the specific intention of being used to fabricate a fenestration product or exterior door. Field-fabricated does not include site-built fenestration.

FENESTRATION PRODUCT, SITE-BUILT. A fenestration designed to be made up of field-glazed or field-assembled units using specific factory cut or otherwise factory-formed framing and glazing units. Examples of site-built fenestration include storefront systems, curtain walls and atrium roof systems.

F-**FACTOR.** The perimeter heat loss factor for slab-on-grade floors (Btu/h × ft × °F) [W/(m × K)].

FLOOR AREA, NET. The actual occupied area not including unoccupied accessory areas such as corridors, stairways, toilet rooms, mechanical rooms and closets.

GENERAL LIGHTING. Interior lighting that provides a substantially uniform level of illumination throughout a space.

GREENHOUSE. A structure or a thermally isolated area of a building that maintains a specialized sunlit environment exclusively used for, and essential to, the cultivation, protection or maintenance of plants. *Greenhouses* are those that are erected for a period of 180 days or more.

GROUP R. Buildings or portions of buildings that contain any of the following occupancies as established in the *International Building Code*:

1. *Group R*-1.
2. *Group R*-2 where located more than three stories in height above grade plane.
3. *Group R*-4 where located more than three stories in height above grade plane.

HEAT TRAP. An arrangement of piping and fittings, such as elbows, or a commercially available heat trap that prevents thermosyphoning of hot water during standby periods.

HEATED SLAB. Slab-on-grade construction in which the heating elements, hydronic tubing or hot air distribution system is in contact with, or placed within or under, the slab.

HIGH SPEED DOOR. A nonswinging door used primarily to facilitate vehicular access or material transportation, with a minimum opening rate of 32 inches (813 mm) per second, a minimum closing rate of 24 inches (610 mm) per second and that includes an automatic-closing device.

HISTORIC BUILDING. Any building or structure that is one or more of the following:

1. Listed, or certified as eligible for listing, by the State Historic Preservation Officer or the Keeper of the National Register of Historic Places, in the National Register of Historic Places.
2. Designated as historic under an applicable state or local law.
3. Certified as a contributing resource within a National Register-listed, state-designated or locally designated historic district.

IEC DESIGN H MOTOR. An electric motor that meets all of the following:

1. It is an induction motor designed for use with three-phase power.
2. It contains a cage rotor.
3. It is capable of direct-on-line starting.
4. It has four, six or eight poles.
5. It is rated from 0.4 kW to 1600 kW at a frequency of 60 hertz.

IEC DESIGN N MOTOR. An electric motor that meets all of the following:

1. It is an induction motor designed for use with three-phase power.
2. It contains a cage rotor.
3. It is capable of direct-on-line starting.
4. It has two, four, six or eight poles.
5. It is rated from 0.4 kW to 1600 kW at a frequency of 60 hertz.

INFILTRATION. The uncontrolled inward air leakage into a building caused by the pressure effects of wind or the effect of differences in the indoor and outdoor air density or both.

INFORMATION TECHNOLOGY EQUIPMENT (ITE). Items including computers, data storage devices, servers and network and communication equipment.

INTEGRATED PART LOAD VALUE (IPLV). A single-number figure of merit based on part-load EER, COP or kW/ton expressing part-load efficiency for air-conditioning and heat pump equipment on the basis of weighted operation at various load capacities for equipment.

INTERNAL CURTAIN SYSTEM. A system consisting of movable panels of fabric or plastic film used to cover and uncover the space enclosed in a *greenhouse* on a daily basis.

ISOLATION DEVICES. Devices that isolate HVAC zones so that they can be operated independently of one another. *Isolation devices* include separate systems, isolation dampers and controls providing shutoff at terminal boxes.

LABELED. Equipment, materials or products to which have been affixed a label, seal, symbol or other identifying mark of a nationally recognized testing laboratory, *approved agency* or other organization concerned with product evaluation that maintains periodic inspection of the production of the labeled items and whose labeling indicates either that the equipment, material or product meets identified standards or has been tested and found suitable for a specified purpose.

LARGE-DIAMETER CEILING FAN. A ceiling fan that is greater than 7 feet (2134 mm) in diameter. These fans are sometimes referred to as High-Volume, Low-Speed (HVLS) fans.

LINER SYSTEM (Ls). A system that includes the following:

1. A continuous vapor barrier liner membrane that is installed below the purlins and that is uninterrupted by framing members.
2. An uncompressed, unfaced insulation resting on top of the liner membrane and located between the purlins.

For multilayer installations, the last rated *R*-value of insulation is for unfaced insulation draped over purlins and then compressed when the metal roof panels are attached.

LISTED. Equipment, materials, products or services included in a list published by an organization acceptable to the *code official* and concerned with evaluation of products or services that maintains periodic inspection of production of *listed* equipment or materials or periodic evaluation of services and whose listing states either that the equipment, material, product or service meets identified standards or has been tested and found suitable for a specified purpose.

LOW-SLOPED ROOF. A roof having a slope less than 2 units vertical in 12 units horizontal.

LOW-VOLTAGE DRY-TYPE DISTRIBUTION TRANSFORMER. A transformer that is air-cooled, does not use oil as a coolant, has an input voltage less than or equal to 600 volts and is rated for operation at a frequency of 60 hertz.

LUMINAIRE-LEVEL LIGHTING CONTROLS. A lighting system consisting of one or more luminaires with embedded lighting control logic, occupancy and ambient light sensors, wireless networking capabilities and local override switching capability, where required.

MANUAL. Capable of being operated by personal intervention (see "*Automatic*").

NAMEPLATE HORSEPOWER. The nominal motor output power rating stamped on the motor nameplate.

NEMA DESIGN A MOTOR. A squirrel-cage motor that meets all of the following:

1. It is designed to withstand full-voltage starting and develop locked-rotor torque as shown in paragraph 12.38.1 of NEMA MG 1.
2. It has pull-up torque not less than the values shown in paragraph 12.40.1 of NEMA MG 1.
3. It has breakdown torque not less than the values shown in paragraph 12.39.1 of NEMA MG 1.
4. It has a locked-rotor current higher than the values shown in paragraph 12.35.1 of NEMA MG 1 for 60 hertz and paragraph 12.35.2 of NEMA MG 1 for 50 hertz.
5. It has a slip at rated load of less than 5 percent for motors with fewer than 10 poles.

NEMA DESIGN B MOTOR. A squirrel-cage motor that meets all of the following:

1. It is designed to withstand full-voltage starting.
2. It develops locked-rotor, breakdown and pull-up torques adequate for general application as specified in Sections 12.38, 12.39 and 12.40 of NEMA MG1.
3. It draws locked-rotor current not to exceed the values shown in Section 12.35.1 for 60 hertz and Section 12.35.2 for 50 hertz of NEMA MG1.
4. It has a slip at rated load of less than 5 percent for motors with fewer than 10 poles.

NEMA DESIGN C MOTOR. A squirrel-cage motor that meets all of the following:

1. Designed to withstand full-voltage starting and develop locked-rotor torque for high-torque applications up to the values shown in paragraph 12.38.2 of NEMA MG1 (incorporated by reference, see A§431.15).
2. It has pull-up torque not less than the values shown in paragraph 12.40.2 of NEMA MG1.
3. It has breakdown torque not less than the values shown in paragraph 12.39.2 of NEMA MG1.
4. It has a locked-rotor current not to exceed the values shown in paragraph 12.35.1 of NEMA MG1 for 60 hertz and paragraph 12.35.2 for 50 hertz.
5. It has a slip at rated load of less than 5 percent.

NETWORKED GUESTROOM CONTROL SYSTEM. A control system, with access from the front desk or other central location associated with a *Group R*-1 building, that is capable of identifying the rented and unrented status of each guestroom according to a timed schedule, and is capable of controlling HVAC in each hotel and motel guestroom separately.

NONSTANDARD PART LOAD VALUE (NPLV). A single-number part-load efficiency figure of merit calculated and referenced to conditions other than IPLV conditions, for units that are not designed to operate at AHRI standard rating conditions.

OCCUPANT SENSOR CONTROL. An automatic control device or system that detects the presence or absence of people within an area and causes lighting, equipment or appliances to be regulated accordingly.

ON-SITE RENEWABLE ENERGY. Energy from renewable energy resources harvested at the building project site.

OPAQUE DOOR. A door that is not less than 50-percent opaque in surface area.

POWERED ROOF/WALL VENTILATORS. A fan consisting of a centrifugal or axial impeller with an integral driver in a weather-resistant housing and with a base designed to fit, usually by means of a curb, over a wall or roof opening.

PROPOSED DESIGN. A description of the proposed building used to estimate annual energy use for determining compliance based on total building performance.

RADIANT HEATING SYSTEM. A heating system that transfers heat to objects and surfaces within a conditioned space, primarily by infrared radiation.

DEFINITIONS

READY ACCESS (TO). That which enables a device, appliance or equipment to be directly reached without requiring the removal or movement of any panel or similar obstruction.

REFRIGERANT DEW POINT. The refrigerant vapor saturation temperature at a specified pressure.

REFRIGERATED WAREHOUSE COOLER. An enclosed storage space capable of being refrigerated to temperatures above 32°F (0°C) that can be walked into and has a total chilled storage area of not less than 3,000 square feet (279 m^2).

REFRIGERATED WAREHOUSE FREEZER. An enclosed storage space capable of being refrigerated to temperatures at or below 32°F (0°C) that can be walked into and has a total chilled storage area of not less than 3,000 square feet (279 m^2).

REFRIGERATION SYSTEM, LOW TEMPERATURE. Systems for maintaining food product in a frozen state in refrigeration applications.

REFRIGERATION SYSTEM, MEDIUM TEMPERATURE. Systems for maintaining food product above freezing in refrigeration applications.

REGISTERED DESIGN PROFESSIONAL. An individual who is registered or licensed to practice their respective design profession as defined by the statutory requirements of the professional registration laws of the state or jurisdiction in which the project is to be constructed.

RENEWABLE ENERGY RESOURCES. Energy derived from solar radiation, wind, waves, tides, landfill gas, biogas, biomass or extracted from hot fluid or steam heated within the earth.

REPAIR. The reconstruction or renewal of any part of an existing building for the purpose of its maintenance or to correct damage.

REROOFING. The process of recovering or replacing an existing roof covering. See "*Roof recover*" and "*Roof replacement*."

RESIDENTIAL BUILDING. For this code, includes detached one- and two-family dwellings and multiple single-family dwellings (townhouses) and *Group R*-2, R-3 and R-4 buildings three stories or less in height above grade plane.

ROOF ASSEMBLY. A system designed to provide weather protection and resistance to design loads. The system consists of a roof covering and roof deck or a single component serving as both the roof covering and the roof deck. A roof assembly includes the roof covering, underlayment, roof deck, insulation, vapor retarder and interior finish.

ROOF RECOVER. The process of installing an additional roof covering over an existing roof covering without removing the existing roof covering.

ROOF REPAIR. Reconstruction or renewal of any part of an existing roof for the purpose of its maintenance.

ROOF REPLACEMENT. The process of removing the existing roof covering, repairing any damaged substrate and installing a new roof covering.

ROOFTOP MONITOR. A raised section of a roof containing vertical fenestration along one or more sides.

R-VALUE (THERMAL RESISTANCE). The inverse of the time rate of heat flow through a body from one of its bounding surfaces to the other surface for a unit temperature difference between the two surfaces, under steady state conditions, per unit area (h × ft^2 × °F/Btu) [(m^2 × K)/W].

SATURATED CONDENSING TEMPERATURE. The saturation temperature corresponding to the measured refrigerant pressure at the condenser inlet for single component and azeotropic refrigerants, and the arithmetic average of the dew point and *bubble point* temperatures corresponding to the refrigerant pressure at the condenser entrance for zeotropic refrigerants.

SERVICE WATER HEATING. Supply of hot water for purposes other than comfort heating.

SLEEPING UNIT. A room or space in which people sleep that can include permanent provisions for living, eating, and either sanitation or kitchen facilities but not both. Such rooms and spaces that are part of a dwelling unit are not *sleeping units*.

SMALL ELECTRIC MOTOR. A general purpose alternating-current single-speed induction motor.

SOLAR HEAT GAIN COEFFICIENT (SHGC). The ratio of the solar heat gain entering the space through the fenestration assembly to the incident solar radiation. Solar heat gain includes directly transmitted solar heat and absorbed solar radiation that is then reradiated, conducted or convected into the space.

STANDARD REFERENCE DESIGN. A version of the *proposed design* that meets the minimum requirements of this code and is used to determine the maximum annual energy use requirement for compliance based on total building performance.

STOREFRONT. A system of doors and windows mulled as a composite fenestration structure that has been designed to resist heavy use. *Storefront* systems include, but are not limited to, exterior fenestration systems that span from the floor level or above to the ceiling of the same story on commercial buildings, with or without mulled windows and doors.

TESTING UNIT ENCLOSURE AREA. The area sum of all the boundary surfaces that define the *dwelling unit, sleeping unit* or occupiable *conditioned space* including top/ceiling, bottom/floor and all side walls. This does not include interior partition walls within the *dwelling unit, sleeping unit*, or occupiable *conditioned space*. Wall height shall be measured from the finished floor of the *conditioned space* to the finished floor or roof/ceiling air barrier above.

THERMAL DISTRIBUTION EFFICIENCY (TDE). The resistance to changes in air heat as air is conveyed through a distance of air duct. TDE is a heat loss calculation evaluating the difference in the heat of the air between the air duct inlet and outlet caused by differences in temperatures between the air in the duct and the duct material. TDE is expressed as a percent difference between the inlet and outlet heat in the duct.

THERMOSTAT. An automatic control device used to maintain temperature at a fixed or adjustable setpoint.

TIME SWITCH CONTROL. An automatic control device or system that controls lighting or other loads, including switching off, based on time schedules.

U-FACTOR (THERMAL TRANSMITTANCE). The coefficient of heat transmission (air to air) through a building component or assembly, equal to the time rate of heat flow per unit area and unit temperature difference between the warm side and cold side air films (Btu/h × ft^2 × °F) [W/(m^2 × K)].

VARIABLE REFRIGERANT FLOW SYSTEM. An engineered direct-expansion (DX) refrigerant system that incorporates a common condensing unit, at least one variable-capacity compressor, a distributed refrigerant piping network to multiple indoor fan heating and cooling units each capable of individual zone temperature control, through integral zone temperature control devices and a common communications network. Variable refrigerant flow utilizes three or more steps of control on common interconnecting piping.

VEGETATIVE ROOF. An assembly of interacting components designed to waterproof a building's top surface that includes, by design, vegetation and related landscape elements.

VENTILATION. The natural or mechanical process of supplying conditioned or unconditioned air to, or removing such air from, any space.

VENTILATION AIR. That portion of supply air that comes from outside (outdoors) plus any recirculated air that has been treated to maintain the desired quality of air within a designated space.

VISIBLE TRANSMITTANCE (VT). The ratio of visible light entering the space through the fenestration product assembly to the incident visible light. Visible transmittance includes the effects of glazing material and frame and is expressed as a number between 0 and 1.

VISIBLE TRANSMITTANCE, ANNUAL (VT$_{annual}$). The ratio of visible light entering the space through the fenestration product assembly to the incident visible light during the course of a year, which includes the effects of glazing material, frame, and light well or tubular conduit, and is expressed as a number between 0 and 1.

VOLTAGE DROP. A decrease in voltage caused by losses in the wiring systems that connect the power source to the load.

WALK-IN COOLER. An enclosed storage space capable of being refrigerated to temperatures above 32°F (0°C) and less than 55°F (12.8°C) that can be walked into, has a ceiling height of not less than 7 feet (2134 mm) and has a total chilled storage area of less than 3,000 square feet (279 m^2).

WALK-IN FREEZER. An enclosed storage space capable of being refrigerated to temperatures at or below 32°F (0°C) that can be walked into, has a ceiling height of not less than 7 feet (2134 mm) and has a total chilled storage area of less than 3,000 square feet (279 m^2).

WALL, ABOVE-GRADE. A wall associated with the *building thermal envelope* that is more than 15 percent above grade and is on the exterior of the building or any wall that is associated with the *building thermal envelope* that is not on the exterior of the building. This includes, but is not limited to, between-floor spandrels, peripheral edges of floors, roof knee walls, dormer walls, gable end walls, walls enclosing a mansard roof and skylight shafts.

WALL, BELOW-GRADE. A wall associated with the basement or first story of the building that is part of the *building thermal envelope*, is not less than 85 percent below grade and is on the exterior of the building.

WATER HEATER. Any heating appliance or equipment that heats potable water and supplies such water to the potable hot water distribution system.

ZONE. A space or group of spaces within a building with heating or cooling requirements that are sufficiently similar so that desired conditions can be maintained throughout using a single controlling device.

CHAPTER 3 [CE]
GENERAL REQUIREMENTS

User note:

About this chapter: Chapter 3 addresses broadly applicable requirements that would not be at home in other chapters having more specific coverage of subject matter. This chapter establishes climate zone by US counties and territories and includes methodology for determining climate zones elsewhere. It also contains product rating, marking and installation requirements for materials such as insulation, windows, doors and siding.

SECTION C301
CLIMATE ZONES

C301.1 General. *Climate zones* from Figure C301.1 or Table C301.1 shall be used for determining the applicable requirements from Chapter 4. Locations not indicated in Table C301.1 shall be assigned a *climate zone* in accordance with Section C301.3.

C301.2 Warm Humid counties. In Table C301.1, Warm Humid counties are identified by an asterisk.

GENERAL REQUIREMENTS

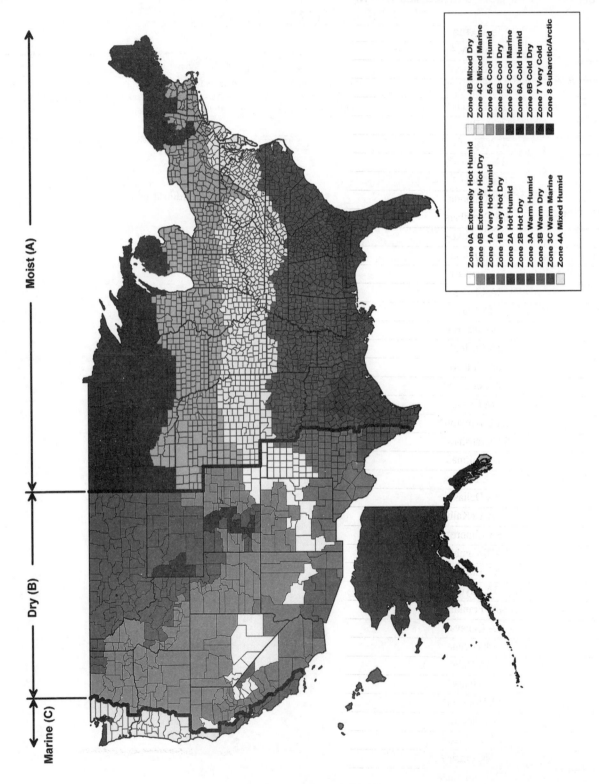

**FIGURE C301.1
CLIMATE ZONES**

TABLE C301.1
CLIMATE ZONES, MOISTURE REGIMES, AND WARM HUMID DESIGNATIONS BY STATE, COUNTY AND TERRITORY[a]

US STATES	
ALABAMA	3A Lowndes*
3A Autauga*	3A Macon*
2A Baldwin*	3A Madison
3A Barbour*	3A Marengo*
3A Bibb	3A Marion
3A Blount	3A Marshall
3A Bullock*	2A Mobile*
3A Butler*	3A Monroe*
3A Calhoun	3A Montgomery*
3A Chambers	3A Morgan
3A Cherokee	3A Perry*
3A Chilton	3A Pickens
3A Choctaw*	3A Pike*
3A Clarke*	3A Randolph
3A Clay	3A Russell*
3A Cleburne	3A Shelby
2A Coffee*	3A St. Clair
3A Colbert	3A Sumter
3A Conecuh*	3A Talladega
3A Coosa	3A Tallapoosa
2A Covington*	3A Tuscaloosa
3A Crenshaw*	3A Walker
3A Cullman	3A Washington*
2A Dale*	3A Wilcox*
3A Dallas*	3A Winston
3A DeKalb	**ALASKA**
3A Elmore*	7 Aleutians East
2A Escambia*	7 Aleutians West
3A Etowah	7 Anchorage
3A Fayette	7 Bethel
3A Franklin	7 Bristol Bay
2A Geneva*	8 Denali
3A Greene	7 Dillingham
3A Hale	8 Fairbanks North Star
2A Henry*	6A Haines
2A Houston*	6A Juneau
3A Jackson	7 Kenai Peninsula
3A Jefferson	5C Ketchikan Gateway
3A Lamar	6A Kodiak Island
3A Lauderdale	7 Lake and Peninsula
3A Lawrence	7 Matanuska-Susitna
3A Lee	8 Nome
3A Limestone	8 North Slope
	8 Northwest Arctic

(continued)

TABLE C301.1—continued
CLIMATE ZONES, MOISTURE REGIMES, AND WARM HUMID DESIGNATIONS BY STATE, COUNTY AND TERRITORY[a]

US STATES—continued
ALASKA *(continued)*
5C Prince of Wales Outer Ketchikan
5C Sitka
6A Skagway-Hoonah-Angoon
8 Southeast Fairbanks
7 Valdez-Cordova
8 Wade Hampton
6A Wrangell-Petersburg
7 Yakutat
8 Yukon-Koyukuk
ARIZONA
5B Apache
3B Cochise
5B Coconino
4B Gila
3B Graham
3B Greenlee
2B La Paz
2B Maricopa
3B Mohave
5B Navajo
2B Pima
2B Pinal
3B Santa Cruz
4B Yavapai
2B Yuma
ARKANSAS
3A Arkansas
3A Ashley
4A Baxter
4A Benton
4A Boone
3A Bradley
3A Calhoun
4A Carroll
3A Chicot
3A Clark
3A Clay
3A Cleburne
3A Cleveland
3A Columbia*
3A Conway
3A Craighead
3A Crawford

3A Crittenden
3A Cross
3A Dallas
3A Desha
3A Drew
3A Faulkner
3A Franklin
4A Fulton
3A Garland
3A Grant
3A Greene
3A Hempstead*
3A Hot Spring
3A Howard
3A Independence
4A Izard
3A Jackson
3A Jefferson
3A Johnson
3A Lafayette*
3A Lawrence
3A Lee
3A Lincoln
3A Little River*
3A Logan
3A Lonoke
4A Madison
4A Marion
3A Miller*
3A Mississippi
3A Monroe
3A Montgomery
3A Nevada
4A Newton
3A Ouachita
3A Perry
3A Phillips
3A Pike
3A Poinsett
3A Polk
3A Pope
3A Prairie
3A Pulaski
3A Randolph

(continued)

TABLE C301.1—continued
CLIMATE ZONES, MOISTURE REGIMES, AND WARM HUMID DESIGNATIONS BY STATE, COUNTY AND TERRITORY[a]

US STATES—continued
ARKANSAS *(continued)*
3A Saline
3A Scott
4A Searcy
3A Sebastian
3A Sevier*
3A Sharp
3A St. Francis
4A Stone
3A Union*
3A Van Buren
4A Washington
3A White
3A Woodruff
3A Yell
CALIFORNIA
3C Alameda
6B Alpine
4B Amador
3B Butte
4B Calaveras
3B Colusa
3B Contra Costa
4C Del Norte
4B El Dorado
3B Fresno
3B Glenn
4C Humboldt
2B Imperial
4B Inyo
3B Kern
3B Kings
4B Lake
5B Lassen
3B Los Angeles
3B Madera
3C Marin
4B Mariposa
3C Mendocino
3B Merced
5B Modoc
6B Mono
3C Monterey
3C Napa
5B Nevada
3B Orange
3B Placer
5B Plumas
3B Riverside
3B Sacramento
3C San Benito
3B San Bernardino
3B San Diego
3C San Francisco
3B San Joaquin
3C San Luis Obispo
3C San Mateo
3C Santa Barbara
3C Santa Clara
3C Santa Cruz
3B Shasta
5B Sierra
5B Siskiyou
3B Solano
3C Sonoma
3B Stanislaus
3B Sutter
3B Tehama
4B Trinity
3B Tulare
4B Tuolumne
3C Ventura
3B Yolo
3B Yuba
COLORADO
5B Adams
6B Alamosa
5B Arapahoe
6B Archuleta
4B Baca
4B Bent
5B Boulder
5B Broomfield
6B Chaffee
5B Cheyenne
7 Clear Creek
6B Conejos
6B Costilla
5B Crowley

(continued)

GENERAL REQUIREMENTS

TABLE C301.1—continued
CLIMATE ZONES, MOISTURE REGIMES, AND WARM HUMID DESIGNATIONS BY STATE, COUNTY AND TERRITORY[a]

US STATES—continued
COLORADO *(continued)*
5B Custer
5B Delta
5B Denver
6B Dolores
5B Douglas
6B Eagle
5B Elbert
5B El Paso
5B Fremont
5B Garfield
5B Gilpin
7 Grand
7 Gunnison
7 Hinsdale
5B Huerfano
7 Jackson
5B Jefferson
5B Kiowa
5B Kit Carson
7 Lake
5B La Plata
5B Larimer
4B Las Animas
5B Lincoln
5B Logan
5B Mesa
7 Mineral
6B Moffat
5B Montezuma
5B Montrose
5B Morgan
4B Otero
6B Ouray
7 Park
5B Phillips
7 Pitkin
4B Prowers
5B Pueblo
6B Rio Blanco
7 Rio Grande
7 Routt
6B Saguache
7 San Juan

6B San Miguel
5B Sedgwick
7 Summit
5B Teller
5B Washington
5B Weld
5B Yuma
CONNECTICUT
5A (all)
DELAWARE
4A (all)
DISTRICT OF COLUMBIA
4A (all)
FLORIDA
2A Alachua*
2A Baker*
2A Bay*
2A Bradford*
2A Brevard*
1A Broward*
2A Calhoun*
2A Charlotte*
2A Citrus*
2A Clay*
2A Collier*
2A Columbia*
2A DeSoto*
2A Dixie*
2A Duval*
2A Escambia*
2A Flagler*
2A Franklin*
2A Gadsden*
2A Gilchrist*
2A Glades*
2A Gulf*
2A Hamilton*
2A Hardee*
2A Hendry*
2A Hernando*
2A Highlands*
2A Hillsborough*
2A Holmes*
2A Indian River*
2A Jackson*

(continued)

TABLE C301.1—continued
CLIMATE ZONES, MOISTURE REGIMES, AND WARM HUMID DESIGNATIONS BY STATE, COUNTY AND TERRITORY[a]

US STATES—continued	
FLORIDA (continued)	3A Barrow
2A Jefferson*	3A Bartow
2A Lafayette*	3A Ben Hill*
2A Lake*	2A Berrien*
2A Lee*	3A Bibb
2A Leon*	3A Bleckley*
2A Levy*	2A Brantley*
2A Liberty*	2A Brooks*
2A Madison*	2A Bryan*
2A Manatee*	3A Bulloch*
2A Marion*	3A Burke
2A Martin*	3A Butts
1A Miami-Dade*	2A Calhoun*
1A Monroe*	2A Camden*
2A Nassau*	3A Candler*
2A Okaloosa*	3A Carroll
2A Okeechobee*	3A Catoosa
2A Orange*	2A Charlton*
2A Osceola*	2A Chatham*
1A Palm Beach*	3A Chattahoochee*
2A Pasco*	3A Chattooga
2A Pinellas*	3A Cherokee
2A Polk*	3A Clarke
2A Putnam*	3A Clay*
2A Santa Rosa*	3A Clayton
2A Sarasota*	2A Clinch*
2A Seminole*	3A Cobb
2A St. Johns*	2A Coffee*
2A St. Lucie*	2A Colquitt*
2A Sumter*	3A Columbia
2A Suwannee*	2A Cook*
2A Taylor*	3A Coweta
2A Union*	3A Crawford
2A Volusia*	3A Crisp*
2A Wakulla*	3A Dade
2A Walton*	3A Dawson
2A Washington*	2A Decatur*
GEORGIA	3A DeKalb
2A Appling*	3A Dodge*
2A Atkinson*	3A Dooly*
2A Bacon*	2A Dougherty*
2A Baker*	3A Douglas
3A Baldwin	2A Early*
3A Banks	2A Echols*
	2A Effingham*

(continued)

GENERAL REQUIREMENTS

TABLE C301.1—continued
CLIMATE ZONES, MOISTURE REGIMES, AND WARM HUMID DESIGNATIONS BY STATE, COUNTY AND TERRITORY[a]

US STATES—continued
GEORGIA (continued)
3A Elbert
3A Emanuel*
2A Evans*
3A Fannin
3A Fayette
3A Floyd
3A Forsyth
3A Franklin
3A Fulton
3A Gilmer
3A Glascock
2A Glynn*
3A Gordon
2A Grady*
3A Greene
3A Gwinnett
3A Habersham
3A Hall
3A Hancock
3A Haralson
3A Harris
3A Hart
3A Heard
3A Henry
3A Houston*
3A Irwin*
3A Jackson
3A Jasper
2A Jeff Davis*
3A Jefferson
3A Jenkins*
3A Johnson*
3A Jones
3A Lamar
2A Lanier*
3A Laurens*
3A Lee*
2A Liberty*
3A Lincoln
2A Long*
2A Lowndes*
3A Lumpkin
3A Macon*
3A Madison
3A Marion*
3A McDuffie
2A McIntosh*
3A Meriwether
2A Miller*
2A Mitchell*
3A Monroe
3A Montgomery*
3A Morgan
3A Murray
3A Muscogee
3A Newton
3A Oconee
3A Oglethorpe
3A Paulding
3A Peach*
3A Pickens
2A Pierce*
3A Pike
3A Polk
3A Pulaski*
3A Putnam
3A Quitman*
3A Rabun
3A Randolph*
3A Richmond
3A Rockdale
3A Schley*
3A Screven*
2A Seminole*
3A Spalding
3A Stephens
3A Stewart*
3A Sumter*
3A Talbot
3A Taliaferro
2A Tattnall*
3A Taylor*
3A Telfair*
3A Terrell*
2A Thomas*
2A Tift*
2A Toombs*
3A Towns

(continued)

GENERAL REQUIREMENTS

TABLE C301.1—continued
CLIMATE ZONES, MOISTURE REGIMES, AND WARM HUMID DESIGNATIONS BY STATE, COUNTY AND TERRITORY[a]

US STATES—continued	
GEORGIA (continued)	6B Franklin
3A Treutlen*	6B Fremont
3A Troup	5B Gem
3A Turner*	5B Gooding
3A Twiggs*	5B Idaho
3A Union	6B Jefferson
3A Upson	5B Jerome
3A Walker	5B Kootenai
3A Walton	5B Latah
2A Ware*	6B Lemhi
3A Warren	5B Lewis
3A Washington	5B Lincoln
2A Wayne*	6B Madison
3A Webster*	5B Minidoka
3A Wheeler*	5B Nez Perce
3A White	6B Oneida
3A Whitfield	5B Owyhee
3A Wilcox*	5B Payette
3A Wilkes	5B Power
3A Wilkinson	5B Shoshone
2A Worth*	6B Teton
HAWAII	5B Twin Falls
1A (all)*	6B Valley
IDAHO	5B Washington
5B Ada	**ILLINOIS**
6B Adams	5A Adams
6B Bannock	4A Alexander
6B Bear Lake	4A Bond
5B Benewah	5A Boone
6B Bingham	5A Brown
6B Blaine	5A Bureau
6B Boise	4A Calhoun
6B Bonner	5A Carroll
6B Bonneville	5A Cass
6B Boundary	5A Champaign
6B Butte	4A Christian
6B Camas	4A Clark
5B Canyon	4A Clay
6B Caribou	4A Clinton
5B Cassia	4A Coles
6B Clark	5A Cook
5B Clearwater	4A Crawford
6B Custer	4A Cumberland
5B Elmore	5A DeKalb
	5A De Witt

(continued)

GENERAL REQUIREMENTS

TABLE C301.1—continued
CLIMATE ZONES, MOISTURE REGIMES, AND WARM HUMID DESIGNATIONS BY STATE, COUNTY AND TERRITORY[a]

US STATES—continued
ILLINOIS *(continued)*
5A Douglas
5A DuPage
5A Edgar
4A Edwards
4A Effingham
4A Fayette
5A Ford
4A Franklin
5A Fulton
4A Gallatin
4A Greene
5A Grundy
4A Hamilton
5A Hancock
4A Hardin
5A Henderson
5A Henry
5A Iroquois
4A Jackson
4A Jasper
4A Jefferson
4A Jersey
5A Jo Daviess
4A Johnson
5A Kane
5A Kankakee
5A Kendall
5A Knox
5A Lake
5A La Salle
4A Lawrence
5A Lee
5A Livingston
5A Logan
5A Macon
4A Macoupin
4A Madison
4A Marion
5A Marshall
5A Mason
4A Massac
5A McDonough
5A McHenry
5A McLean
5A Menard
5A Mercer
4A Monroe
4A Montgomery
5A Morgan
5A Moultrie
5A Ogle
5A Peoria
4A Perry
5A Piatt
5A Pike
4A Pope
4A Pulaski
5A Putnam
4A Randolph
4A Richland
5A Rock Island
4A Saline
5A Sangamon
5A Schuyler
5A Scott
4A Shelby
5A Stark
4A St. Clair
5A Stephenson
5A Tazewell
4A Union
5A Vermilion
4A Wabash
5A Warren
4A Washington
4A Wayne
4A White
5A Whiteside
5A Will
4A Williamson
5A Winnebago
5A Woodford
INDIANA
5A Adams
5A Allen
4A Bartholomew
5A Benton
5A Blackford
5A Boone
4A Brown

(continued)

TABLE C301.1—continued
CLIMATE ZONES, MOISTURE REGIMES, AND WARM HUMID DESIGNATIONS BY STATE, COUNTY AND TERRITORY[a]

US STATES—continued	
INDIANA *(continued)*	4A Martin
5A Carroll	5A Miami
5A Cass	4A Monroe
4A Clark	5A Montgomery
4A Clay	4A Morgan
5A Clinton	5A Newton
4A Crawford	5A Noble
4A Daviess	4A Ohio
4A Dearborn	4A Orange
4A Decatur	4A Owen
5A De Kalb	5A Parke
5A Delaware	4A Perry
4A Dubois	4A Pike
5A Elkhart	5A Porter
4A Fayette	4A Posey
4A Floyd	5A Pulaski
5A Fountain	4A Putnam
4A Franklin	5A Randolph
5A Fulton	4A Ripley
4A Gibson	4A Rush
5A Grant	4A Scott
4A Greene	4A Shelby
5A Hamilton	4A Spencer
5A Hancock	5A Starke
4A Harrison	5A Steuben
4A Hendricks	5A St. Joseph
5A Henry	4A Sullivan
5A Howard	4A Switzerland
5A Huntington	5A Tippecanoe
4A Jackson	5A Tipton
5A Jasper	4A Union
5A Jay	4A Vanderburgh
4A Jefferson	5A Vermillion
4A Jennings	4A Vigo
4A Johnson	5A Wabash
4A Knox	5A Warren
5A Kosciusko	4A Warrick
5A LaGrange	4A Washington
5A Lake	5A Wayne
5A LaPorte	5A Wells
4A Lawrence	5A White
5A Madison	5A Whitley
4A Marion	**IOWA**
5A Marshall	5A Adair
	5A Adams

(continued)

GENERAL REQUIREMENTS

TABLE C301.1—continued
CLIMATE ZONES, MOISTURE REGIMES, AND WARM HUMID DESIGNATIONS BY STATE, COUNTY AND TERRITORY[a]

US STATES—continued	
IOWA *(continued)*	5A Humboldt
5A Allamakee	5A Ida
5A Appanoose	5A Iowa
5A Audubon	5A Jackson
5A Benton	5A Jasper
6A Black Hawk	5A Jefferson
5A Boone	5A Johnson
5A Bremer	5A Jones
5A Buchanan	5A Keokuk
5A Buena Vista	6A Kossuth
5A Butler	5A Lee
5A Calhoun	5A Linn
5A Carroll	5A Louisa
5A Cass	5A Lucas
5A Cedar	6A Lyon
6A Cerro Gordo	5A Madison
5A Cherokee	5A Mahaska
5A Chickasaw	5A Marion
5A Clarke	5A Marshall
6A Clay	5A Mills
5A Clayton	6A Mitchell
5A Clinton	5A Monona
5A Crawford	5A Monroe
5A Dallas	5A Montgomery
5A Davis	5A Muscatine
5A Decatur	6A O'Brien
5A Delaware	6A Osceola
5A Des Moines	5A Page
6A Dickinson	6A Palo Alto
5A Dubuque	5A Plymouth
6A Emmet	5A Pocahontas
5A Fayette	5A Polk
5A Floyd	5A Pottawattamie
5A Franklin	5A Poweshiek
5A Fremont	5A Ringgold
5A Greene	5A Sac
5A Grundy	5A Scott
5A Guthrie	5A Shelby
5A Hamilton	6A Sioux
6A Hancock	5A Story
5A Hardin	5A Tama
5A Harrison	5A Taylor
5A Henry	5A Union
5A Howard	5A Van Buren
	5A Wapello

(continued)

TABLE C301.1—continued
CLIMATE ZONES, MOISTURE REGIMES, AND WARM HUMID DESIGNATIONS BY STATE, COUNTY AND TERRITORY[a]

US STATES—continued
IOWA *(continued)*
5A Warren
5A Washington
5A Wayne
5A Webster
6A Winnebago
5A Winneshiek
5A Woodbury
6A Worth
5A Wright
KANSAS
4A Allen
4A Anderson
4A Atchison
4A Barber
4A Barton
4A Bourbon
4A Brown
4A Butler
4A Chase
4A Chautauqua
4A Cherokee
5A Cheyenne
4A Clark
4A Clay
4A Cloud
4A Coffey
4A Comanche
4A Cowley
4A Crawford
5A Decatur
4A Dickinson
4A Doniphan
4A Douglas
4A Edwards
4A Elk
4A Ellis
4A Ellsworth
4A Finney
4A Ford
4A Franklin
4A Geary
5A Gove
4A Graham

4A Grant
4A Gray
5A Greeley
4A Greenwood
4A Hamilton
4A Harper
4A Harvey
4A Haskell
4A Hodgeman
4A Jackson
4A Jefferson
5A Jewell
4A Johnson
4A Kearny
4A Kingman
4A Kiowa
4A Labette
4A Lane
4A Leavenworth
4A Lincoln
4A Linn
5A Logan
4A Lyon
4A Marion
4A Marshall
4A McPherson
4A Meade
4A Miami
4A Mitchell
4A Montgomery
4A Morris
4A Morton
4A Nemaha
4A Neosho
4A Ness
5A Norton
4A Osage
4A Osborne
4A Ottawa
4A Pawnee
5A Phillips
4A Pottawatomie
4A Pratt
5A Rawlins
4A Reno

(continued)

GENERAL REQUIREMENTS

TABLE C301.1—continued
CLIMATE ZONES, MOISTURE REGIMES, AND WARM HUMID DESIGNATIONS BY STATE, COUNTY AND TERRITORY[a]

US STATES—continued
KANSAS (continued)
5A Republic
4A Rice
4A Riley
4A Rooks
4A Rush
4A Russell
4A Saline
5A Scott
4A Sedgwick
4A Seward
4A Shawnee
5A Sheridan
5A Sherman
5A Smith
4A Stafford
4A Stanton
4A Stevens
4A Sumner
5A Thomas
4A Trego
4A Wabaunsee
5A Wallace
4A Washington
5A Wichita
4A Wilson
4A Woodson
4A Wyandotte
KENTUCKY
4A (all)
LOUISIANA
2A Acadia*
2A Allen*
2A Ascension*
2A Assumption*
2A Avoyelles*
2A Beauregard*
3A Bienville*
3A Bossier*
3A Caddo*
2A Calcasieu*
3A Caldwell*
2A Cameron*
3A Catahoula*

3A Claiborne*
3A Concordia*
3A De Soto*
2A East Baton Rouge*
3A East Carroll
2A East Feliciana*
2A Evangeline*
3A Franklin*
3A Grant*
2A Iberia*
2A Iberville*
3A Jackson*
2A Jefferson*
2A Jefferson Davis*
2A Lafayette*
2A Lafourche*
3A La Salle*
3A Lincoln*
2A Livingston*
3A Madison*
3A Morehouse
3A Natchitoches*
2A Orleans*
3A Ouachita*
2A Plaquemines*
2A Pointe Coupee*
2A Rapides*
3A Red River*
3A Richland*
3A Sabine*
2A St. Bernard*
2A St. Charles*
2A St. Helena*
2A St. James*
2A St. John the Baptist*
2A St. Landry*
2A St. Martin*
2A St. Mary*
2A St. Tammany*
2A Tangipahoa*
3A Tensas*
2A Terrebonne*
3A Union*
2A Vermilion*
3A Vernon*

(continued)

TABLE C301.1—continued
CLIMATE ZONES, MOISTURE REGIMES, AND WARM HUMID DESIGNATIONS BY STATE, COUNTY AND TERRITORY[a]

US STATES—continued	
LOUISIANA *(continued)*	4A St. Mary's
2A Washington*	4A Talbot
3A Webster*	4A Washington
2A West Baton Rouge*	4A Wicomico
3A West Carroll	4A Worcester
2A West Feliciana*	**MASSACHUSETTS**
3A Winn*	5A (all)
MAINE	**MICHIGAN**
6A Androscoggin	6A Alcona
7 Aroostook	6A Alger
6A Cumberland	5A Allegan
6A Franklin	6A Alpena
6A Hancock	6A Antrim
6A Kennebec	6A Arenac
6A Knox	6A Baraga
6A Lincoln	5A Barry
6A Oxford	5A Bay
6A Penobscot	6A Benzie
6A Piscataquis	5A Berrien
6A Sagadahoc	5A Branch
6A Somerset	5A Calhoun
6A Waldo	5A Cass
6A Washington	6A Charlevoix
6A York	6A Cheboygan
MARYLAND	6A Chippewa
5A Allegany	6A Clare
4A Anne Arundel	5A Clinton
4A Baltimore	6A Crawford
4A Baltimore (city)	6A Delta
4A Calvert	6A Dickinson
4A Caroline	5A Eaton
4A Carroll	6A Emmet
4A Cecil	5A Genesee
4A Charles	6A Gladwin
4A Dorchester	6A Gogebic
4A Frederick	6A Grand Traverse
5A Garrett	5A Gratiot
4A Harford	5A Hillsdale
4A Howard	6A Houghton
4A Kent	5A Huron
4A Montgomery	5A Ingham
4A Prince George's	5A Ionia
4A Queen Anne's	6A Iosco
4A Somerset	6A Iron
	6A Isabella

(continued)

TABLE C301.1—continued
CLIMATE ZONES, MOISTURE REGIMES, AND WARM HUMID DESIGNATIONS BY STATE, COUNTY AND TERRITORY[a]

US STATES—continued	
MICHIGAN *(continued)*	5A Washtenaw
5A Jackson	5A Wayne
5A Kalamazoo	6A Wexford
6A Kalkaska	**MINNESOTA**
5A Kent	7 Aitkin
7 Keweenaw	6A Anoka
6A Lake	6A Becker
5A Lapeer	7 Beltrami
6A Leelanau	6A Benton
5A Lenawee	6A Big Stone
5A Livingston	6A Blue Earth
6A Luce	6A Brown
6A Mackinac	7 Carlton
5A Macomb	6A Carver
6A Manistee	7 Cass
7 Marquette	6A Chippewa
6A Mason	6A Chisago
6A Mecosta	6A Clay
6A Menominee	7 Clearwater
5A Midland	7 Cook
6A Missaukee	6A Cottonwood
5A Monroe	7 Crow Wing
5A Montcalm	6A Dakota
6A Montmorency	6A Dodge
5A Muskegon	6A Douglas
6A Newaygo	6A Faribault
5A Oakland	5A Fillmore
6A Oceana	6A Freeborn
6A Ogemaw	6A Goodhue
6A Ontonagon	6A Grant
6A Osceola	6A Hennepin
6A Oscoda	5A Houston
6A Otsego	7 Hubbard
5A Ottawa	6A Isanti
6A Presque Isle	7 Itasca
6A Roscommon	6A Jackson
5A Saginaw	6A Kanabec
5A Sanilac	6A Kandiyohi
6A Schoolcraft	7 Kittson
5A Shiawassee	7 Koochiching
5A St. Clair	6A Lac qui Parle
5A St. Joseph	7 Lake
5A Tuscola	7 Lake of the Woods
5A Van Buren	6A Le Sueur
	6A Lincoln

(continued)

GENERAL REQUIREMENTS

TABLE C301.1—continued
CLIMATE ZONES, MOISTURE REGIMES, AND WARM HUMID DESIGNATIONS BY STATE, COUNTY AND TERRITORY[a]

US STATES—continued	
MINNESOTA *(continued)*	5A Winona
6A Lyon	6A Wright
7 Mahnomen	6A Yellow Medicine
7 Marshall	**MISSISSIPPI**
6A Martin	3A Adams*
6A McLeod	3A Alcorn
6A Meeker	3A Amite*
6A Mille Lacs	3A Attala
6A Morrison	3A Benton
6A Mower	3A Bolivar
6A Murray	3A Calhoun
6A Nicollet	3A Carroll
6A Nobles	3A Chickasaw
7 Norman	3A Choctaw
6A Olmsted	3A Claiborne*
6A Otter Tail	3A Clarke
7 Pennington	3A Clay
7 Pine	3A Coahoma
6A Pipestone	3A Copiah*
7 Polk	3A Covington*
6A Pope	3A DeSoto
6A Ramsey	3A Forrest*
7 Red Lake	3A Franklin*
6A Redwood	2A George*
6A Renville	3A Greene*
6A Rice	3A Grenada
6A Rock	2A Hancock*
7 Roseau	2A Harrison*
6A Scott	3A Hinds*
6A Sherburne	3A Holmes
6A Sibley	3A Humphreys
6A Stearns	3A Issaquena
6A Steele	3A Itawamba
6A Stevens	2A Jackson*
7 St. Louis	3A Jasper
6A Swift	3A Jefferson*
6A Todd	3A Jefferson Davis*
6A Traverse	3A Jones*
6A Wabasha	3A Kemper
7 Wadena	3A Lafayette
6A Waseca	3A Lamar*
6A Washington	3A Lauderdale
6A Watonwan	3A Lawrence*
6A Wilkin	3A Leake
	3A Lee

(continued)

GENERAL REQUIREMENTS

TABLE C301.1—continued
CLIMATE ZONES, MOISTURE REGIMES, AND WARM HUMID DESIGNATIONS BY STATE, COUNTY AND TERRITORY[a]

US STATES—continued	
MISSISSIPPI *(continued)*	5A Andrew
3A Leflore	5A Atchison
3A Lincoln*	4A Audrain
3A Lowndes	4A Barry
3A Madison	4A Barton
3A Marion*	4A Bates
3A Marshall	4A Benton
3A Monroe	4A Bollinger
3A Montgomery	4A Boone
3A Neshoba	4A Buchanan
3A Newton	4A Butler
3A Noxubee	4A Caldwell
3A Oktibbeha	4A Callaway
3A Panola	4A Camden
2A Pearl River*	4A Cape Girardeau
3A Perry*	4A Carroll
3A Pike*	4A Carter
3A Pontotoc	4A Cass
3A Prentiss	4A Cedar
3A Quitman	4A Chariton
3A Rankin*	4A Christian
3A Scott	5A Clark
3A Sharkey	4A Clay
3A Simpson*	4A Clinton
3A Smith*	4A Cole
2A Stone*	4A Cooper
3A Sunflower	4A Crawford
3A Tallahatchie	4A Dade
3A Tate	4A Dallas
3A Tippah	5A Daviess
3A Tishomingo	5A DeKalb
3A Tunica	4A Dent
3A Union	4A Douglas
3A Walthall*	3A Dunklin
3A Warren*	4A Franklin
3A Washington	4A Gasconade
3A Wayne*	5A Gentry
3A Webster	4A Greene
3A Wilkinson*	5A Grundy
3A Winston	5A Harrison
3A Yalobusha	4A Henry
3A Yazoo	4A Hickory
MISSOURI	5A Holt
5A Adair	4A Howard
	4A Howell

(continued)

C3-18   2021 INTERNATIONAL ENERGY CONSERVATION CODE®

TABLE C301.1—continued
CLIMATE ZONES, MOISTURE REGIMES, AND WARM HUMID DESIGNATIONS BY STATE, COUNTY AND TERRITORY[a]

US STATES—continued	
MISSOURI (continued)	4A Reynolds
4A Iron	4A Ripley
4A Jackson	4A Saline
4A Jasper	5A Schuyler
4A Jefferson	5A Scotland
4A Johnson	4A Scott
5A Knox	4A Shannon
4A Laclede	5A Shelby
4A Lafayette	4A St. Charles
4A Lawrence	4A St. Clair
5A Lewis	4A St. Francois
4A Lincoln	4A St. Louis
5A Linn	4A St. Louis (city)
5A Livingston	4A Ste. Genevieve
5A Macon	4A Stoddard
4A Madison	4A Stone
4A Maries	5A Sullivan
5A Marion	4A Taney
4A McDonald	4A Texas
5A Mercer	4A Vernon
4A Miller	4A Warren
4A Mississippi	4A Washington
4A Moniteau	4A Wayne
4A Monroe	4A Webster
4A Montgomery	5A Worth
4A Morgan	4A Wright
4A New Madrid	**MONTANA**
4A Newton	6B (all)
5A Nodaway	**NEBRASKA**
4A Oregon	5A (all)
4A Osage	**NEVADA**
4A Ozark	4B Carson City (city)
3A Pemiscot	5B Churchill
4A Perry	3B Clark
4A Pettis	4B Douglas
4A Phelps	5B Elko
5A Pike	4B Esmeralda
4A Platte	5B Eureka
4A Polk	5B Humboldt
4A Pulaski	5B Lander
5A Putnam	4B Lincoln
5A Ralls	4B Lyon
4A Randolph	4B Mineral
4A Ray	4B Nye
	5B Pershing

(continued)

GENERAL REQUIREMENTS

TABLE C301.1—continued
CLIMATE ZONES, MOISTURE REGIMES, AND WARM HUMID DESIGNATIONS BY STATE, COUNTY AND TERRITORY[a]

US STATES—continued	
NEVADA *(continued)*	4B DeBaca
5B Storey	3B Doña Ana
5B Washoe	3B Eddy
5B White Pine	4B Grant
NEW HAMPSHIRE	4B Guadalupe
6A Belknap	5B Harding
6A Carroll	3B Hidalgo
5A Cheshire	3B Lea
6A Coos	4B Lincoln
6A Grafton	5B Los Alamos
5A Hillsborough	3B Luna
5A Merrimack	5B McKinley
5A Rockingham	5B Mora
5A Strafford	3B Otero
6A Sullivan	4B Quay
NEW JERSEY	5B Rio Arriba
4A Atlantic	4B Roosevelt
5A Bergen	5B Sandoval
4A Burlington	5B San Juan
4A Camden	5B San Miguel
4A Cape May	5B Santa Fe
4A Cumberland	3B Sierra
4A Essex	4B Socorro
4A Gloucester	5B Taos
4A Hudson	5B Torrance
5A Hunterdon	4B Union
4A Mercer	4B Valencia
4A Middlesex	**NEW YORK**
4A Monmouth	5A Albany
5A Morris	5A Allegany
4A Ocean	4A Bronx
5A Passaic	5A Broome
4A Salem	5A Cattaraugus
5A Somerset	5A Cayuga
5A Sussex	5A Chautauqua
4A Union	5A Chemung
5A Warren	6A Chenango
NEW MEXICO	6A Clinton
4B Bernalillo	5A Columbia
4B Catron	5A Cortland
3B Chaves	6A Delaware
4B Cibola	5A Dutchess
5B Colfax	5A Erie
4B Curry	6A Essex
	6A Franklin

(continued)

TABLE C301.1—continued
CLIMATE ZONES, MOISTURE REGIMES, AND WARM HUMID DESIGNATIONS BY STATE, COUNTY AND TERRITORY[a]

US STATES—continued	
NEW YORK *(continued)*	5A Wyoming
6A Fulton	5A Yates
5A Genesee	**NORTH CAROLINA**
5A Greene	3A Alamance
6A Hamilton	3A Alexander
6A Herkimer	5A Alleghany
6A Jefferson	3A Anson
4A Kings	5A Ashe
6A Lewis	5A Avery
5A Livingston	3A Beaufort
6A Madison	3A Bertie
5A Monroe	3A Bladen
6A Montgomery	3A Brunswick*
4A Nassau	4A Buncombe
4A New York	4A Burke
5A Niagara	3A Cabarrus
6A Oneida	4A Caldwell
5A Onondaga	3A Camden
5A Ontario	3A Carteret*
5A Orange	3A Caswell
5A Orleans	3A Catawba
5A Oswego	3A Chatham
6A Otsego	3A Cherokee
5A Putnam	3A Chowan
4A Queens	3A Clay
5A Rensselaer	3A Cleveland
4A Richmond	3A Columbus*
5A Rockland	3A Craven
5A Saratoga	3A Cumberland
5A Schenectady	3A Currituck
5A Schoharie	3A Dare
5A Schuyler	3A Davidson
5A Seneca	3A Davie
5A Steuben	3A Duplin
6A St. Lawrence	3A Durham
4A Suffolk	3A Edgecombe
6A Sullivan	3A Forsyth
5A Tioga	3A Franklin
5A Tompkins	3A Gaston
6A Ulster	3A Gates
6A Warren	4A Graham
5A Washington	3A Granville
5A Wayne	3A Greene
4A Westchester	3A Guilford
	3A Halifax

(continued)

GENERAL REQUIREMENTS

TABLE C301.1—continued
CLIMATE ZONES, MOISTURE REGIMES, AND WARM HUMID DESIGNATIONS BY STATE, COUNTY AND TERRITORY[a]

US STATES—continued	
NORTH CAROLINA (continued)	4A Surry
3A Harnett	4A Swain
4A Haywood	4A Transylvania
4A Henderson	3A Tyrrell
3A Hertford	3A Union
3A Hoke	3A Vance
3A Hyde	3A Wake
3A Iredell	3A Warren
4A Jackson	3A Washington
3A Johnston	5A Watauga
3A Jones	3A Wayne
3A Lee	3A Wilkes
3A Lenoir	3A Wilson
3A Lincoln	4A Yadkin
4A Macon	5A Yancey
4A Madison	**NORTH DAKOTA**
3A Martin	6A Adams
4A McDowell	6A Barnes
3A Mecklenburg	7 Benson
4A Mitchell	6A Billings
3A Montgomery	7 Bottineau
3A Moore	6A Bowman
3A Nash	7 Burke
3A New Hanover*	6A Burleigh
3A Northampton	6A Cass
3A Onslow*	7 Cavalier
3A Orange	6A Dickey
3A Pamlico	7 Divide
3A Pasquotank	6A Dunn
3A Pender*	6A Eddy
3A Perquimans	6A Emmons
3A Person	6A Foster
3A Pitt	6A Golden Valley
3A Polk	7 Grand Forks
3A Randolph	6A Grant
3A Richmond	6A Griggs
3A Robeson	6A Hettinger
3A Rockingham	6A Kidder
3A Rowan	6A LaMoure
3A Rutherford	6A Logan
3A Sampson	7 McHenry
3A Scotland	6A McIntosh
3A Stanly	6A McKenzie
4A Stokes	6A McLean
	6A Mercer

(continued)

TABLE C301.1—continued
CLIMATE ZONES, MOISTURE REGIMES, AND WARM HUMID DESIGNATIONS BY STATE, COUNTY AND TERRITORY[a]

US STATES—continued	
NORTH DAKOTA *(continued)*	5A Darke
6A Morton	5A Defiance
6A Mountrail	5A Delaware
7 Nelson	5A Erie
6A Oliver	5A Fairfield
7 Pembina	4A Fayette
7 Pierce	4A Franklin
7 Ramsey	5A Fulton
6A Ransom	4A Gallia
7 Renville	5A Geauga
6A Richland	4A Greene
7 Rolette	5A Guernsey
6A Sargent	4A Hamilton
6A Sheridan	5A Hancock
6A Sioux	5A Hardin
6A Slope	5A Harrison
6A Stark	5A Henry
6A Steele	4A Highland
6A Stutsman	4A Hocking
7 Towner	5A Holmes
6A Traill	5A Huron
7 Walsh	4A Jackson
7 Ward	5A Jefferson
6A Wells	5A Knox
6A Williams	5A Lake
OHIO	4A Lawrence
4A Adams	5A Licking
5A Allen	5A Logan
5A Ashland	5A Lorain
5A Ashtabula	5A Lucas
4A Athens	4A Madison
5A Auglaize	5A Mahoning
5A Belmont	5A Marion
4A Brown	5A Medina
4A Butler	4A Meigs
5A Carroll	5A Mercer
5A Champaign	5A Miami
5A Clark	5A Monroe
4A Clermont	5A Montgomery
4A Clinton	5A Morgan
5A Columbiana	5A Morrow
5A Coshocton	5A Muskingum
5A Crawford	5A Noble
5A Cuyahoga	5A Ottawa
	5A Paulding

(continued)

GENERAL REQUIREMENTS

TABLE C301.1—continued
CLIMATE ZONES, MOISTURE REGIMES, AND WARM HUMID DESIGNATIONS BY STATE, COUNTY AND TERRITORY[a]

US STATES—continued	
OHIO *(continued)*	4A Craig
5A Perry	3A Creek
4A Pickaway	3A Custer
4A Pike	4A Delaware
5A Portage	3A Dewey
5A Preble	4A Ellis
5A Putnam	4A Garfield
5A Richland	3A Garvin
4A Ross	3A Grady
5A Sandusky	4A Grant
4A Scioto	3A Greer
5A Seneca	3A Harmon
5A Shelby	4A Harper
5A Stark	3A Haskell
5A Summit	3A Hughes
5A Trumbull	3A Jackson
5A Tuscarawas	3A Jefferson
5A Union	3A Johnston
5A Van Wert	4A Kay
4A Vinton	3A Kingfisher
4A Warren	3A Kiowa
4A Washington	3A Latimer
5A Wayne	3A Le Flore
5A Williams	3A Lincoln
5A Wood	3A Logan
5A Wyandot	3A Love
OKLAHOMA	4A Major
3A Adair	3A Marshall
4A Alfalfa	3A Mayes
3A Atoka	3A McClain
4B Beaver	3A McCurtain
3A Beckham	3A McIntosh
3A Blaine	3A Murray
3A Bryan	3A Muskogee
3A Caddo	3A Noble
3A Canadian	4A Nowata
3A Carter	3A Okfuskee
3A Cherokee	3A Oklahoma
3A Choctaw	3A Okmulgee
4B Cimarron	4A Osage
3A Cleveland	4A Ottawa
3A Coal	3A Pawnee
3A Comanche	3A Payne
3A Cotton	3A Pittsburg
	3A Pontotoc

(continued)

GENERAL REQUIREMENTS

TABLE C301.1—continued
CLIMATE ZONES, MOISTURE REGIMES, AND WARM HUMID DESIGNATIONS BY STATE, COUNTY AND TERRITORY[a]

US STATES—continued	
OKLAHOMA *(continued)*	5B Sherman
3A Pottawatomie	4C Tillamook
3A Pushmataha	5B Umatilla
3A Roger Mills	5B Union
3A Rogers	5B Wallowa
3A Seminole	5B Wasco
3A Sequoyah	4C Washington
3A Stephens	5B Wheeler
4B Texas	4C Yamhill
3A Tillman	**PENNSYLVANIA**
3A Tulsa	4A Adams
3A Wagoner	5A Allegheny
4A Washington	5A Armstrong
3A Washita	5A Beaver
4A Woods	5A Bedford
4A Woodward	4A Berks
OREGON	5A Blair
5B Baker	5A Bradford
4C Benton	4A Bucks
4C Clackamas	5A Butler
4C Clatsop	5A Cambria
4C Columbia	5A Cameron
4C Coos	5A Carbon
5B Crook	5A Centre
4C Curry	4A Chester
5B Deschutes	5A Clarion
4C Douglas	5A Clearfield
5B Gilliam	5A Clinton
5B Grant	5A Columbia
5B Harney	5A Crawford
5B Hood River	4A Cumberland
4C Jackson	4A Dauphin
5B Jefferson	4A Delaware
4C Josephine	5A Elk
5B Klamath	5A Erie
5B Lake	5A Fayette
4C Lane	5A Forest
4C Lincoln	4A Franklin
4C Linn	5A Fulton
5B Malheur	5A Greene
4C Marion	5A Huntingdon
5B Morrow	5A Indiana
4C Multnomah	5A Jefferson
4C Polk	5A Juniata
	5A Lackawanna

(continued)

GENERAL REQUIREMENTS

TABLE C301.1—continued
CLIMATE ZONES, MOISTURE REGIMES, AND WARM HUMID DESIGNATIONS BY STATE, COUNTY AND TERRITORY[a]

US STATES—continued	
PENNSYLVANIA (continued)	3A Calhoun
4A Lancaster	3A Charleston*
5A Lawrence	3A Cherokee
4A Lebanon	3A Chester
5A Lehigh	3A Chesterfield
5A Luzerne	3A Clarendon
5A Lycoming	3A Colleton*
5A McKean	3A Darlington
5A Mercer	3A Dillon
5A Mifflin	3A Dorchester*
5A Monroe	3A Edgefield
4A Montgomery	3A Fairfield
5A Montour	3A Florence
5A Northampton	3A Georgetown*
5A Northumberland	3A Greenville
4A Perry	3A Greenwood
4A Philadelphia	3A Hampton*
5A Pike	3A Horry*
5A Potter	2A Jasper*
5A Schuylkill	3A Kershaw
5A Snyder	3A Lancaster
5A Somerset	3A Laurens
5A Sullivan	3A Lee
5A Susquehanna	3A Lexington
5A Tioga	3A Marion
5A Union	3A Marlboro
5A Venango	3A McCormick
5A Warren	3A Newberry
5A Washington	3A Oconee
5A Wayne	3A Orangeburg
5A Westmoreland	3A Pickens
5A Wyoming	3A Richland
4A York	3A Saluda
RHODE ISLAND	3A Spartanburg
5A (all)	3A Sumter
SOUTH CAROLINA	3A Union
3A Abbeville	3A Williamsburg
3A Aiken	3A York
3A Allendale*	**SOUTH DAKOTA**
3A Anderson	6A Aurora
3A Bamberg*	6A Beadle
3A Barnwell*	5A Bennett
2A Beaufort*	5A Bon Homme
3A Berkeley*	6A Brookings
	6A Brown

(continued)

GENERAL REQUIREMENTS

TABLE C301.1—continued
CLIMATE ZONES, MOISTURE REGIMES, AND WARM HUMID DESIGNATIONS BY STATE, COUNTY AND TERRITORY[a]

US STATES—continued	
SOUTH DAKOTA (continued)	6A Moody
5A Brule	6A Pennington
6A Buffalo	6A Perkins
6A Butte	6A Potter
6A Campbell	6A Roberts
5A Charles Mix	6A Sanborn
6A Clark	6A Shannon
5A Clay	6A Spink
6A Codington	5A Stanley
6A Corson	6A Sully
6A Custer	5A Todd
6A Davison	5A Tripp
6A Day	6A Turner
6A Deuel	5A Union
6A Dewey	6A Walworth
5A Douglas	5A Yankton
6A Edmunds	6A Ziebach
6A Fall River	**TENNESSEE**
6A Faulk	4A Anderson
6A Grant	3A Bedford
5A Gregory	4A Benton
5A Haakon	4A Bledsoe
6A Hamlin	4A Blount
6A Hand	4A Bradley
6A Hanson	4A Campbell
6A Harding	4A Cannon
6A Hughes	4A Carroll
5A Hutchinson	4A Carter
6A Hyde	4A Cheatham
5A Jackson	3A Chester
6A Jerauld	4A Claiborne
5A Jones	4A Clay
6A Kingsbury	4A Cocke
6A Lake	3A Coffee
6A Lawrence	3A Crockett
6A Lincoln	4A Cumberland
5A Lyman	3A Davidson
6A Marshall	3A Decatur
6A McCook	4A DeKalb
6A McPherson	4A Dickson
6A Meade	3A Dyer
5A Mellette	3A Fayette
6A Miner	4A Fentress
6A Minnehaha	3A Franklin
	3A Gibson

(continued)

2021 INTERNATIONAL ENERGY CONSERVATION CODE®

GENERAL REQUIREMENTS

TABLE C301.1—continued
CLIMATE ZONES, MOISTURE REGIMES, AND WARM HUMID DESIGNATIONS BY STATE, COUNTY AND TERRITORY[a]

US STATES—continued	
TENNESSEE *(continued)*	4A Putnam
3A Giles	4A Rhea
4A Grainger	4A Roane
4A Greene	4A Robertson
3A Grundy	3A Rutherford
4A Hamblen	4A Scott
3A Hamilton	4A Sequatchie
4A Hancock	4A Sevier
3A Hardeman	3A Shelby
3A Hardin	4A Smith
4A Hawkins	4A Stewart
3A Haywood	4A Sullivan
3A Henderson	4A Sumner
4A Henry	3A Tipton
3A Hickman	4A Trousdale
4A Houston	4A Unicoi
4A Humphreys	4A Union
4A Jackson	4A Van Buren
4A Jefferson	4A Warren
4A Johnson	4A Washington
4A Knox	3A Wayne
4A Lake	4A Weakley
3A Lauderdale	4A White
3A Lawrence	3A Williamson
3A Lewis	4A Wilson
3A Lincoln	**TEXAS**
4A Loudon	2A Anderson*
4A Macon	3B Andrews
3A Madison	2A Angelina*
3A Marion	2A Aransas*
3A Marshall	3A Archer
3A Maury	4B Armstrong
4A McMinn	2A Atascosa*
3A McNairy	2A Austin*
4A Meigs	4B Bailey
4A Monroe	2B Bandera
4A Montgomery	2A Bastrop*
3A Moore	3B Baylor
4A Morgan	2A Bee*
4A Obion	2A Bell*
4A Overton	2A Bexar*
3A Perry	3A Blanco*
4A Pickett	3B Borden
4A Polk	2A Bosque*
	3A Bowie*

(continued)

TABLE C301.1—continued
CLIMATE ZONES, MOISTURE REGIMES, AND WARM HUMID DESIGNATIONS BY STATE, COUNTY AND TERRITORY[a]

US STATES—continued
TEXAS *(continued)*
2A Brazoria*
2A Brazos*
3B Brewster
4B Briscoe
2A Brooks*
3A Brown*
2A Burleson*
3A Burnet*
2A Caldwell*
2A Calhoun*
3B Callahan
1A Cameron*
3A Camp*
4B Carson
3A Cass*
4B Castro
2A Chambers*
2A Cherokee*
3B Childress
3A Clay
4B Cochran
3B Coke
3B Coleman
3A Collin*
3B Collingsworth
2A Colorado*
2A Comal*
3A Comanche*
3B Concho
3A Cooke
2A Coryell*
3B Cottle
3B Crane
3B Crockett
3B Crosby
3B Culberson
4B Dallam
2A Dallas*
3B Dawson
4B Deaf Smith
3A Delta
3A Denton*
2A DeWitt*

3B Dickens
2B Dimmit
4B Donley
2A Duval*
3A Eastland
3B Ector
2B Edwards
2A Ellis*
3B El Paso
3A Erath*
2A Falls*
3A Fannin
2A Fayette*
3B Fisher
4B Floyd
3B Foard
2A Fort Bend*
3A Franklin*
2A Freestone*
2B Frio
3B Gaines
2A Galveston*
3B Garza
3A Gillespie*
3B Glasscock
2A Goliad*
2A Gonzales*
4B Gray
3A Grayson
3A Gregg*
2A Grimes*
2A Guadalupe*
4B Hale
3B Hall
3A Hamilton*
4B Hansford
3B Hardeman
2A Hardin*
2A Harris*
3A Harrison*
4B Hartley
3B Haskell
2A Hays*
3B Hemphill
3A Henderson*

(continued)

GENERAL REQUIREMENTS

TABLE C301.1—continued
CLIMATE ZONES, MOISTURE REGIMES, AND WARM HUMID DESIGNATIONS BY STATE, COUNTY AND TERRITORY[a]

US STATES—continued	
TEXAS *(continued)*	3B Loving
1A Hidalgo*	3B Lubbock
2A Hill*	3B Lynn
4B Hockley	2A Madison*
3A Hood*	3A Marion*
3A Hopkins*	3B Martin
2A Houston*	3B Mason
3B Howard	2A Matagorda*
3B Hudspeth	2B Maverick
3A Hunt*	3B McCulloch
4B Hutchinson	2A McLennan*
3B Irion	2A McMullen*
3A Jack	2B Medina
2A Jackson*	3B Menard
2A Jasper*	3B Midland
3B Jeff Davis	2A Milam*
2A Jefferson*	3A Mills*
2A Jim Hogg*	3B Mitchell
2A Jim Wells*	3A Montague
2A Johnson*	2A Montgomery*
3B Jones	4B Moore
2A Karnes*	3A Morris*
3A Kaufman*	3B Motley
3A Kendall*	3A Nacogdoches*
2A Kenedy*	2A Navarro*
3B Kent	2A Newton*
3B Kerr	3B Nolan
3B Kimble	2A Nueces*
3B King	4B Ochiltree
2B Kinney	4B Oldham
2A Kleberg*	2A Orange*
3B Knox	3A Palo Pinto*
3A Lamar*	3A Panola*
4B Lamb	3A Parker*
3A Lampasas*	4B Parmer
2B La Salle	3B Pecos
2A Lavaca*	2A Polk*
2A Lee*	4B Potter
2A Leon*	3B Presidio
2A Liberty*	3A Rains*
2A Limestone*	4B Randall
4B Lipscomb	3B Reagan
2A Live Oak*	2B Real
3A Llano*	3A Red River*
	3B Reeves

(continued)

GENERAL REQUIREMENTS

TABLE C301.1—continued
CLIMATE ZONES, MOISTURE REGIMES, AND WARM HUMID DESIGNATIONS BY STATE, COUNTY AND TERRITORY[a]

US STATES—continued	
TEXAS *(continued)*	2A Washington*
2A Refugio*	2B Webb
4B Roberts	2A Wharton*
2A Robertson*	3B Wheeler
3A Rockwall*	3A Wichita
3B Runnels	3B Wilbarger
3A Rusk*	1A Willacy*
3A Sabine*	2A Williamson*
3A San Augustine*	2A Wilson*
2A San Jacinto*	3B Winkler
2A San Patricio*	3A Wise
3A San Saba*	3A Wood*
3B Schleicher	4B Yoakum
3B Scurry	3A Young
3B Shackelford	2B Zapata
3A Shelby*	2B Zavala
4B Sherman	**UTAH**
3A Smith*	5B Beaver
3A Somervell*	5B Box Elder
2A Starr*	5B Cache
3A Stephens	5B Carbon
3B Sterling	6B Daggett
3B Stonewall	5B Davis
3B Sutton	6B Duchesne
4B Swisher	5B Emery
2A Tarrant*	5B Garfield
3B Taylor	5B Grand
3B Terrell	5B Iron
3B Terry	5B Juab
3B Throckmorton	5B Kane
3A Titus*	5B Millard
3B Tom Green	6B Morgan
2A Travis*	5B Piute
2A Trinity*	6B Rich
2A Tyler*	5B Salt Lake
3A Upshur*	5B San Juan
3B Upton	5B Sanpete
2B Uvalde	5B Sevier
2B Val Verde	6B Summit
3A Van Zandt*	5B Tooele
2A Victoria*	6B Uintah
2A Walker*	5B Utah
2A Waller*	6B Wasatch
3B Ward	3B Washington
	5B Wayne

(continued)

GENERAL REQUIREMENTS

TABLE C301.1—continued
CLIMATE ZONES, MOISTURE REGIMES, AND WARM HUMID DESIGNATIONS BY STATE, COUNTY AND TERRITORY[a]

US STATES—continued	
UTAH *(continued)*	4C Grays Harbor
5B Weber	5C Island
VERMONT	4C Jefferson
6A (all)	4C King
VIRGINIA	5C Kitsap
4A (all except as follows:)	5B Kittitas
5A Alleghany	5B Klickitat
5A Bath	4C Lewis
3A Brunswick	5B Lincoln
3A Chesapeake	4C Mason
5A Clifton Forge	5B Okanogan
5A Covington	4C Pacific
3A Emporia	6B Pend Oreille
3A Franklin	4C Pierce
3A Greensville	5C San Juan
3A Halifax	4C Skagit
3A Hampton	5B Skamania
5A Highland	4C Snohomish
3A Isle of Wight	5B Spokane
3A Mecklenburg	6B Stevens
3A Newport News	4C Thurston
3A Norfolk	4C Wahkiakum
3A Pittsylvania	5B Walla Walla
3A Portsmouth	4C Whatcom
3A South Boston	5B Whitman
3A Southampton	5B Yakima
3A Suffolk	**WEST VIRGINIA**
3A Surry	5A Barbour
3A Sussex	4A Berkeley
3A Virginia Beach	4A Boone
WASHINGTON	4A Braxton
5B Adams	5A Brooke
5B Asotin	4A Cabell
5B Benton	4A Calhoun
5B Chelan	4A Clay
5C Clallam	4A Doddridge
4C Clark	4A Fayette
5B Columbia	4A Gilmer
4C Cowlitz	5A Grant
5B Douglas	4A Greenbrier
6B Ferry	5A Hampshire
5B Franklin	5A Hancock
5B Garfield	5A Hardy
5B Grant	5A Harrison
	4A Jackson

(continued)

TABLE C301.1—continued
CLIMATE ZONES, MOISTURE REGIMES, AND WARM HUMID DESIGNATIONS BY STATE, COUNTY AND TERRITORY[a]

US STATES—continued	
WEST VIRGINIA *(continued)*	6A Buffalo
4A Jefferson	6A Burnett
4A Kanawha	5A Calumet
4A Lewis	6A Chippewa
4A Lincoln	6A Clark
4A Logan	5A Columbia
5A Marion	5A Crawford
5A Marshall	5A Dane
4A Mason	5A Dodge
4A McDowell	6A Door
4A Mercer	6A Douglas
5A Mineral	6A Dunn
4A Mingo	6A Eau Claire
5A Monongalia	6A Florence
4A Monroe	5A Fond du Lac
4A Morgan	6A Forest
4A Nicholas	5A Grant
5A Ohio	5A Green
5A Pendleton	5A Green Lake
4A Pleasants	5A Iowa
5A Pocahontas	6A Iron
5A Preston	6A Jackson
4A Putnam	5A Jefferson
4A Raleigh	5A Juneau
5A Randolph	5A Kenosha
4A Ritchie	6A Kewaunee
4A Roane	5A La Crosse
4A Summers	5A Lafayette
5A Taylor	6A Langlade
5A Tucker	6A Lincoln
4A Tyler	6A Manitowoc
4A Upshur	6A Marathon
4A Wayne	6A Marinette
4A Webster	6A Marquette
5A Wetzel	6A Menominee
4A Wirt	5A Milwaukee
4A Wood	5A Monroe
4A Wyoming	6A Oconto
WISCONSIN	6A Oneida
5A Adams	5A Outagamie
6A Ashland	5A Ozaukee
6A Barron	6A Pepin
6A Bayfield	6A Pierce
6A Brown	6A Polk
	6A Portage

(continued)

TABLE C301.1—continued
CLIMATE ZONES, MOISTURE REGIMES, AND WARM HUMID DESIGNATIONS BY STATE, COUNTY AND TERRITORY[a]

US STATES—continued
WISCONSIN *(continued)*
6A Price
5A Racine
5A Richland
5A Rock
6A Rusk
5A Sauk
6A Sawyer
6A Shawano
6A Sheboygan
6A St. Croix
6A Taylor
6A Trempealeau
5A Vernon
6A Vilas
5A Walworth
6A Washburn
5A Washington
5A Waukesha
6A Waupaca
5A Waushara
5A Winnebago
6A Wood
WYOMING
6B Albany
6B Big Horn
6B Campbell
6B Carbon
6B Converse
6B Crook

6B Fremont
5B Goshen
6B Hot Springs
6B Johnson
5B Laramie
7 Lincoln
6B Natrona
6B Niobrara
6B Park
5B Platte
6B Sheridan
7 Sublette
6B Sweetwater
7 Teton
6B Uinta
6B Washakie
6B Weston
US TERRITORIES
AMERICAN SAMOA
1A (all)*
GUAM
1A (all)*
NORTHERN MARIANA ISLANDS
1A (all)*
PUERTO RICO
1A (all except as follows:)*
2B Barraquitas
2B Cayey
VIRGIN ISLANDS
1A (all)*

a. Key: A – Moist, B – Dry, C – Marine. Absence of moisture designation indicates moisture regime is irrelevant. Asterisk (*) indicates a Warm Humid location.

C301.3 Climate zone definitions. To determine the climate zones for locations not listed in this code, use the following information to determine climate zone numbers and letters in accordance with Items 1 through 5.

1. Determine the thermal climate zone, 0 through 8, from Table C301.3 using the heating (HDD) and cooling degree-days (CDD) for the location.
2. Determine the moisture zone (Marine, Dry or Humid) in accordance with Items 2.1 through 2.3.
    2.1. If monthly average temperature and precipitation data are available, use the Marine, Dry and Humid definitions to determine the moisture zone (C, B or A).
    2.2. If annual average temperature information (including degree-days) and annual precipitation (i.e., annual mean) are available, use Items 2.2.1 through 2.2.3 to determine the moisture zone. If the moisture zone is not Marine, then use the Dry definition to determine whether Dry or Humid.
        2.2.1. If thermal climate zone is 3 and CDD50°F ≤ 4,500 (CDD10°C ≤ 2500), climate zone is Marine (3C).
        2.2.2. If thermal climate zone is 4 and CDD50°F ≤ 2,700 (CDD10°C ≤ 1500), climate zone is Marine (4C).
        2.2.3. If thermal climate zone is 5 and CDD50°F ≤ 1,800 (CDD10°C ≤ 1000), climate zone is Marine (5C).
    2.3. If only degree-day information is available, use Items 2.3.1 through 2.3.3 to determine the moisture zone. If the moisture zone is not Marine, then it is not possible to assign Humid or Dry moisture zone for this location.
        2.3.1. If thermal climate zone is 3 and CDD50°F ≤ 4,500 (CDD10°C ≤ 2500), climate zone is Marine (3C).
        2.3.2. If thermal climate zone is 4 and CDD50°F ≤ 2,700 (CDD10°C ≤ 1500), climate zone is Marine (4C).
        2.3.3. If thermal climate zone is 5 and CDD50°F ≤ 1,800 (CDD10°C ≤ 1000), climate zone is Marine (5C).
3. Marine (C) Zone definition: Locations meeting all the criteria in Items 3.1 through 3.4.
    3.1. Mean temperature of coldest month between 27°F (-3°C) and 65°F (18°C).
    3.2. Warmest month mean < 72°F (22°C).
    3.3. Not fewer than four months with mean temperatures over 50°F (10°C).
    3.4. Dry season in summer. The month with the heaviest precipitation in the cold season has at least three times as much precipitation as the month with the least precipitation in the rest of the year. The cold season is October through March in the Northern Hemisphere and April through September in the Southern Hemisphere.
4. Dry (B) definition: Locations meeting the criteria in Items 4.1 through 4.4.
    4.1. Not Marine (C).
    4.2. If 70 percent or more of the precipitation, P, occurs during the high sun period, defined as April through September in the Northern Hemisphere and October through March in the Southern Hemisphere, then the dry/humid threshold is in accordance with Equation 3-1.

    $P < 0.44 \times (T - 7)$
    $[P < 20.0 \times (T + 14)$ in SI units]
    (Equation 3-1)
    where:
    P = Annual precipitation, inches (mm).
    T = Annual mean temperature, °F (°C).

    4.3. If between 30 and 70 percent of the precipitation, P, occurs during the high sun period, defined as April through September in the Northern Hemisphere and October through March in the Southern Hemisphere, then the dry/humid threshold is in accordance with Equation 3-2.

    $P < 0.44 \times (T - 19.5)$
    $[P < 20.0 \times (T + 7)$ in SI units]
    (Equation 3-2)
    where:
    P = Annual precipitation, inches (mm).
    T = Annual mean temperature, °F (°C).

    4.4. If 30 percent or less of the precipitation, P, occurs during the high sun period, defined as April through September in the Northern Hemisphere and October through March in the Southern Hemisphere, then the dry/humid threshold is in accordance with Equation 3-3.

    $P < 0.44 \times (T - 32)$
    $[P < 20.0 \times T$ in SI units]
    (Equation 3-3)
    where:
    P = Annual precipitation, inches (mm).
    T = Annual mean temperature, °F (°C).

5. Humid (A) definition: Locations that are not Marine (C) or Dry (B).

GENERAL REQUIREMENTS

TABLE C301.3
THERMAL CLIMATE ZONE DEFINITIONS

ZONE NUMBER	THERMAL CRITERIA	
	IP Units	SI Units
0	10,800 < CDD50°F	6000 < CDD10°C
1	9,000 < CDD50°F ≤ 10,800	5000 < CDD10°C ≤ 6000
2	6,300 < CDD50°F ≤ 9,000	3500 < CDD10°C ≤ 5000
3	CDD50°F ≤ 6,300 AND HDD65°F ≤ 3,600	CDD10°C < 3500 AND HDD18°C ≤ 2000
4	CDD50°F ≤ 6,300 AND 3,600 < HDD65°F ≤ 5,400	CDD10°C < 3500 AND 2000 < HDD18°C ≤ 3000
5	CDD50°F < 6,300 AND 5,400 < HDD65°F ≤ 7,200	CDD10°C < 3500 AND 3000 < HDD18°C ≤ 4000
6	7,200 < HDD65°F ≤ 9,000	4000 < HDD18°C ≤ 5000
7	9,000 < HDD65°F ≤ 12,600	5000 < HDD18°C ≤ 7000
8	12,600 < HDD65°F	7000 < HDD18°C

For SI: °C = [(°F) − 32]/1.8.

C301.4 Tropical climate region. The tropical climate region shall be defined as:

1. Hawaii, Puerto Rico, Guam, American Samoa, US Virgin Islands, Commonwealth of Northern Mariana Islands; and
2. Islands in the area between the Tropic of Cancer and the Tropic of Capricorn.

SECTION C302
DESIGN CONDITIONS

C302.1 Interior design conditions. The interior design temperatures used for heating and cooling load calculations shall be a maximum of 72°F (22°C) for heating and minimum of 75°F (24°C) for cooling.

SECTION C303
MATERIALS, SYSTEMS AND EQUIPMENT

C303.1 Identification. Materials, systems and equipment shall be identified in a manner that will allow a determination of compliance with the applicable provisions of this code.

C303.1.1 Building thermal envelope insulation. An *R*-value identification mark shall be applied by the manufacturer to each piece of *building thermal envelope* insulation 12 inches (305 mm) or greater in width. Alternatively, the insulation installers shall provide a certification listing the type, manufacturer and *R*-value of insulation installed in each element of the *building thermal envelope*. For blown-in or sprayed fiberglass and cellulose insulation, the initial installed thickness, settled thickness, settled *R*-value, installed density, coverage area and number of bags installed shall be indicated on the certification. For sprayed polyurethane foam (SPF) insulation, the installed thickness of the areas covered and *R*-value of installed thickness shall be indicated on the certification. For insulated siding, the *R*-value shall be labeled on the product's package and shall be indicated on the certification. The insulation installer shall sign, date and post the certification in a conspicuous location on the job site.

Exception: For roof insulation installed above the deck, the *R*-value shall be labeled as required by the material standards specified in Table 1508.2 of the *International Building Code*.

C303.1.1.1 Blown-in or sprayed roof/ceiling insulation. The thickness of blown-in or sprayed fiberglass and cellulose roof/ceiling insulation shall be written in inches (mm) on markers and one or more of such markers shall be installed for every 300 square feet (28 m^2) of attic area throughout the attic space. The markers shall be affixed to the trusses or joists and marked with the minimum initial installed thickness with numbers not less than 1 inch (25 mm) in height. Each marker shall face the attic *access* opening. Spray polyurethane foam thickness and installed *R*-value shall be indicated on certification provided by the insulation installer.

C303.1.2 Insulation mark installation. Insulating materials shall be installed such that the manufacturer's *R*-value mark is readily observable upon inspection. For insulation materials that are installed without an observable manufacturer's *R*-value mark, such as blown or draped products, an insulation certificate complying with Section C303.1.1 shall be left immediately after installation by the installer, in a conspicuous location within the building, to certify the installed *R*-value of the insulation material.

C303.1.3 Fenestration product rating. *U*-factors of fenestration products shall be determined as follows:

1. For windows, doors and skylights, *U*-factor ratings shall be determined in accordance with NFRC 100.
2. Where required for garage doors and rolling doors, *U*-factor ratings shall be determined in accordance with either NFRC 100 or ANSI/DASMA 105.

U-factors shall be determined by an accredited, independent laboratory, and *labeled* and certified by the manufacturer.

Products lacking such a *labeled U*-factor shall be assigned a default *U*-factor from Table C303.1.3(1) or Table C303.1.3(2). The *solar heat gain coefficient* (SHGC) and *visible transmittance* (VT) of glazed fenestration products (windows, glazed doors and skylights) shall be determined in accordance with NFRC 200 by an accredited, independent laboratory, and *labeled* and certified by the manufacturer. Products lacking such a *labeled* SHGC or VT shall be assigned a default SHGC or VT from Table C303.1.3(3). For Tubular Daylighting Devices, VT_{annual} shall be measured and rated in accordance with NFRC 203.

TABLE C303.1.3(1)
DEFAULT GLAZED WINDOW, GLASS DOOR AND SKYLIGHT U-FACTORS

FRAME TYPE	WINDOW AND GLASS DOOR		SKYLIGHT	
	Single	Double	Single	Double
Metal	1.20	0.80	2.00	1.30
Metal with Thermal Break	1.10	0.65	1.90	1.10
Nonmetal or Metal Clad	0.95	0.55	1.75	1.05
Glazed Block	0.60			

TABLE C303.1.3(2)
DEFAULT OPAQUE DOOR U-FACTORS

DOOR TYPE	OPAQUE U-FACTOR
Uninsulated Metal	1.20
Insulated Metal (Rolling)	0.90
Insulated Metal (Other)	0.60
Wood	0.50
Insulated, nonmetal edge, max 45% glazing, any glazing double pane	0.35

TABLE C303.1.3(3)
DEFAULT GLAZED FENESTRATION SHGC AND VT

	SINGLE GLAZED		DOUBLE GLAZED		GLAZED BLOCK
	Clear	Tinted	Clear	Tinted	
SHGC	0.8	0.7	0.7	0.6	0.6
VT	0.6	0.3	0.6	0.3	0.6

C303.1.4 Insulation product rating. The thermal resistance (*R*-value) of insulation shall be determined in accordance with the US Federal Trade Commission *R*-value rule (CFR Title 16, Part 460) in units of h × ft² × °F/Btu at a mean temperature of 75°F (24°C).

C303.1.4.1 Insulated siding. The thermal resistance (*R*-value) of insulated siding shall be determined in accordance with ASTM C1363. Installation for testing shall be in accordance with the manufacturer's instructions.

C303.2 Installation. Materials, systems and equipment shall be installed in accordance with the manufacturer's instructions and the *International Building Code*.

C303.2.1 Protection of exposed foundation insulation. Insulation applied to the exterior of basement walls, crawl space walls and the perimeter of slab-on-grade floors shall have a rigid, opaque and weather-resistant protective covering to prevent the degradation of the insulation's thermal performance. The protective covering shall cover the exposed exterior insulation and extend not less than 6 inches (153 mm) below grade.

C303.2.2 Multiple layers of continuous insulation board. Where two or more layers of continuous insulation board are used in a construction assembly, the continuous insulation boards shall be installed in accordance with Section C303.2. Where the continuous insulation board manufacturer's instructions do not address installation of two or more layers, the edge joints between each layer of continuous insulation boards shall be staggered.

CHAPTER 4 [CE]
COMMERCIAL ENERGY EFFICIENCY

User note:

About this chapter: Chapter 4 presents the paths and options for compliance with the energy efficiency provisions. Chapter 4 contains energy efficiency provisions for the building envelope, mechanical and water heating systems, lighting and additional efficiency requirements. A performance alternative is also provided to allow for energy code compliance other than by the prescriptive method.

SECTION C401
GENERAL

C401.1 Scope. The provisions in this chapter are applicable to commercial *buildings* and their *building sites*.

C401.2 Application. Commercial buildings shall comply with Section C401.2.1 or C401.2.2.

C401.2.1 International Energy Conservation Code. Commercial buildings shall comply with one of the following:

1. Prescriptive Compliance. The Prescriptive Compliance option requires compliance with Sections C402 through C406 and Section C408. Dwelling units and sleeping units in Group R-2 buildings without systems serving multiple units shall be deemed to be in compliance with this chapter, provided that they comply with Section R406.
2. Total Building Performance. The Total Building Performance option requires compliance with Section C407.

Exception: Additions, alterations, repairs and changes of occupancy to existing buildings complying with Chapter 5.

C401.2.2 ASHRAE 90.1. Commercial buildings shall comply with the requirements of ANSI/ASHRAE/IESNA 90.1.

C401.3 Thermal envelope certificate. A permanent thermal envelope certificate shall be completed by an *approved* party. Such certificate shall be posted on a wall in the space where the space conditioning equipment is located, a utility room or other *approved* location. If located on an electrical panel, the certificate shall not cover or obstruct the visibility of the circuit directory label, service disconnect label or other required labels. A copy of the certificate shall also be included in the construction files for the project. The certificate shall include the following:

1. *R*-values of insulation installed in or on ceilings, roofs, walls, foundations and slabs, *basement walls*, crawl space walls and floors and ducts outside *conditioned spaces*.
2. *U*-factors and *solar heat gain coefficients* (SHGC) of fenestrations.
3. Results from any *building* envelope air leakage testing performed on the *building*.

Where there is more than one value for any component of the building envelope, the certificate shall indicate the area-weighted average value where available. If the area-weighted average is not available, the certificate shall list each value that applies to 10 percent or more of the total component area.

SECTION C402
BUILDING ENVELOPE REQUIREMENTS

C402.1 General. *Building thermal envelope* assemblies for buildings that are intended to comply with the code on a prescriptive basis in accordance with the compliance path described in Item 1 of Section C401.2.1 shall comply with the following:

1. The opaque portions of the *building thermal envelope* shall comply with the specific insulation requirements of Section C402.2 and the thermal requirements of either the *R*-value-based method of Section C402.1.3; the *U*-, *C*- and *F*-factor-based method of Section C402.1.4; or the component performance alternative of Section C402.1.5.
2. Roof solar reflectance and thermal emittance shall comply with Section C402.3.
3. Fenestration in building envelope assemblies shall comply with Section C402.4.
4. Air leakage of building envelope assemblies shall comply with Section C402.5.

Alternatively, where buildings have a vertical fenestration area or skylight area exceeding that allowed in Section C402.4, the building and *building thermal envelope* shall comply with Item 2 of Section C401.2.1 or Section C401.2.2.

Walk-in coolers, walk-in freezers, refrigerated warehouse coolers and refrigerated warehouse freezers shall comply with Section C403.11.

C402.1.1 Low-energy buildings and greenhouses. The following low-energy buildings, or portions thereof separated from the remainder of the building by *building thermal envelope* assemblies complying with this section, shall be exempt from the *building thermal envelope* provisions of Section C402.

1. Those with a peak design rate of energy usage less than 3.4 Btu/h × ft^2 (10.7 W/m^2) or 1.0 watt per

square foot (10.7 W/m²) of floor area for space conditioning purposes.

2. Those that do not contain *conditioned space*.

C402.1.1.1 Greenhouses. Greenhouse structures or areas that are mechanically heated or cooled and that comply with all of the following shall be exempt from the building envelope requirements of this code:

1. Exterior opaque envelope assemblies comply with Sections C402.2 and C402.4.5.

    Exception: Low energy greenhouses that comply with Section C402.1.1.

2. Interior partition *building thermal envelope* assemblies that separate the greenhouse from *conditioned space* comply with Sections C402.2, C402.4.3 and C402.4.5.

3. Fenestration assemblies that comply with the thermal envelope requirements in Table C402.1.1.1. The *U*-factor for a roof shall be for the roof assembly or a roof that includes the assembly and an *internal curtain system*.

    Exception: Unconditioned greenhouses.

TABLE C402.1.1.1
FENESTRATION THERMAL ENVELOPE MAXIMUM REQUIREMENTS

COMPONENT	*U*-FACTOR (BTU/h × ft² × °F)
Skylight	0.5
Vertical fenestration	0.7

C402.1.2 Equipment buildings. Buildings that comply with the following shall be exempt from the *building thermal envelope* provisions of this code:

1. Are separate buildings with floor area not more than 1,200 square feet (110 m²).

2. Are intended to house electric equipment with installed equipment power totaling not less than 7 watts per square foot (75 W/m²) and not intended for human occupancy.

3. Have a heating system capacity not greater than (17,000 Btu/hr) (5 kW) and a heating thermostat setpoint that is restricted to not more than 50°F (10°C).

4. Have an average wall and roof *U*-factor less than 0.200 in *Climate Zones* 1 through 5 and less than 0.120 in *Climate Zones* 6 through 8.

5. Comply with the roof solar reflectance and thermal emittance provisions for *Climate Zone* 1.

C402.1.3 Insulation component *R*-value-based method. *Building thermal envelope* opaque assemblies shall comply with the requirements of Sections C402.2 and C402.4 based on the *climate zone* specified in Chapter 3. For opaque portions of the *building thermal envelope* intended to comply on an insulation component

R-value basis, the *R*-values for cavity insulation and continuous insulation shall be not less than that specified in Table C402.1.3. Where cavity insulation is installed in multiple layers, the cavity insulation *R*-values shall be summed to determine compliance with the cavity insulation *R*-value requirements. Where continuous insulation is installed in multiple layers, the continuous insulation *R*-values shall be summed to determine compliance with the continuous insulation *R*-value requirements. Cavity insulation *R*-values shall not be used to determine compliance with the continuous insulation *R*-value requirements in Table C402.1.3. Commercial buildings or portions of commercial buildings enclosing Group R occupancies shall use the *R*-values from the "*Group R*" column of Table C402.1.3. Commercial buildings or portions of commercial buildings enclosing occupancies other than *Group R* shall use the *R*-values from the "All other" column of Table C402.1.3.

C402.1.4 Assembly *U*-factor, *C*-factor or *F*-factor-based method. *Building thermal envelope* opaque assemblies shall meet the requirements of Sections C402.2 and C402.4 based on the climate zone specified in Chapter 3. *Building thermal envelope* opaque assemblies intended to comply on an assembly *U*-, *C*- or *F*-factor basis shall have a *U*-, *C*- or *F*-factor not greater than that specified in Table C402.1.4. Commercial buildings or portions of commercial buildings enclosing Group R occupancies shall use the *U*-, *C*- or *F*-factor from the "*Group R*" column of Table C402.1.4. Commercial buildings or portions of commercial buildings enclosing occupancies other than *Group R* shall use the *U*-, *C*- or *F*-factor from the "All other" column of Table C402.1.4

C402.1.4.1 Roof/ceiling assembly. The maximum roof/ceiling assembly *U*-factor shall not exceed that specified in Table C402.1.4 based on construction materials used in the roof/ceiling assembly.

C402.1.4.1.1 Tapered, above-deck insulation based on thickness. Where used as a component of a maximum roof/ceiling assembly *U*-factor calculation, the sloped roof insulation *R*-value contribution to that calculation shall use the average thickness in inches (mm) along with the material *R*-value-per-inch (per-mm) solely for *U*-factor compliance as prescribed in Section C402.1.4.

C402.1.4.1.2 Suspended ceilings. Insulation installed on suspended ceilings having removable ceiling tiles shall not be considered part of the assembly *U*-factor of the roof/ceiling construction.

C402.1.4.1.3 Joints staggered. Continuous insulation board shall be installed in not less than two layers, and the edge joints between each layer of insulation shall be staggered, except where insulation tapers to the roof deck at a gutter edge, roof drain or scupper.

COMMERCIAL ENERGY EFFICIENCY

TABLE C402.1.3
OPAQUE THERMAL ENVELOPE INSULATION COMPONENT MINIMUM REQUIREMENTS, R-VALUE METHOD[a]

CLIMATE ZONE	0 AND 1		2		3		4 EXCEPT MARINE		5 AND MARINE 4		6		7		8	
	All other	Group R	All other	Group R	All other	Group R	All other	Group R	All other	Group R	All other	Group R	All other	Group R	All other	Group R
Roofs																
Insulation entirely above roof deck	R-20ci	R-25ci	R-25ci	R-25ci	R-25ci	R-25ci	R-30ci	R-30ci	R-30ci	R-30ci	R-30ci	R-30ci	R-35ci	R-35ci	R-35ci	R-35ci
Metal buildings[b]	R-19 + R-11 LS	R-19 + R-11 LS	R-19 + R11 LS	R-19 + R-11 LS	R-19 + R-11 LS	R-19 + R-11 LS	R-19 + R-11 LS	R-19 + R-11 LS	R-19 + R-11 LS	R-19 + R-11 LS	R-25 + R-11 LS	R-30 + R-11 LS	R-30 + R-11 LS	R-30 + R-11 LS	R-25 + R-11 + R-11 LS	R-25 + R-11 + R-11 LS
Attic and other	R-38	R-38	R-38	R-38	R-38	R-38	R-49	R-49	R-49	R-49	R-49	R-49	R-60	R-60	R-60	R-60
Walls, above grade																
Mass[f]	R-5.7ci[c]	R-5.7ci[c]	R-5.7ci[f]	R-5.7ci[f]	R-7.6ci	R-7.6ci	R-9.5ci	R-11.4ci	R-11.4ci	R-13.3ci	R-13.3ci	R-15.2ci	R-15.2ci	R-15.2ci	R-25ci	R-25ci
Metal building	R-13 + R-6.5ci	R-13 + R-6.5ci	R-13 + R-6.5ci	R-13 + R-13ci	R-13 + R-6.5ci	R-13 + R-13ci	R-13 + R-13ci	R-13 + R-14ci	R-13 + R-14ci	R-13 + R-14ci	R-13 + R-14ci	R-13 + R-14ci	R-13 + R-17ci	R-13 + R-19.5ci	R-13 + R-19.5ci	R-13 + R-19.5ci
Metal framed	R-13 + R-5ci	R-13 + R-5ci	R-13 + R-5ci	R-13 + R-7.5ci	R-13 + R-7.5ci	R-13 + R-7.5ci	R-13 + R-7.5ci	R-13 + R-10ci	R-13 + R-10ci	R-13 + R-10ci	R-13 + R-12.5ci	R-13 + R-12.5ci	R-13 + R-12.5ci	R-13 + R-15.6ci	R-13 + R-18.8ci	R-13 + R-18.8ci
Wood framed and other	R-13 + R-3.8ci or R-20	R-13 + R-3.8ci or R-20	R-13 + R-3.8ci or R-20	R-13 + R-3.8ci or R-20	R-13 + R-3.8ci or R-20	R-13 + R-3.8ci or R-20	R-13 + R-3.8ci or R-20	R-13 + R-7.5ci or R-20 + R-3.8ci	R-13 + R-7.5ci or R20 + R3.8ci	R-13 + R-7.5ci or R-20 + R-3.8ci	R-13 + R-7.5ci or R-20 + R-3.8ci	R-13 + R-7.5ci or R-20 + R-3.8ci	R-13 + R-7.5ci or R-20 + R-3.8ci	R-13 + R-7.5ci or R-20 + R-3.8ci	R-13 + R-18.8ci	R-13 + R-18.8ci
Walls, below grade																
Below-grade wall[d]	NR	NR	NR	NR	NR	NR	R-7.5ci	R-10ci	R-10ci	R-10ci	R-10ci	R-15ci	R-15ci	R-15ci	R-15ci	R-15ci
Floors																
Mass[e]	NR	NR	R-6.3ci	R-8.3ci	R-10ci	R-10ci	R-14.6ci	R-16.7ci	R-16.7ci	R-16.7ci	R-16.7ci	R-16.7ci	R-20.9ci	R-20.9ci	R-23ci	R-23ci
Joist/framing	R-13	R-13	R-30	R-30	R-30	R-30	R-30	R-30	R-30	R-30	R-38	R-38	R-38	R-38	R-38	R-38
Slab-on-grade floors																
Unheated slabs	NR	NR	NR	NR	NR	NR	R-15 for 24" below	R-15 for 24" below	R-15 for 24" below	R-20 for 24" below	R-20 for 24" below	R-20 for 24" below	R-20 for 24" below	R-20 for 48" below	R-20 for 48" below	R-20 for 48" below
Heated slabs[g]	R-7.5 for 12" below+ R-5 full slab	R-7.5 for 12" below+ R-5 full slab	R-7.5 for 12" below+ R-5 full slab	R-7.5 for 12" below+ R-5 full slab	R-10 for 24" below+ R-5 full slab	R-10 for 24" below+ R-5 full slab	R-15 for 24" below+ R-5 full slab	R-15 for 24" below+ R-5 full slab	R-15 for 36" below+ R-5 full slab	R-15 for 36" below+ R-5 full slab	R-15 for 36" below+ R-5 full slab	R-20 for 48" below+ R-5 full slab	R-20 for 48" below+ R-5 full slab	R-20 for 48" below+ R-5 full slab	R-25 for 48" below+ R-5 full slab	R-20 for 48" below+ R-5 full slab

For SI: 1 inch = 25.4 mm, 1 pound per square foot = 4.88 kg/m^2, 1 pound per cubic foot = 16 kg/m^3.

ci = Continuous Insulation, NR = No Requirement, LS = Liner System.

a. Assembly descriptions can be found in ANSI/ASHRAE/IESNA 90.1 Appendix A.
b. Where using R-value compliance method, a thermal spacer block shall be provided, otherwise use the U-factor compliance method in Table C402.1.4.
c. R-5.7ci is allowed to be substituted with concrete block walls complying with ASTM C90, ungrouted or partially grouted at 32 inches or less on center vertically and 48 inches or less on center horizontally, with ungrouted cores filled with materials having a maximum thermal conductivity of 0.44 Btu-in/h-ft^2·°F.
d. Where heated slabs are below grade, below-grade walls shall comply with the exterior insulation requirements for heated slabs.
e. "Mass floors" shall be in accordance with Section C402.2.3.
f. "Mass walls" shall be in accordance with Section C402.2.2.
g. The first value is for perimeter insulation and the second value is for full, under-slab insulation. Perimeter insulation is not required to extend below the bottom of the slab.

TABLE C402.1.4
OPAQUE THERMAL ENVELOPE ASSEMBLY MAXIMUM REQUIREMENTS, U-FACTOR METHOD[a, b]

CLIMATE ZONE	0 AND 1		2		3		4 EXCEPT MARINE		5 AND MARINE 4		6		7		8	
	All other	Group R	All other	Group R	All other	Group R	All other	Group R	All other	Group R	All other	Group R	All other	Group R	All other	Group R
Roofs																
Insulation entirely above roof deck	U-0.048	U-0.039	U-0.039	U-0.039	U-0.039	U-0.039	U-0.032	U-0.032	U-0.032	U-0.032	U-0.032	U-0.032	U-0.028	U-0.028	U-0.028	U-0.028
Metal buildings	U-0.035	U-0.035	U-0.035	U-0.035	U-0.035	U-0.035	U-0.035	U-0.035	U-0.035	U-0.035	U-0.031	U-0.029	U-0.029	U-0.029	U-0.026	U-0.026
Attic and other	U-0.027	U-0.027	U-0.027	U-0.027	U-0.027	U-0.027	U-0.021	U-0.021	U-0.021	U-0.021	U-0.021	U-0.021	U-0.017	U-0.017	U-0.017	U-0.017
Walls, above grade																
Mass[f]	U-0.151	U-0.151	U-0.151	U-0.123	U-0.123	U-0.104	U-0.104	U-0.090	U-0.090	U-0.080	U-0.080	U-0.071	U-0.071	U-0.071	U-0.037	U-0.037
Metal building	U-0.079	U-0.079	U-0.079	U-0.079	U-0.079	U-0.052	U-0.052	U-0.050	U-0.050	U-0.050	U-0.050	U-0.050	U-0.044	U-0.039	U-0.039	U-0.039
Metal framed	U-0.077	U-0.077	U-0.077	U-0.064	U-0.064	U-0.064	U-0.064	U-0.055	U-0.055	U-0.049	U-0.049	U-0.049	U-0.049	U-0.042	U-0.037	U-0.037
Wood framed and other[c]	U-0.064	U-0.064	U-0.064	U-0.064	U-0.064	U-0.064	U-0.064	U-0.051	U-0.051	U-0.051	U-0.051	U-0.051	U-0.051	U-0.051	U-0.032	U-0.032
Walls, below grade																
Below-grade wall[c]	C-1.140[e]	C-1.140[e]	C-1.140[e]	C-1.140[e]	C-1.140[e]	C-1.140[e]	C-0.119	C-0.119	C-0.092	C-0.092	C-0.092	C-0.063	C-0.063	C-0.063	C-0.063	C-0.063
Floors																
Mass[d]	U-0.322[e]	U-0.322[e]	U-0.107	U-0.087	U-0.074	U-0.074	U-0.057	U-0.057	U-0.051	U-0.051	U-0.051	U-0.051	U-0.042	U-0.042	U-0.038	U-0.038
Joist/framing	U-0.066[e]	U-0.066[e]	U-0.033	U-0.033	U-0.033	U-0.033	U-0.033	U-0.033	U-0.033	U-0.033	U-0.027	U-0.027	U-0.027	U-0.027	U-0.027	U-0.027
Slab-on-grade floors																
Unheated slabs	F-0.73[e]	F-0.73[e]	F-0.73[e]	F-0.73[e]	F-0.73[e]	F-0.54	F-0.52	F-0.52	F-0.52	F-0.52	F-0.51	F-0.434	F-0.51	F-0.434	F-0.424	F-0.424
Heated slabs	F-0.69	F-0.69	F-0.69	F-0.69	F-0.66	F-0.66	F-0.62	F-0.62	F-0.62	F-0.62	F-0.62	F-0.602	F-0.602	F-0.602	F-0.602	F-0.602
Opaque doors																
Nonswinging door	U-0.31	U-0.31	U-0.31	U-0.31	U-0.31	U-0.31	U-0.31	U-0.31	U-0.31	U-0.31	U-0.31	U-0.31	U-0.31	U-0.31	U-0.31	U-0.31
Swinging door[g]	U-0.37	U-0.37	U-0.37	U-0.37	U-0.37	U-0.37	U-0.37	U-0.37	U-0.37	U-0.37	U-0.37	U-0.37	U-0.37	U-0.37	U-0.37	U-0.37
Garage door < 14% glazing[h]	U-0.31	U-0.31	U-0.31	U-0.31	U-0.31	U-0.31	U-0.31	U-0.31	U-0.31	U-0.31	U-0.31	U-0.31	U-0.31	U-0.31	U-0.31	U-0.31

For SI: 1 pound per square foot = 4.88 kg/m², 1 pound per cubic foot = 16 kg/m³.
ci = Continuous Insulation, NR = No Requirement, LS = Liner System.

a. Where assembly U-factors, C-factors and F-factors are established in ANSI/ASHRAE/IESNA 90.1 Appendix A, such opaque assemblies shall be a compliance alternative where those values meet the criteria of this table, and provided that the construction, excluding the cladding system on walls, complies with the appropriate construction details from ANSI/ASHRAE/ISNEA 90.1 Appendix A.
b. Where U-factors have been established by testing in accordance with ASTM C1363, such opaque assemblies shall be a compliance alternative where those values meet the criteria of this table. The R-value of continuous insulation shall be permitted to be added to or subtracted from the original tested design.
c. Where heated slabs are below grade, below-grade walls shall comply with the U-factor requirements for above-grade mass walls.
d. "Mass floors" shall be in accordance with Section C402.2.3.
e. These C-, F- and U-factors are based on assemblies that are not required to contain insulation.
f. "Mass walls" shall be in accordance with Section C402.2.2.
g. Swinging door U-factors shall be determined in accordance with NFRC-100.
h. Garage doors having a single row of fenestration shall have an assembly U-factor less than or equal to 0.44 in Climate Zones 0 through 6 and less than or equal to 0.36 in Climate Zones 7 and 8, provided that the fenestration area is not less than 14 percent and not more than 25 percent of the total door area.

C402.1.4.2 Thermal resistance of cold-formed steel walls. *U*-factors of walls with cold-formed steel studs shall be permitted to be determined in accordance with Equation 4-1.

$$U = 1/[R_s + (ER)] \quad \text{(Equation 4-1)}$$

where:

R_s = The cumulative *R*-value of the wall components along the path of heat transfer, excluding the *cavity insulation* and steel studs.

ER = The effective *R*-value of the *cavity insulation* with steel studs as specified in Table C402.1.4.2.

TABLE C402.1.4.2
EFFECTIVE *R*-VALUES FOR STEEL STUD WALL ASSEMBLIES

NOMINAL STUD DEPTH (inches)	SPACING OF FRAMING (inches)	CAVITY *R*-VALUE (insulation)	CORRECTION FACTOR (F_c)	EFFECTIVE *R*-VALUE (ER) (Cavity *R*-Value × F_c)
3½	16	13	0.46	5.98
		15	0.43	6.45
3½	24	13	0.55	7.15
		15	0.52	7.80
6	16	19	0.37	7.03
		21	0.35	7.35
6	24	19	0.45	8.55
		21	0.43	9.03
8	16	25	0.31	7.75
	24	25	0.38	9.50

For SI: 1 inch = 25.4 mm.

C402.1.5 Component performance alternative. Building envelope values and fenestration areas determined in accordance with Equation 4-2 shall be an alternative to compliance with the *U*-, *F*- and *C*-factors in Tables C402.1.4 and C402.4 and the maximum allowable fenestration areas in Section C402.4.1. *Fenestration* shall meet the applicable SHGC requirements of Section C402.4.3.

$$A + B + C + D + E \leq \text{Zero} \quad \text{(Equation 4-2)}$$

where:

A = Sum of the (UA Dif) values for each distinct assembly type of the *building thermal envelope*, other than slabs on grade and below-grade walls.

UA Dif = UA Proposed − UA Table.

UA Proposed = Proposed *U*-value × Area.

UA Table = (*U*-factor from Table C402.1.3, C402.1.4 or C402.4) × Area.

B = Sum of the (FL Dif) values for each distinct slab-on-grade perimeter condition of the *building thermal envelope*.

FL Dif = FL Proposed − FL Table.

FL Proposed = Proposed *F*-value × Perimeter length.

FL Table = (*F*-factor specified in Table C402.1.4) × Perimeter length.

C = Sum of the (CA Dif) values for each distinct *below-grade wall* assembly type of the *building thermal envelope*.

CA Dif = CA Proposed − CA Table.

CA Proposed = Proposed *C*-value × Area.

CA Table = (Maximum allowable *C*-factor specified in Table C402.1.4) × Area.

Where the proposed vertical glazing area is less than or equal to the maximum vertical glazing area allowed by Section C402.4.1, the value of D (Excess Vertical Glazing Value) shall be zero. Otherwise:

D = (DA × UV) − (DA × U Wall), but not less than zero.

DA = (Proposed Vertical Glazing Area) − (Vertical Glazing Area allowed by Section C402.4.1).

UA Wall = Sum of the (UA Proposed) values for each opaque assembly of the exterior wall.

U Wall = Area-weighted average *U*-value of all above-grade wall assemblies.

UAV = Sum of the (UA Proposed) values for each vertical glazing assembly.

UV = UAV/total vertical glazing area.

Where the proposed skylight area is less than or equal to the skylight area allowed by Section C402.4.1, the value of E (Excess Skylight Value) shall be zero. Otherwise:

E = (EA × US) − (EA × U Roof), but not less than zero.

EA = (Proposed Skylight Area) − (Allowable Skylight Area as specified in Section C402.4.1).

U Roof = Area-weighted average *U*-value of all roof assemblies.

UAS = Sum of the (UA Proposed) values for each skylight assembly.

US = UAS/total skylight area.

C402.2 Specific building thermal envelope insulation requirements. Insulation in *building thermal envelope* opaque assemblies shall comply with Sections C402.2.1 through C402.2.7 and Table C402.1.3.

C402.2.1 Roof assembly. The minimum thermal resistance (*R*-value) of the insulating material installed either between the roof framing or continuously on the roof assembly shall be as specified in Table C402.1.3, based on construction materials used in the roof assembly.

C402.2.1.1 Tapered, above-deck insulation based on thickness. Where used as a component of a roof/ceiling assembly *R*-value calculation, the sloped roof insulation *R*-value contribution to that calculation shall use the average thickness in inches (mm) along with the material *R*-value-per-inch (per-mm) solely for *R*-value compliance as prescribed in Section 402.1.3.

C402.2.1.2 Minimum thickness, lowest point. The minimum thickness of above-deck roof insulation at its lowest point, gutter edge, roof drain or scupper, shall be not less than 1 inch (25 mm).

C402.2.1.3 Suspended ceilings. Insulation installed on suspended ceilings having removable ceiling tiles shall not be considered part of the minimum thermal resistance (*R*-value) of roof insulation in roof/ceiling construction.

C402.2.1.4 Joints staggered. Continuous insulation board shall be installed in not less than two layers and the edge joints between each layer of insulation shall be staggered, except where insulation tapers to the roof deck at a gutter edge, roof drain or scupper.

C402.2.1.5 Skylight curbs. Skylight curbs shall be insulated to the level of roofs with insulation entirely above the deck or R-5, whichever is less.

> **Exception:** Unit skylight curbs included as a component of a skylight listed and labeled in accordance with NFRC 100 shall not be required to be insulated.

C402.2.2 Above-grade walls. The minimum thermal resistance (*R*-value) of materials installed in the wall cavity between framing members and continuously on the walls shall be as specified in Table C402.1.3, based on framing type and construction materials used in the wall assembly. The *R*-value of integral insulation installed in concrete masonry units shall not be used in determining compliance with Table C402.1.3 except as otherwise noted in the table. In determining compliance with Table C402.1.4, the use of the *U*-factor of concrete masonry units with integral insulation shall be permitted.

"Mass walls" where used as a component in the thermal envelope of a building shall comply with one of the following:

1. Weigh not less than 35 pounds per square foot (171 kg/m^2) of wall surface area.
2. Weigh not less than 25 pounds per square foot (122 kg/m^2) of wall surface area where the material weight is not more than 120 pcf (1900 kg/m^3).
3. Have a heat capacity exceeding 7 Btu/ft^2 × °F (144 kJ/m^2 × K).
4. Have a heat capacity exceeding 5 Btu/ft^2 × °F (103 kJ/m^2 × K), where the material weight is not more than 120 pcf (1900 kg/m^3).

C402.2.3 Floors. The thermal properties (component *R*-values or assembly *U*-, *C*- or *F*-factors) of floor assemblies over outdoor air or unconditioned space shall be as specified in Table C402.1.3 or C402.1.4 based on the construction materials used in the floor assembly. Floor framing *cavity insulation* or structural slab insulation shall be installed to maintain permanent contact with the underside of the subfloor decking or structural slabs.

"Mass floors" where used as a component of the thermal envelope of a building shall provide one of the following weights:

1. 35 pounds per square foot (171 kg/m^2) of floor surface area.
2. 25 pounds per square foot (122 kg/m^2) of floor surface area where the material weight is not more than 120 pounds per cubic foot (1923 kg/m^3).

Exceptions:

1. The floor framing *cavity insulation* or structural slab insulation shall be permitted to be in contact with the top side of sheathing or continuous insulation installed on the bottom side of floor assemblies where combined with insulation that meets or exceeds the minimum *R*-value in Table C402.1.3 for "Metal framed" or "Wood framed and other" values for "Walls, above grade" and extends from the bottom to the top of all perimeter floor framing or floor assembly members.
2. Insulation applied to the underside of concrete floor slabs shall be permitted an airspace of not more than 1 inch (25 mm) where it turns up and is in contact with the underside of the floor under walls associated with the *building thermal envelope*.

C402.2.4 Slabs-on-grade. The minimum thermal resistance (*R*-value) of the insulation for unheated or heated slab-on-grade floors designed in accordance with the *R*-value method of Section C402.1.3 shall be as specified in Table C402.1.3.

C402.2.4.1 Insulation installation. Where installed, the perimeter insulation shall be placed on the outside of the foundation or on the inside of the foundation wall. The perimeter insulation shall extend downward from the top of the slab for the minimum distance shown in the table or to the top of the footing, whichever is less, or downward to not less than the bottom of the slab and then horizontally to the interior or exterior for the total distance shown in the table. Insulation extending away from the building shall be protected by pavement or by not less than 10 inches (254 mm) of soil. Where installed, full slab insulation shall be continuous under the entire area of the slab-on-grade floor, except at structural column locations and service penetrations. Insulation required at the heated slab perimeter shall not be required to extend below the bottom of the heated slab and shall be continuous with the full slab insulation.

> **Exception:** Where the slab-on-grade floor is greater than 24 inches (61 mm) below the finished exterior grade, perimeter insulation is not required.

C402.2.5 Below-grade walls. The *C*-factor for the below-grade exterior walls shall be in accordance with Table C402.1.4. The *R*-value of the insulating material installed continuously within or on the below-grade exterior walls of the building envelope shall be in accordance with Table C402.1.3. The *C*-factor or *R*-value required shall extend to a depth of not less than 10 feet (3048 mm) below the outside finished ground level, or to the level of the lowest floor of the conditioned space enclosed by the below-grade wall, whichever is less.

C402.2.6 Insulation of radiant heating systems. *Radiant heating system* panels, and their associated components that are installed in interior or exterior assemblies, shall be insulated to an *R*-value of not less than R-3.5 on all surfaces not facing the space being heated. *Radiant heating system* panels that are installed in the *building thermal envelope* shall be separated from the exterior of the building or unconditioned or exempt spaces by not less than the *R*-value of insulation installed in the opaque assembly in which they are installed or the assembly shall comply with Section C402.1.4.

> **Exception:** Heated slabs on grade insulated in accordance with Section C402.2.4.

C402.2.7 Airspaces. Where the *R*-value of an airspace is used for compliance in accordance with Section C402.1, the airspace shall be enclosed in an unventilated cavity constructed to minimize airflow into and out of the enclosed airspace. Airflow shall be deemed minimized where the enclosed airspace is located on the interior side of the continuous air barrier and is bounded on all sides by building components.

> **Exception:** The thermal resistance of airspaces located on the exterior side of the continuous air barrier and adjacent to and behind the exterior wall-covering material shall be determined in accordance with ASTM C1363 modified with an airflow entering the bottom and exiting the top of the airspace at an air movement rate of not less than 70 mm/second.

C402.3 Roof solar reflectance and thermal emittance. Low-sloped roofs directly above cooled conditioned spaces in *Climate Zones* 0 through 3 shall comply with one or more of the options in Table C402.3.

> **Exceptions:** The following roofs and portions of roofs are exempt from the requirements of Table C402.3:
> 1. Portions of the roof that include or are covered by the following:
>   1.1. Photovoltaic systems or components.
>   1.2. Solar air or water-heating systems or components.
>   1.3. Vegetative roofs or landscaped roofs.
>   1.4. Above-roof decks or walkways.
>   1.5. Skylights.
>   1.6. HVAC systems and components, and other opaque objects mounted above the roof.
> 2. Portions of the roof shaded during the peak sun angle on the summer solstice by permanent features of the building or by permanent features of adjacent buildings.
> 3. Portions of roofs that are ballasted with a minimum stone ballast of 17 pounds per square foot (74 kg/m²) or 23 psf (117 kg/m²) pavers.
> 4. Roofs where not less than 75 percent of the roof area complies with one or more of the exceptions to this section.

TABLE C402.3
MINIMUM ROOF REFLECTANCE AND EMITTANCE OPTIONS[a]

Three-year-aged solar reflectance[b] of 0.55 and 3-year aged thermal emittance[c] of 0.75
Three-year-aged solar reflectance index[d] of 64

a. The use of area-weighted averages to comply with these requirements shall be permitted. Materials lacking 3-year-aged tested values for either solar reflectance or thermal emittance shall be assigned both a 3-year-aged solar reflectance in accordance with Section C402.3.1 and a 3-year-aged thermal emittance of 0.90.
b. Aged solar reflectance tested in accordance with ASTM C1549, ASTM E903 or ASTM E1918 or CRRC-S100.
c. Aged thermal emittance tested in accordance with ASTM C1371 or ASTM E408 or CRRC-S100.
d. Solar reflectance index (SRI) shall be determined in accordance with ASTM E1980 using a convection coefficient of 2.1 Btu/h × ft² × °F (12 W/m² × K). Calculation of aged SRI shall be based on aged tested values of solar reflectance and thermal emittance.

C402.3.1 Aged roof solar reflectance. Where an aged solar reflectance required by Section C402.3 is not available, it shall be determined in accordance with Equation 4-3.

$$R_{aged} = [0.2 + 0.7(R_{initial} - 0.2)] \quad \text{(Equation 4-3)}$$

where:

R_{aged} = The aged solar reflectance.

$R_{initial}$ = The initial solar reflectance determined in accordance with CRRC-S100.

C402.4 Fenestration. Fenestration shall comply with Sections C402.4.1 through C402.4.5 and Table C402.4. Daylight responsive controls shall comply with this section and Section C405.2.4.

TABLE C402.4
BUILDING ENVELOPE FENESTRATION MAXIMUM U-FACTOR AND SHGC REQUIREMENTS

CLIMATE ZONE	0 AND 1		2		3		4 EXCEPT MARINE		5 AND MARINE 4		6		7		8	
Vertical fenestration																
U-factor																
Fixed fenestration	0.50		0.45		0.42		0.36		0.36		0.34		0.29		0.26	
Operable fenestration	0.62		0.60		0.54		0.45		0.45		0.42		0.36		0.32	
Entrance doors	0.83		0.77		0.68		0.63		0.63		0.63		0.63		0.63	
SHGC																
	Fixed	Operable	Fixed	Operable	Fixed	Operable	Fixed	Operable	Fixed	Operable	Fixed	Operable	Fixed	Operable	Fixed	Operable
PF < 0.2	0.23	0.21	0.25	0.23	0.25	0.23	0.36	0.33	0.38	0.33	0.38	0.34	0.40	0.36	0.40	0.36
$0.2 \leq PF < 0.5$	0.28	0.25	0.30	0.28	0.30	0.28	0.43	0.40	0.46	0.40	0.46	0.41	0.48	0.43	0.48	0.43
$PF \geq 0.5$	0.37	0.34	0.40	0.37	0.40	0.37	0.58	0.53	0.61	0.53	0.61	0.54	0.64	0.58	0.64	0.58
Skylights																
U-factor	0.70		0.65		0.55		0.50		0.50		0.50		0.44		0.41	
SHGC	0.30		0.30		0.30		0.40		0.40		0.40		NR		NR	

NR = No Requirement, PF = Projection Factor.

C402.4.1 Maximum area. The vertical fenestration area, not including opaque doors and opaque spandrel panels, shall be not greater than 30 percent of the gross above-grade wall area. The skylight area shall be not greater than 3 percent of the gross roof area.

C402.4.1.1 Increased vertical fenestration area with daylight responsive controls. In *Climate Zones* 0 through 6, not more than 40 percent of the gross above-grade wall area shall be vertical fenestration, provided that all of the following requirements are met:

1. In buildings not greater than two stories above grade, not less than 50 percent of the net floor area is within a *daylight zone*.
2. In buildings three or more stories above grade, not less than 25 percent of the net floor area is within a *daylight zone*.
3. *Daylight responsive controls* are installed in *daylight zones*.
4. Visible transmittance (VT) of vertical fenestration is not less than 1.1 times solar heat gain coefficient (SHGC).

   Exception: Fenestration that is outside the scope of NFRC 200 is not required to comply with Item 4.

C402.4.1.2 Increased skylight area with daylight responsive controls. The skylight area shall be not more than 6 percent of the roof area provided that *daylight responsive controls* are installed in *toplit daylight zones*.

C402.4.2 Minimum skylight fenestration area. Skylights shall be provided in enclosed spaces greater than 2,500 square feet (232 m²) in floor area, directly under a roof with not less than 75 percent of the ceiling area with a ceiling height greater than 15 feet (4572 mm), and used as an office, lobby, atrium, concourse, corridor, storage space, gymnasium/exercise center, convention center, automotive service area, space where manufacturing occurs, nonrefrigerated warehouse, retail store, distribution/sorting area, transportation depot or workshop. The total *toplit daylight zone* shall be not less than half the floor area and shall comply with one of the following:

1. A minimum skylight area to *toplit daylight zone* of not less than 3 percent where all skylights have a VT of not less than 0.40, or VT_{annual} of not less than 0.26, as determined in accordance with Section C303.1.3.
2. A minimum skylight effective aperture, determined in accordance with Equation 4-4, of:

   2.1. Not less than 1 percent using a skylight's VT rating; or

   2.2. Not less than 0.66 percent using a Tubular Daylight Device's VT_{annual} rating.

$$\text{Skylight Effective Aperture} = \frac{0.85 \times \text{Skylight Area} \times \text{Skylight VT} \times \text{WF}}{\text{Toplit Zone}}$$

(Equation 4-4)

where:

Skylight area = Total fenestration area of skylights.

Skylight VT = Area weighted average visible transmittance of skylights.

WF = Area weighted average well factor, where well factor is 0.9 if light well depth is less than 2 feet (610 mm), or 0.7 if light well depth is 2 feet (610 mm) or greater, or 1.0 for Tubular Daylighting Devices with VT_{annual} ratings.

Light well depth = Measure vertically from the underside of the lowest point of the skylight glazing to the ceiling plane under the skylight.

> **Exception:** Skylights above *daylight zones* of enclosed spaces are not required in:
> 1. Buildings in *Climate Zones* 6 through 8.
> 2. Spaces where the designed *general lighting* power densities are less than 0.5 W/ft^2 (5.4 W/m^2).
> 3. Areas where it is documented that existing structures or natural objects block direct beam sunlight on not less than half of the roof over the enclosed area for more than 1,500 daytime hours per year between 8 a.m. and 4 p.m.
> 4. Spaces where the *daylight zone* under rooftop monitors is greater than 50 percent of the enclosed space floor area.
> 5. Spaces where the total area minus the area of *sidelit daylight zones* is less than 2,500 square feet (232 m^2), and where the lighting is controlled in accordance with Section C405.2.3.
> 6. Spaces designed as storm shelters complying with ICC 500.

C402.4.2.1 Lighting controls in toplit daylight zones. *Daylight responsive controls* shall be provided in toplit daylight zones.

C402.4.2.2 Haze factor. Skylights in office, storage, automotive service, manufacturing, nonrefrigerated warehouse, retail store and distribution/sorting area spaces shall have a glazing material or diffuser with a haze factor greater than 90 percent when tested in accordance with ASTM D1003.

> **Exception:** Skylights and tubular daylighting devices designed and installed to exclude direct sunlight entering the occupied space by the use of fixed or automated baffles, the geometry of skylight and light well or the use of optical diffuser components.

C402.4.3 Maximum *U*-factor and SHGC. The maximum *U*-factor and solar heat gain coefficient (SHGC) for fenestration shall be as specified in Table C402.4.

The window projection factor shall be determined in accordance with Equation 4-5.

$$PF = A/B \hspace{2cm} \text{(Equation 4-5)}$$

where:

PF = Projection factor (decimal).

A = Distance measured horizontally from the farthest continuous extremity of any overhang, eave or permanently attached shading device to the vertical surface of the glazing.

B = Distance measured vertically from the bottom of the glazing to the underside of the overhang, eave or permanently attached shading device.

Where different windows or glass doors have different *PF* values, they shall each be evaluated separately.

C402.4.3.1 Increased skylight SHGC. In *Climate Zones* 0 through 6, skylights shall be permitted a maximum SHGC of 0.60 where located above *daylight zones* provided with *daylight responsive controls*.

C402.4.3.2 Increased skylight *U*-factor. Where skylights are installed above *daylight zones* provided with *daylight responsive controls*, a maximum *U*-factor of 0.9 shall be permitted in *Climate Zones* 0 through 3 and a maximum *U*-factor of 0.75 shall be permitted in *Climate Zones* 4 through 8.

C402.4.3.3 Dynamic glazing. Where dynamic glazing is intended to satisfy the SHGC and VT requirements of Table C402.4, the ratio of the higher to lower labeled SHGC shall be greater than or equal to 2.4, and the *dynamic glazing* shall be automatically controlled to modulate the amount of solar gain into the space in multiple steps. Dynamic glazing shall be considered separately from other fenestration, and area-weighted averaging with other fenestration that is not dynamic glazing shall not be permitted.

> **Exception:** Dynamic glazing is not required to comply with this section where both the lower and higher labeled SHGC already comply with the requirements of Table C402.4.

C402.4.3.4 Area-weighted *U*-factor. An area-weighted average shall be permitted to satisfy the *U*-factor requirements for each fenestration product category listed in Table C402.4. Individual fenestration products from different fenestration product categories listed in Table C402.4 shall not be combined in calculating area-weighted average *U*-factor.

C402.4.4 Daylight zones. Daylight zones referenced in Sections C402.4.1.1 through C402.4.3.2 shall comply with Sections C405.2.4.2 and C405.2.4.3, as applicable.

Daylight zones shall include *toplit daylight zones* and sidelit *daylight* zones.

C402.4.5 Doors. Opaque swinging doors shall comply with Table C402.1.4. Opaque nonswinging doors shall comply with Table C402.1.4. Opaque doors shall be considered as part of the gross area of above-grade walls that are part of the *building thermal envelope*. Opaque doors shall comply with Section C402.4.5.1 or C402.4.5.2. Other doors shall comply with the provisions of Section C402.4.3 for vertical fenestration.

C402.4.5.1 Opaque swinging doors. Opaque swinging doors shall comply with Table C402.1.4.

C402.4.5.2 Nonswinging doors. Opaque nonswinging doors that are horizontally hinged sectional doors with a single row of fenestration shall have an assembly *U*-factor less than or equal to 0.440 in Climate Zones 0 through 6 and less than or equal to 0.360 in Climate Zones 7 and 8, provided that the fenestration area is not less than 14 percent and not more than 25 percent of the total door area.

> **Exception:** Other doors shall comply with the provisions of Section C402.4.3 for vertical fenestration.

C402.5 Air leakage—thermal envelope. The *building thermal envelope* shall comply with Sections C402.5.1 through Section C402.5.11.1, or the building *thermal envelope* shall be tested in accordance with Section C402.5.2 or C402.5.3. Where compliance is based on such testing, the building shall also comply with Sections C402.5.7, C402.5.8 and C402.5.9.

C402.5.1 Air barriers. A continuous air barrier shall be provided throughout the *building thermal envelope*. The continuous air barriers shall be located on the inside or outside of the building thermal envelope, located within the assemblies composing the building thermal envelope, or any combination thereof. The air barrier shall comply with Sections C402.5.1.1, and C402.5.1.2.

> **Exception:** Air barriers are not required in buildings located in *Climate Zone* 2B.

C402.5.1.1 Air barrier construction. The *continuous air barrier* shall be constructed to comply with the following:

1. The air barrier shall be continuous for all assemblies that are the thermal envelope of the building and across the joints and assemblies.
2. Air barrier joints and seams shall be sealed, including sealing transitions in places and changes in materials. The joints and seals shall be securely installed in or on the joint for its entire length so as not to dislodge, loosen or otherwise impair its ability to resist positive and negative pressure from wind, stack effect and mechanical ventilation.
3. Penetrations of the air barrier shall be caulked, gasketed or otherwise sealed in a manner compatible with the construction materials and location. Sealing shall allow for expansion, contraction and mechanical vibration. Joints and seams associated with penetrations shall be sealed in the same manner or taped. Sealing materials shall be securely installed around the penetration so as not to dislodge, loosen or otherwise impair the penetrations' ability to resist positive and negative pressure from wind, stack effect and mechanical ventilation. Sealing of concealed fire sprinklers, where required, shall be in a manner that is recommended by the manufacturer. Caulking or other adhesive sealants shall not be used to fill voids between fire sprinkler cover plates and walls or ceilings.
4. Recessed lighting fixtures shall comply with Section C402.5.10. Where similar objects are installed that penetrate the air barrier, provisions shall be made to maintain the integrity of the air barrier.

C402.5.1.2 Air barrier compliance. A continuous air barrier for the opaque building envelope shall comply with the following:

1. Buildings or portions of buildings, including Group R and I occupancies, shall meet the provisions of Section C402.5.2.

    > **Exception:** Buildings in Climate Zones 2B, 3C and 5C.

2. Buildings or portions of buildings other than Group R and I occupancies shall meet the provisions of Section C402.5.3.

    > **Exceptions:**
    > 1. Buildings in Climate Zones 2B, 3B, 3C and 5C.
    > 2. Buildings larger than 5,000 square feet (464.5 m^2) floor area in Climate Zones 0B, 1, 2A, 4B and 4C.
    > 3. Buildings between 5,000 square feet (464.5 m^2) and 50,000 square feet (4645 m^2) floor area in Climate Zones 0A, 3A and 5B.

3. Buildings or portions of buildings that do not complete air barrier testing shall meet the provisions of Section C402.5.1.3 or C402.5.1.4 in addition to Section C402.5.1.5.

C402.5.1.3 Materials. Materials with an air permeability not greater than 0.004 cfm/ft^2 (0.02 L/s × m^2) under a pressure differential of 0.3 inch water gauge (75 Pa) when tested in accordance with ASTM E2178 shall comply with this section. Materials in Items 1 through 16 shall be deemed to comply with this section, provided that joints are sealed and materials are installed as air barriers in accordance with the manufacturer's instructions.

1. Plywood with a thickness of not less than $^3/_8$ inch (10 mm).
2. Oriented strand board having a thickness of not less than $^3/_8$ inch (10 mm).
3. Extruded polystyrene insulation board having a thickness of not less than $^1/_2$ inch (12.7 mm).
4. Foil-back polyisocyanurate insulation board having a thickness of not less than $^1/_2$ inch (12.7 mm).
5. Closed-cell spray foam having a minimum density of 1.5 pcf (2.4 kg/m^3) and having a thickness of not less than $1^1/_2$ inches (38 mm).
6. Open-cell spray foam with a density between 0.4 and 1.5 pcf (0.6 and 2.4 kg/m^3) and having a thickness of not less than 4.5 inches (113 mm).
7. Exterior or interior gypsum board having a thickness of not less than $^1/_2$ inch (12.7 mm).
8. Cement board having a thickness of not less than $^1/_2$ inch (12.7 mm).
9. Built-up roofing membrane.
10. Modified bituminous roof membrane.
11. Single-ply roof membrane.
12. A Portland cement/sand parge, or gypsum plaster having a thickness of not less than $^5/_8$ inch (15.9 mm).
13. Cast-in-place and precast concrete.
14. Fully grouted concrete block masonry.
15. Sheet steel or aluminum.
16. Solid or hollow masonry constructed of clay or shale masonry units.

C402.5.1.4 Assemblies. Assemblies of materials and components with an average air leakage not greater than 0.04 cfm/ft^2 (0.2 L/s × m^2) under a pressure differential of 0.3 inch of water gauge (w.g.)(75 Pa) when tested in accordance with ASTM E2357, ASTM E1677, ASTM D8052 or ASTM E283 shall comply with this section. Assemblies listed in Items 1 through 3 shall be deemed to comply, provided that joints are sealed and the requirements of Section C402.5.1.1 are met.

1. Concrete masonry walls coated with either one application of block filler or two applications of a paint or sealer coating.
2. Masonry walls constructed of clay or shale masonry units with a nominal width of 4 inches (102 mm) or more.
3. A Portland cement/sand parge, stucco or plaster not less than $^1/_2$ inch (12.7 mm) in thickness.

C402.5.1.5 Building envelope performance verification. The installation of the continuous air barrier shall be verified by the *code official*, a *registered design professional* or *approved* agency in accordance with the following:

1. A review of the construction documents and other supporting data shall be conducted to assess compliance with the requirements in Section C402.5.1.
2. Inspection of continuous air barrier components and assemblies shall be conducted during construction while the air barrier is still accessible for inspection and repair to verify compliance with the requirements of Sections C402.5.1.3 and C402.5.1.4.
3. A final commissioning report shall be provided for inspections completed by the *registered design professional* or *approved* agency. The commissioning report shall be provided to the building owner or owner's authorized agent and the *code official*. The report shall identify deficiencies found during the review of the construction documents and inspection and details of corrective measures taken.

C402.5.2 Dwelling and sleeping unit enclosure testing. The *building thermal envelope* shall be tested in accordance with ASTM E779, ANSI/RESNET/ICC 380, ASTM E1827 or an equivalent method approved by the *code official*. The measured air leakage shall not exceed 0.30 cfm/ft^2 (1.5 L/s m^2) of the testing unit enclosure area at a pressure differential of 0.2 inch water gauge (50 Pa). Where multiple dwelling units or sleeping units or other occupiable conditioned spaces are contained within one *building thermal envelope*, each unit shall be considered an individual testing unit, and the building air leakage shall be the weighted average of all testing unit results, weighted by each testing unit's enclosure area. Units shall be tested separately with an unguarded blower door test as follows:

1. Where buildings have fewer than eight testing units, each testing unit shall be tested.
2. For buildings with eight or more testing units, the greater of seven units or 20 percent of the testing units in the building shall be tested, including a top floor unit, a ground floor unit and a unit with the largest testing unit enclosure area. For each tested unit that exceeds the maximum air leakage rate, an additional two units shall be tested, including a mixture of testing unit types and locations.

C402.5.3 Building thermal envelope testing. The *building thermal envelope* shall be tested in accordance with ASTM E779, ANSI/RESNET/ICC 380, ASTM E3158 or ASTM E1827 or an equivalent method approved by the code official. The measured air leakage shall not exceed 0.40 cfm/ft² (2.0 L/s × m²) of the *building thermal envelope* area at a pressure differential of 0.3 inch water gauge (75 Pa). Alternatively, portions of the building shall be tested and the measured air leakages shall be area weighted by the surface areas of the building envelope in each portion. The weighted average test results shall not exceed the whole building leakage limit. In the alternative approach, the following portions of the building shall be tested:

1. The entire envelope area of all stories that have any spaces directly under a roof.
2. The entire envelope area of all stories that have a building entrance, exposed floor, or loading dock, or are below grade.
3. Representative above-grade sections of the building totaling at least 25 percent of the wall area enclosing the remaining conditioned space.

Exception: Where the measured air leakage rate exceeds 0.40 cfm/ft² (2.0 L/s × m²) but does not exceed 0.60 cfm/ft² (3.0 L/s × m²), a diagnostic evaluation using smoke tracer or infrared imaging shall be conducted while the building is pressurized along with a visual inspection of the air barrier. Any leaks noted shall be sealed where such sealing can be made without destruction of existing building components. An additional report identifying the corrective actions taken to seal leaks shall be submitted to the code official and the building owner, and shall be deemed to comply with the requirements of this section.

C402.5.4 Air leakage of fenestration. The air leakage of fenestration assemblies shall meet the provisions of Table C402.5.4. Testing shall be in accordance with the applicable reference test standard in Table C402.5.4 by an accredited, independent testing laboratory and *labeled* by the manufacturer.

Exceptions:

1. Field-fabricated fenestration assemblies that are sealed in accordance with Section C402.5.1.
2. Fenestration in buildings that comply with the testing alternative of Section C402.5 are not required to meet the air leakage requirements in Table C402.5.4.

C402.5.5 Rooms containing fuel-burning appliances. In *Climate Zones* 3 through 8, where combustion air is supplied through openings in an exterior wall to a room or space containing a space-conditioning fuel-burning appliance, one of the following shall apply:

1. The room or space containing the appliance shall be located outside of the *building thermal envelope*.
2. The room or space containing the appliance shall be enclosed and isolated from conditioned spaces inside the *building thermal envelope*. Such rooms shall comply with all of the following:
   2.1. The walls, floors and ceilings that separate the enclosed room or space from conditioned spaces shall be insulated to be not less than equivalent to the insulation requirement of below-grade walls as specified in Table C402.1.3 or Table C402.1.4.
   2.2. The walls, floors and ceilings that separate the enclosed room or space from conditioned spaces shall be sealed in accordance with Section C402.5.1.1.

TABLE C402.5.4
MAXIMUM AIR LEAKAGE RATE FOR FENESTRATION ASSEMBLIES

FENESTRATION ASSEMBLY	MAXIMUM RATE (CFM/FT²)	TEST PROCEDURE
Windows	0.20[a]	AAMA/WDMA/CSA101/I.S.2/A440 or NFRC 400
Sliding doors	0.20[a]	
Swinging doors	0.20[a]	
Skylights—with condensation weepage openings	0.30	
Skylights—all other	0.20[a]	
Curtain walls	0.06	NFRC 400 or ASTM E283 at 1.57 psf (75 Pa)
Storefront glazing	0.06	
Commercial glazed swinging entrance doors	1.00	
Power-operated sliding doors and power operated folding doors	1.00	
Revolving doors	1.00	
Garage doors	0.40	ANSI/DASMA 105, NFRC 400, or ASTM E283 at 1.57 psf (75 Pa)
Rolling doors	1.00	
High-speed doors	1.30	

For SI: 1 cubic foot per minute = 0.47 L/s, 1 square foot = 0.093 m².

a. The maximum rate for windows, sliding and swinging doors, and skylights is permitted to be 0.3 cfm per square foot of fenestration or door area when tested in accordance with AAMA/WDMA/CSA101/I.S.2/A440 at 6.24 psf (300 Pa).

2.3. The doors into the enclosed room or space shall be fully gasketed.

2.4. Water lines and ducts in the enclosed room or space shall be insulated in accordance with Section C403.

2.5. Where an air duct supplying combustion air to the enclosed room or space passes through *conditioned space*, the duct shall be insulated to an *R*-value of not less than R-8.

Exception: Fireplaces and stoves complying with Sections 901 through 905 of the *International Mechanical Code*, and Section 2111.14 of the *International Building Code*.

C402.5.6 Doors and access openings to shafts, chutes, stairways and elevator lobbies. Doors and *access* openings from conditioned space to shafts, chutes stairways and elevator lobbies not within the scope of the fenestration assemblies covered by Section C402.5.4 shall be gasketed, weather-stripped or sealed.

Exceptions:

1. Door openings required to comply with Section 716 of the *International Building Code*.

2. Doors and door openings required to comply with UL 1784 by the *International Building Code*.

C402.5.7 Air intakes, exhaust openings, stairways and shafts. Stairway enclosures, elevator shaft vents and other outdoor air intakes and exhaust openings integral to the building envelope shall be provided with dampers in accordance with Section C403.7.7.

C402.5.8 Loading dock weather seals. Cargo door openings and loading door openings shall be equipped with weather seals that restrict infiltration and provide direct contact along the top and sides of vehicles that are parked in the doorway.

C402.5.9 Vestibules. Building entrances shall be protected with an enclosed vestibule, with all doors opening into and out of the vestibule equipped with self-closing devices. Vestibules shall be designed so that in passing through the vestibule it is not necessary for the interior and exterior doors to open at the same time. The installation of one or more revolving doors in the *building entrance* shall not eliminate the requirement that a vestibule be provided on any doors adjacent to revolving doors.

Exceptions: Vestibules are not required for the following:

1. Buildings in *Climate Zones* 0 through 2.

2. Doors not intended to be used by the public, such as doors to mechanical or electrical equipment rooms, or intended solely for employee use.

3. Doors opening directly from a *sleeping unit* or dwelling unit.

4. Doors that open directly from a space less than 3,000 square feet (298 m^2) in area.

5. Revolving doors.

6. Doors used primarily to facilitate vehicular movement or material handling and adjacent personnel doors.

7. Doors that have an air curtain with a velocity of not less than 6.56 feet per second (2 m/s) at the floor that have been tested in accordance with ANSI/AMCA 220 and installed in accordance with the manufacturer's instructions. Manual or automatic controls shall be provided that will operate the air curtain with the opening and closing of the door. Air curtains and their controls shall comply with Section C408.2.3.

C402.5.10 Recessed lighting. Recessed luminaires installed in the *building thermal envelope* shall be all of the following:

1. IC-rated.

2. Labeled as having an air leakage rate of not more 2.0 cfm (0.944 L/s) when tested in accordance with ASTM E283 at a 1.57 psf (75 Pa) pressure differential.

3. Sealed with a gasket or caulk between the housing and interior wall or ceiling covering.

C402.5.11 Operable openings interlocking. Where occupancies utilize operable openings to the outdoors that are larger than 40 square feet (3.7 m^2) in area, such openings shall be interlocked with the heating and cooling system so as to raise the cooling setpoint to 90°F (32°C) and lower the heating setpoint to 55°F (13°C) whenever the operable opening is open. The change in heating and cooling setpoints shall occur within 10 minutes of opening the operable opening.

Exceptions:

1. Separately zoned areas associated with the preparation of food that contain appliances that contribute to the HVAC loads of a restaurant or similar type of occupancy.

2. Warehouses that utilize overhead doors for the function of the occupancy, where approved by the code official.

3. The first entrance doors where located in the exterior wall and are part of a vestibule system.

C402.5.11.1 Operable controls. Controls shall comply with Section C403.14.

SECTION C403
BUILDING MECHANICAL SYSTEMS

C403.1 General. Mechanical systems and equipment serving the building heating, cooling, ventilating or refrigerating needs shall comply with this section.

> **Exception:** Data center systems are exempt from the requirements of Sections C403.4 and C403.5.

C403.1.1 Calculation of heating and cooling loads. Design loads associated with heating, ventilating and air conditioning of the building shall be determined in accordance with ANSI/ASHRAE/ACCA Standard 183 or by an *approved* equivalent computational procedure using the design parameters specified in Chapter 3. Heating and cooling loads shall be adjusted to account for load reductions that are achieved where energy recovery systems are utilized in the HVAC system in accordance with the ASHRAE HVAC Systems and Equipment Handbook by an approved equivalent computational procedure.

C403.1.2 Data centers. Data center systems shall comply with Sections 6 and 8 of ASHRAE 90.4 with the following changes:

1. Replace design mechanical load component (MLC) values specified in Table 6.2.1.1 of the ASHRAE 90.4 with the values in Table C403.1.2(1) as applicable in each climate zone.

2. Replace annualized MLC values specified in Table 6.2.1.2 of the ASHRAE 90.4 with the values in Table C403.1.2(2) as applicable in each climate zone.

TABLE C403.1.2(1)
MAXIMUM DESIGN MECHANICAL LOAD COMPONENT
(DESIGN MLC)

CLIMATE ZONE	DESIGN MLC AT 100% AND AT 50% ITE LOAD
0A	0.24
0B	0.26
1A	0.23
2A	0.24
3A	0.23
4A	0.23
5A	0.22
6A	0.22
1B	0.28
2B	0.27
3B	0.26
4B	0.23
5B	0.23
6B	0.21
3C	0.19
4C	0.21
5C	0.19
7	0.20
8	0.19

TABLE C403.1.2(2)
MAXIMUM ANNUALIZED MECHANICAL LOAD COMPONENT
(ANNUALIZED MLC)

CLIMATE ZONE	HVAC MAXIMUM ANNUALIZED MLC AT 100% AND AT 50% ITE LOAD
0A	0.19
0B	0.20
1A	0.18
2A	0.19
3A	0.18
4A	0.17
5A	0.17
6A	0.17
1B	0.16
2B	0.18
3B	0.18
4B	0.18
5B	0.16
6B	0.17
3C	0.16
4C	0.16
5C	0.16
7	0.16
8	0.16

C403.2 System design. Mechanical systems shall be designed to comply with Sections C403.2.1 through C403.2.3. Where elements of a building's mechanical systems are addressed in Sections C403.3 through C403.14, such elements shall comply with the applicable provisions of those sections.

C403.2.1 Zone isolation required. HVAC systems serving *zones* that are over 25,000 square feet (2323 m^2) in floor area or that span more than one floor and are designed to operate or be occupied nonsimultaneously shall be divided into isolation areas. Each isolation area shall be equipped with *isolation devices* and controls configured to automatically shut off the supply of conditioned air and outdoor air to and exhaust air from the isolation area. Each isolation area shall be controlled independently by a device meeting the requirements of Section C403.4.2.2. Central systems and plants shall be provided with controls and devices that will allow system and equipment operation for any length of time while serving only the smallest isolation area served by the system or plant.

> **Exceptions:**
>
> 1. Exhaust air and outdoor air connections to isolation areas where the fan system to which they connect is not greater than 5,000 cfm (2360 L/s).

2. Exhaust airflow from a single isolation area of less than 10 percent of the design airflow of the exhaust system to which it connects.

3. Isolation areas intended to operate continuously or intended to be inoperative only when all other isolation areas in a *zone* are inoperative.

C403.2.2 Ventilation. Ventilation, either natural or mechanical, shall be provided in accordance with Chapter 4 of the *International Mechanical Code*. Where mechanical ventilation is provided, the system shall provide the capability to reduce the outdoor air supply to the minimum required by Chapter 4 of the *International Mechanical Code*.

C403.2.3 Fault detection and diagnostics. New buildings with an HVAC system serving a gross conditioned floor area of 100,000 square feet (9290 m^2) or larger shall include a fault detection and diagnostics (FDD) system to monitor the HVAC system's performance and automatically identify faults. The FDD system shall:

1. Include permanently installed sensors and devices to monitor the HVAC system's performance.

2. Sample the HVAC system's performance at least once every 15 minutes.

3. Automatically identify and report HVAC system faults.

4. Automatically notify authorized personnel of identified HVAC system faults.

5. Automatically provide prioritized recommendations for repair of identified faults based on analysis of data collected from the sampling of HVAC system performance.

6. Be capable of transmitting the prioritized fault repair recommendations to remotely located authorized personnel.

Exception: R-1 and R-2 occupancies.

C403.3 Heating and cooling equipment efficiencies. Heating and cooling equipment installed in mechanical systems shall be sized in accordance with Section C403.3.1 and shall be not less efficient in the use of energy than as specified in Section C403.3.2.

C403.3.1 Equipment sizing. The output capacity of heating and cooling equipment shall be not greater than that of the smallest available equipment size that exceeds the loads calculated in accordance with Section C403.1.1. A single piece of equipment providing both heating and cooling shall satisfy this provision for one function with the capacity for the other function as small as possible, within available equipment options.

Exceptions:

1. Required standby equipment and systems provided with controls and devices that allow such systems or equipment to operate automatically only when the primary equipment is not operating.

2. Multiple units of the same equipment type with combined capacities exceeding the design load and provided with controls that are configured to sequence the operation of each unit based on load.

C403.3.2 HVAC equipment performance requirements. Equipment shall meet the minimum efficiency requirements of Tables C403.3.2(1) through C403.3.2(16) when tested and rated in accordance with the applicable test procedure. Plate-type liquid-to-liquid heat exchangers shall meet the minimum requirements of AHRI 400. The efficiency shall be verified through certification under an approved certification program or, where a certification program does not exist, the equipment efficiency ratings shall be supported by data furnished by the manufacturer. Where multiple rating conditions or performance requirements are provided, the equipment shall satisfy all stated requirements. Where components, such as indoor or outdoor coils, from different manufacturers are used, calculations and supporting data shall be furnished by the designer that demonstrates that the combined efficiency of the specified components meets the requirements herein.

C403.3.2.1 Water-cooled centrifugal chilling packages. Equipment not designed for operation at AHRI Standard 550/590 test conditions of 44.00°F leaving and 54.00°F entering chilled-fluid temperatures, and with 85.00°F entering and 94.30°F leaving condenser-fluid temperatures, shall have maximum full-load kW/ton (FL) and part-load rating requirements adjusted using the following equations:

$FL_{adj} = FL/K_{adj}$ \hfill (Equation 4-6)

$PLV_{adj} = IPLV.IP/K_{adj}$ \hfill (Equation 4-7)

where:

K_{adj} = $A \times B$

FL = Full-load kW/ton value from Table C403.3.2(3).

FL_{adj} = Maximum full-load kW/ton rating, adjusted for nonstandard conditions.

$IPLV.IP$ = $IPLV.IP$ value from Table C403.3.2(3).

PLV_{adj} = Maximum *NPLV* rating, adjusted for nonstandard conditions.

A = $0.00000014592 \times (LIFT)^4 - 0.0000346496 \times (LIFT)^3 + 0.00314196 \times (LIFT)^2 - 0.147199 \times (LIFT) + 3.93073$

B = $0.0015 \times L_{vg}E_{vap} + 0.934$

$LIFT$ = $L_{vg}Cond - L_{vg}E_{vap}$

$L_{vg}Cond$ = Full-load condenser leaving fluid temperature (°F).

$L_{vg}E_{vap}$ = Full-load evaporator leaving temperature (°F).

The FL_{adj} and PLV_{adj} values are applicable only for centrifugal chillers meeting all of the following full-load design ranges:

- $36.00°F \leq L_{vg}E_{vap} \leq 60.00°F$
- $L_{vg}Cond \leq 115.00°F$
- $20.00°F \leq LIFT \leq 80.00°F$

Manufacturers shall calculate the FL_{adj} and PLV_{adj} before determining whether to label the chiller. Centrifugal chillers designed to operate outside of these ranges are not covered by this code.

C403.3.2.2 Positive displacement (air- and water-cooled) chilling packages. Equipment with a leaving fluid temperature higher than 32°F (0°C) and water-cooled positive displacement chilling packages with a condenser leaving fluid temperature below 115°F (46°C) shall meet the requirements of the tables in Section C403.3.2 when tested or certified with water at standard rating conditions, in accordance with the referenced test procedure.

C403.3.3 Hot gas bypass limitation. Cooling systems shall not use hot gas bypass or other evaporator pressure control systems unless the system is designed with multiple steps of unloading or continuous capacity modulation. The capacity of the hot gas bypass shall be limited as indicated in Table C403.3.3, as limited by Section C403.5.1.

TABLE C403.3.2(1)
ELECTRICALLY OPERATED UNITARY AIR CONDITIONERS AND CONDENSING UNITS—MINIMUM EFFICIENCY REQUIREMENTS[c, d]

EQUIPMENT TYPE	SIZE CATEGORY	HEADING SECTION TYPE	SUBCATEGORY OR RATING CONDITION	MINIMUM EFFICIENCY	TEST PROCEDURE[a]
Air conditioners, air cooled	< 65,000 Btu/h[b]	All	Split system, three phase and applications outside US single phase[b]	13.0 SEER before 1/1/2023 13.4 SEER2 after 1/1/2023	AHRI 210/240—2017 before 1/1/2023 AHRI 210/240—2023 after 1/1/2023
			Single-package, three phase and applications outside US single phase[b]	14.0 SEER before 1/1/2023 13.4 SEER2 after 1/1/2023	
Space constrained, air cooled	≤ 30,000 Btu/h[b]	All	Split system, three phase and applications outside US single phase[b]	12.0 SEER before 1/1/2023 11.7 SEER2 after 1/1/2023	AHRI 210/240—2017 before 1/1/2023 AHRI 210/240—2023 after 1/1/2023
			Single package, three phase and applications outside US single phase[b]	12.0 SEER before 1/1/2023 11.7 SEER2 after 1/1/2023	
Small duct, high velocity, air cooled	< 65,000 Btu/h[b]	All	Split system, three phase and applications outside US single phase[b]	12.0 SEER before 1/1/2023 12.1 SEER2 after 1/1/2023	AHRI 210/240—2017 before 1/1/2023 AHRI 210/240—2023 after 1/1/2023
Air conditioners, air cooled	≥ 65,000 Btu/h and < 135,000 Btu/h	Electric resistance (or none)	Split system and single package	11.2 EER 12.9 IEER before 1/1/2023 14.8 IEER after 1/1/2023	AHRI 340/360
		All other		11.0 EER 12.7 IEER before 1/1/2023 14.6 IEER after 1/1/2023	
	≥ 135,000 Btu/h and < 240,000 Btu/h	Electric resistance (or none)		11.0 EER 12.4 IEER before 1/1/2023 14.2 IEER after 1/1/2023	
		All other		10.8 EER 12.2 IEER before 1/1/2023 14.0 IEER after 1/1/2023	

(continued)

COMMERCIAL ENERGY EFFICIENCY

TABLE C403.3.2(1)—continued
ELECTRICALLY OPERATED UNITARY AIR CONDITIONERS AND CONDENSING UNITS—MINIMUM EFFICIENCY REQUIREMENTS[c, d]

EQUIPMENT TYPE	SIZE CATEGORY	HEADING SECTION TYPE	SUBCATEGORY OR RATING CONDITION	MINIMUM EFFICIENCY	TEST PROCEDURE[a]
Air conditioners, air cooled *(continued)*	≥ 240,000 Btu/h and < 760,000 Btu/h	Electric resistance (or none)	Split system and single package	10.0 EER 11.6 IEER before 1/1/2023 13.2 IEER after 1/1/2023	AHRI 340/360
		All other		9.8 EER 11.4 IEER before 1/1/2023 13.0 IEER after 1/1/2023	
	≥ 760,000 Btu/h	Electric resistance (or none)		9.7 EER 11.2 IEER before 1/1/2023 12.5 IEER after 1/1/2023	
		All other		9.5 EER 11.0 IEER before 1/1/2023 12.3 IEER after 1/1/2023	
Air conditioners, water cooled	< 65,000 Btu/h	All	Split system and single package	12.1 EER 12.3 IEER	AHRI 210/240
	≥ 65,000 Btu/h and < 135,000 Btu/h	Electric resistance (or none)		12.1 EER 13.9 IEER	AHRI 340/360
		All other		11.9 EER 13.7 IEER	
	≥ 135,000 Btu/h and < 240,000 Btu/h	Electric resistance (or none)		12.5 EER 13.9 IEER	
		All other		12.3 EER 13.7 IEER	
	≥ 240,000 Btu/h and < 760,000 Btu/h	Electric resistance (or none)		12.4 EER 13.6 IEER	
		All other		12.2 EER 13.4 IEER	
	≥ 760,000 Btu/h	Electric resistance (or none)		12.2 EER 13.5 IEER	
		All other		12.0 EER 13.3 IEER	

(continued)

TABLE C403.3.2(1)—continued
ELECTRICALLY OPERATED UNITARY AIR CONDITIONERS AND CONDENSING UNITS—MINIMUM EFFICIENCY REQUIREMENTS[c, d]

EQUIPMENT TYPE	SIZE CATEGORY	HEADING SECTION TYPE	SUBCATEGORY OR RATING CONDITION	MINIMUM EFFICIENCY	TEST PROCEDURE[a]
Air conditioners, evaporatively cooled	< 65,000 Btu/h[b]	All	Split system and single package	12.1 EER 12.3 IEER	AHRI 210/240
	≥ 65,000 Btu/h and < 135,000 Btu/h	Electric resistance (or none)		12.1 EER 12.3 IEER	AHRI 340/360
		All other		11.9 EER 12.1 IEER	
	≥ 135,000 Btu/h and < 240,000 Btu/h	Electric resistance (or none)		12.0 EER 12.2 IEER	
		All other		11.8 EER 12.0 IEER	
	≥ 240,000 Btu/h and < 760,000 Btu/h	Electric resistance (or none)		11.9 EER 12.1 IEER	
		All other		11.7 EER 11.9 IEER	
	≥ 760,000 Btu/h	Electric resistance (or none)		11.7 EER 11.9 IEER	
		All other		11.5 EER 11.7 IEER	
Condensing units, air cooled	≥ 135,000 Btu/h	—	—	10.5 EER 11.8 IEER	AHRI 365
Condensing units, water cooled	≥ 135,000 Btu/h	—	—	13.5 EER 14.0 IEER	AHRI 365
Condensing units, evaporatively cooled	≥ 135,000 Btu/h	—	—	13.5 EER 14.0 IEER	AHRI 365

For SI: 1 British thermal unit per hour = 0.2931 W.

a. Chapter 6 contains a complete specification of the referenced standards, which include test procedures, including the reference year version of the test procedure.
b. Single-phase, US air-cooled air conditioners less than 65,000 Btu/h are regulated as consumer products by the US Department of Energy Code of Federal Regulations DOE 10 CFR 430. SEER and SEER2 values for single-phase products are set by the US Department of Energy.
c. DOE 10 CFR 430 Subpart B Appendix M1 includes the test procedure updates effective 1/1/2023 that will be incorporated in AHRI 210/240—2023.
d. This table is a replica of ASHRAE 90.1 Table 6.8.1-1 Electrically Operated Unitary Air Conditioners and Condensing Units—Minimum Efficiency Requirements.

COMMERCIAL ENERGY EFFICIENCY

TABLE C403.3.2(2)
ELECTRICALLY OPERATED AIR-COOLED UNITARY HEAT PUMPS—MINIMUM EFFICIENCY REQUIREMENTS[c, d]

EQUIPMENT TYPE	SIZE CATEGORY	HEADING SECTION TYPE	SUBCATEGORY OR RATING CONDITION	MINIMUM EFFICIENCY	TEST PROCEDURE[a]
Air cooled (cooling mode)	< 66,000 Btu/h	All	Split system, three phase and applications outside US single phase[b]	14.0 SEER before 1/1/2023 14.3 SEER2 after 1/1/2023	AHRI 210/240—2017 before 1/1/2023 AHRI 210/240—2023 after 1/1/2023
			Single package, three phase and applications outside US single phase[b]	14.0 SEER before 1/1/2023 13.4 SEER2 after 1/1/2023	
Space constrained, air cooled (cooling mode)	≤ 30,000 Btu/h	All	Split system, three phase and applications outside US single phase[b]	12.0 SEER before 1/1/2023 11.7 SEER2 after 1/1/2023	AHRI 210/240—2017 before 1/1/2023 AHRI 210/240—2023 after 1/1/2023
			Single package, three phase and applications outside US single phase[b]	12.0 SEER before 1/1/2023 11.7 SEER2 after 1/1/2023	
Single duct, high velocity, air cooled (cooling mode)	< 65,000	All	Split system, three phase and applications outside US single phase[b]	12.0 SEER before 1/1/2023 12.0 SEER2 after 1/1/2023	AHRI 210/240—2017 before 1/1/2023 AHRI 210/240—2023 after 1/1/2023
Air cooled (cooling mode)	≥ 65,000 Btu/h and < 135,000 Btu/h	Electric resistance (or none)	Split system and single package	11.0 EER 12.2 IEER before 1/1/2023 14.1 IEER after 1/1/2023	AHRI 340/360
		All other		10.8 EER 12.0 IEER before 1/1/2023 13.9 IEER after 1/1/2023	
	≥ 135,000 Btu/h and < 240,000 Btu/h	Electric resistance (or none)		10.6 EER 11.6 IEER before 1/1/2023 13.5 IEER after 1/1/2023	
		All other		10.4 EER 11.4 IEER before 1/1/2023 13.3 IEER after 1/1/2023	
	≥ 240,000 Btu/h	Electric resistance (or none)		9.5 EER 10.6 IEER before 1/1/2023 12.5 IEER after 1/1/2023	
		All other		9.3 EER 10.4 IEER before 1/1/2023 12.3 IEER after 1/1/2023	
Air cooled (heating mode)	< 65,000 Btu/h	All	Split system, three phase and applications outside US single phase[b]	8.2 HSPF before 1/1/2023 7.5 HSPF2 after 1/1/2023	AHRI 210/240—2017 before 1/1/2023 AHRI 210/240—2023 after 1/1/2023
			Single package, three phase and applications outside US single phase[b]	8.0 HSPF before 1/1/2023 6.7 HSPF2 after 1/1/2023	

(continued)

TABLE C403.3.2(2)—continued
ELECTRICALLY OPERATED AIR-COOLED UNITARY HEAT PUMPS—MINIMUM EFFICIENCY REQUIREMENTS[c, d]

EQUIPMENT TYPE	SIZE CATEGORY	HEADING SECTION TYPE	SUBCATEGORY OR RATING CONDITION	MINIMUM EFFICIENCY	TEST PROCEDURE[a]
Space constrained, air cooled (heating mode)	≤ 30,000 Btu/h	All	Split system, three phase and applications outside US single phase[b]	7.4 HSPF before 1/1/2023 6.3 HSPF2 after 1/1/2023	AHRI 210/240—2017 before 1/1/2023 AHRI 210/240—2023 after 1/1/2023
			Single package, three phase and applications outside US single phase[b]	7.4 HSPF before 1/1/2023 6.3 HSPF2 after 1/1/2023	
Small duct, high velocity, air cooled (heating mode)	< 65,000 Btu/h	All	Split system, three phase and applications outside US single phase[b]	7.2 HSPF before 1/1/2023 6.1 HSPF2 after 1/1/2023	AHRI 210/240—2017 before 1/1/2023 AHRI 210/240—2023 after 1/1/2023
Air cooled (heating mode)	≥ 65,000 Btu/h and < 135,000 Btu/h (cooling capacity)	All	47°F db/43°F wb outdoor air	3.30 COP_H before 1/1/2023 3.40 COP_H after 1/1/2023	AHRI 340/360
			17°F db/15°F wb outdoor air	2.25 COP_H	
	≥ 135,000 Btu/h and < 240,000 Btu/h (cooling capacity)		47°F db/43°F wb outdoor air	3.20 COP_H before 1/1/2023 3.30 SOP_H after 1/1/2023	
			17°F db/15°F wb outdoor air	2.05 COP_H	
	≥ 240,000 Btu/h (cooling capacity)		47°F db/43°F wb outdoor air	3.20 COP_H	
			17°F db/15°F wb outdoor air	2.05 COP_H	

For SI: 1 British thermal unit per hour = 0.2931 W, °C = [(°F) − 32]/1.8, wb = wet bulb, db = dry bulb.

a. Chapter 6 contains a complete specification of the referenced standards, which include test procedures, including the reference year version of the test procedure.
b. Single-phase, US air-cooled heat pumps less than 65,000 Btu/h are regulated as consumer products by the US Department of Energy Code of Federal Regulations DOE 10 CFR 430. SEER, SEER2 and HSPF values for single-phase products are set by the US Department of Energy.
c. DOE 10 CFR 430 Subpart B Appendix M1 includes the test procedure updates effective 1/1/2023 that will be incorporated in AHRI 210/240—2023.
d. This table is a replica of ASHRAE 90.1 Table 6.8.1-2 Electrically Operated Air-Cooled Unitary Heat Pumps—Minimum Efficiency Requirements.

TABLE C403.3.2(3)
WATER-CHILLING PACKAGES—MINIMUM EFFICIENCY REQUIREMENTS[a, b, e, f]

EQUIPMENT TYPE	SIZE CATEGORY	UNITS	PATH A	PATH B	TEST PROCEDURE[c]
Air cooled chillers	< 150 tons	EER (Btu/Wh)	≥ 10.100 FL	≥ 9.700 FL	AHRI 550/590
			≥ 13.700 IPLV.IP	≥ 15.800 IPLV.IP	
	≥ 150 tons		≥ 10.100 FL	≥ 9.700 FL	
			≥ 14.000 IPLV.IP	≥ 16.100 IPLV.IP	
Air cooled without condenser, electrically operated	All capacities	EER (Btu/Wh)	Air-cooled chillers without condenser must be rated with matching condensers and comply with air-cooled chiller efficiency requirements		AHRI 550/590
Water cooled, electrically operated positive displacement	< 75 tons	kW/ton	≤ 0.750 FL	≤ 0.780 FL	AHRI 550/590
			≤ 0.600 IPLV.IP	≤ 0.500 IPLV.IP	
	≥ 75 tons and < 150 tons		≤ 0.720 FL	≤ 0.750 FL	
			≤ 0.560 IPLV.IP	≤ 0.490 IPLV.IP	
	≥ 150 tons and < 300 tons		≤ 0.660 FL	≤ 0.680 FL	
			≤ 0.540 IPLV.IP	≤ 0.440 IPLV.IP	
	≥ 300 tons and < 600 tons		≤ 0.610 FL	≤ 0.625 FL	
			≤ 0.520 IPLV.IP	≤ 0.410 IPLV.IP	
	≥ 600 tons		≤ 0.560 FL	≤ 0.585 FL	
			≤ 0.500 IPLV.IP	≤ 0.380 IPLV.IP	
Water cooled, electrically operated centrifugal	< 150 tons	kW/ton	≤ 0.610 FL	≤ 0.695 FL	AHRI 550/590
			≤ 0.550 IPLV.IP	≤ 0.440 IPLV.IP	
			≤ 0.610 FL	≤ 0.635 FL	
			≤ 0.550 IPLV.IP	≤ 0.400 IPLV.IP	
	≥ 300 tons and < 400 tons		≤ 0.560 FL	≤ 0.595 FL	
			≤ 0.520 IPLV.IP	≤ 0.390 IPLV.IP	
	≥ 400 tons and < 600 tons		≤ 0.560 FL	≤ 0.585 FL	
			≤ 0.500 IPLV.IP	≤ 0.380 IPLV.IP	
	≥ 600 tons		≤ 0.560 FL	≤ 0.585 FL	
			≤ 0.500 IPLV.IP	≤ 0.380 IPLV.IP	
Air cooled absorption, single effect	All capacities	COP (W/W)	≥ 0.600 FL	NA[d]	AHRI 560
Water cooled absorption, single effect	All capacities	COP (W/W)	≥ 0.700 FL	NA[d]	AHRI 560
Absorption double effect, indirect fired	All capacities	COP (W/W)	≥ 1.000 FL	NA[d]	AHRI 560
			≥ 0.150 IPLV.IP		
Absorption double effect, direct fired	All capacities	COP (W/W)	≥ 1.000 FL	NA[d]	AHRI 560
			≥ 1.000 IPLV		

a. Chapter 6 contains a complete specification of the referenced standards, which include test procedures, including the reference year version of the test procedure.
b. The requirements for centrifugal chillers shall be adjusted for nonstandard rating conditions per Section C403.3.2.1 and are applicable only for the range of conditions listed there. The requirements for air-cooled, water-cooled positive displacement and absorption chillers are at standard rating conditions defined in the reference test procedure.
c. Both the full-load and IPLV.IP requirements must be met or exceeded to comply with this standard. When there is a Path B, compliance can be with either Path A or Path B for any application.
d. NA means the requirements are not applicable for Path B, and only Path A can be used for compliance.
e. FL is the full-load performance requirements, and IPLV.IP is for the part-load performance requirements.
f. This table is a replica of ASHRAE 90.1 Table 6.8.1-3 Water-Chilling Packages—Minimum Efficiency Requirements.

COMMERCIAL ENERGY EFFICIENCY

TABLE C403.3.2(4)
ELECTRICALLY OPERATED PACKAGED TERMINAL AIR CONDITIONERS, PACKAGED TERMINAL HEAT PUMPS, SINGLE-PACKAGE VERTICAL AIR CONDITIONERS, SINGLE-PACKAGE VERTICAL HEAT PUMPS, ROOM AIR CONDITIONERS AND ROOM AIR-CONDITIONER HEAT PUMPS—MINIMUM EFFICIENCY REQUIREMENTS[e]

EQUIPMENT TYPE	SIZE CATEGORY (INPUT)	SUBCATEGORY OR RATING CONDITION	MINIMUM EFFICIENCY[d]	TEST PROCEDURE[a]
PTAC (cooling mode) standard size	< 7,000 Btu/h	95°F db/75°F wb outdoor air[c]	11.9 EER	AHRI 310/380
	≥ 7,000 Btu/h and ≤ 15,000 Btu/h		$14.0 - (0.300 \times Cap/1{,}000)$ EER[d]	
	> 15,000 Btu/h		9.5 EER	
PTAC (cooling mode) nonstandard size[a]	< 7,000 Btu/h	95°F db/75°F wb outdoor air[c]	9.4 EER	AHRI 310/380
	≥ 7,000 Btu/h and ≤ 15,000 Btu/h		$10.9 - (0.213 \times Cap/1{,}000)$ EER[d]	
	> 15,000 Btu/h		7.7 EER	
PTHP (cooling mode) standard size	< 7,000 Btu/h	95°F db/75°F wb outdoor air[c]	11.9 EER	AHRI 310/380
	≥ 7,000 Btu/h and ≤ 15,000 Btu/h		$14.0 - (0.300 \times Cap/1{,}000)$ EER[d]	
	> 15,000 Btu/h		9.5 EER	
PTHP (cooling mode) nonstandard size[b]	< 7,000 Btu/h	95°F db/75°F wb outdoor air[c]	9.3 EER	AHRI 310/380
	≥ 7,000 Btu/h and ≤ 15,000 Btu/h		$10.8 - (0.213 \times Cap/1{,}000)$ EER[d]	
	> 15,000 Btu/h		7.6 EER	
PTHP (heating mode) standard size	< 7,000 Btu/h	47°F db/43°F wb outdoor air	3.3 COP_H	AHRI 310/380
	≥ 7,000 Btu/h and ≤ 15,000 Btu/h		$3.7 - (0.052 \times Cap/1{,}000)$ COP_H[d]	
	> 15,000 Btu/h		2.90 COP_H	
PTHP (heating mode) nonstandard size[b]	< 7,000 Btu/h	47°F db/43°F wb outdoor air	2.7 COP_H	AHRI 310/380
	≥ 7,000 Btu/h and ≤ 15,000 Btu/h		$2.9 - (0.026 \times Cap/1000)$ COP_H[d]	
	> 15,000 Btu/h		2.5 COP_H	
SPVAC (cooling mode) single and three phase	< 65,000 Btu/h	95°F db/75°F wb outdoor air[c]	11.0 EER	AHRI 390
	≥ 65,000 Btu/h and ≤ 135,000 Btu/h		10.0 EER	
	≥ 135,000 Btu/h and ≤ 240,000 Btu/h		10.0 EER	
SPVHP (cooling mode)	< 65,000 Btu/h	95°F db/75°F wb outdoor air[c]	11.0 EER	AHRI 390
	≥ 65,000 Btu/h and ≤ 135,000 Btu/h		10.0 EER	
	≥ 135,000 Btu/h and ≤ 240,000 Btu/h		10.1 EER	
SPVHP (heating mode)	< 65,000 Btu/h	47°F db/43°F wb outdoor air	3.3 COP_H	AHRI 390
	≥ 65,000 Btu/h and ≤ 135,000 Btu/h		3.0 COP_H	
	≥ 135,000 Btu/h and ≤ 240,000 Btu/h		3.0 COP_H	

(continued)

TABLE C403.3.2(4)—continued
ELECTRICALLY OPERATED PACKAGED TERMINAL AIR CONDITIONERS, PACKAGED TERMINAL HEAT PUMPS, SINGLE-PACKAGE VERTICAL AIR CONDITIONERS, SINGLE-PACKAGE VERTICAL HEAT PUMPS, ROOM AIR CONDITIONERS AND ROOM AIR-CONDITIONER HEAT PUMPS—MINIMUM EFFICIENCY REQUIREMENTS[e]

EQUIPMENT TYPE	SIZE CATEGORY (INPUT)	SUBCATEGORY OR RATING CONDITION	MINIMUM EFFICIENCY[d]	TEST PROCEDURE[a]
Room air conditioners without reverse cycle with louvered sides for applications outside US	< 6,000 Btu/h	—	11.0 CEER	ANSI/AHAM RAC-1
	≥ 6,000 Btu/h and < 8,000 Btu/h	—	11.0 CEER	
	≥ 8,000 Btu/h and < 14,000 Btu/h	—	10.9 CEER	
	≥ 14,000 Btu/h and < 20,000 Btu/h	—	10.7 CEER	
	≥ 20,000 Btu/h and < 28,000 Btu/h	—	9.4 CEER	
	≥ 28,000 Btu/h	—	9.0 CEER	
Room air conditioners without louvered sides	< 6,000 Btu/h	—	10.0 CEER	ANSI/AHAM RAC-1
	≥ 6,000 Btu/h and < 8,000 Btu/h	—	10.0 CEER	
	≥ 8,000 Btu/h and < 11,000 Btu/h	—	9.6 CEER	
	≥ 11,000 Btu/h and < 14,000 Btu/h	—	9.5 CEER	
	≥ 14,000 Btu/h and < 20,000 Btu/h	—	9.3 CEER	
	≥ 20,000 Btu/h	—	9.4 CEER	
Room air conditioners with reverse cycle, with louvered sides for applications outside US	< 20,000 Btu/h	—	9.8 CEER	ANSI/AHAM RAC-1
	≥ 20,000 Btu/h	—	9.3 CEER	
Room air conditioners with reverse cycle without louvered sides for applications outside US	< 14,000 Btu/h	—	9.3 CEER	ANSI/AHAM RAC-1
	≥ 14,000 Btu/h	—	8.7 CEER	
Room air conditioners, casement only for applications outside US	All	—	9.5 CEER	ANSI/AHAM RAC-1
Room air conditioners, casement slider for applications outside US	All	—	10.4 CEER	ANSI/AHAM RAC-1

For SI: 1 British thermal unit per hour = 0.2931 W, °C = [(°F) – 32]/1.8, wb = wet bulb, db = dry bulb.

"Cap" = The rated cooling capacity of the project in Btu/h. Where the unit's capacity is less than 7,000 Btu/h, use 7,000 Btu/h in the calculation. Where the unit's capacity is greater than 15,000 Btu/h, use 15,000 Btu/h in the calculations.

a. Chapter 6 contains a complete specification of the referenced standards, which include test procedures, including the reference year version of the test procedure.
b. Nonstandard size units must be factory labeled as follows: "MANUFACTURED FOR NONSTANDARD SIZE APPLICATIONS ONLY; NOT TO BE INSTALLED IN NEW STANDARD PROJECTS." Nonstandard size efficiencies apply only to units being installed in existing sleeves having an external wall opening of less than 16 inches (406 mm) high or less than 42 inches (1067 mm) wide and having a cross-sectional area less than 670 square inches (0.43 m²).
c. The cooling-mode wet bulb temperature requirement only applies for units that reject condensate to the condenser coil.
d. "Cap" in EER and COPH equations for PTACs and PTHPs means cooling capacity in Btu/h at 95°F outdoor dry-bulb temperature.
e. This table is a replica of ASHRAE 90.1 Table 6.8.1-4 Electrically Operated Packaged Terminal Air Conditioners, Packaged Terminal Heat Pumps, Single-Package Vertical Air Conditioners, Single-Package Vertical Heat Pumps, Room Air Conditioners, and Room Air-Conditioner Heat Pumps—Minimum Efficiency Requirements.

TABLE C403.3.2(5)
WARM-AIR FURNACES AND COMBINATION WARM-AIR FURNACES/AIR-CONDITIONING UNITS, WARM-AIR DUCT FURNACES AND UNIT HEATERS—MINIMUM EFFICIENCY REQUIREMENTS[g]

EQUIPMENT TYPE	SIZE CATEGORY (INPUT)	SUBCATEGORY OR RATING CONDITION	MINIMUM EFFICIENCY	TEST PROCEDURE[a]
Warm-air furnace, gas fired for application outside the US	< 225,000 Btu/h	Maximum capacity[c]	80% AFUE (nonweatherized) or 81% AFUE (weatherized) or 80% $E_t^{b,d}$	DOE 10 CFR 430 Appendix N or Section 2.39, Thermal Efficiency, ANSI Z21.47
Warm-air furnace, gas fired	< 225,000 Btu/h	Maximum capacity[c]	80% $E_t^{b,d}$ before 1/1/2023 81% E_t^d after 1/1/2023	Section 2.39, Thermal Efficiency, ANSI Z21.47
Warm-air furnace, oil fired for application outside the US	< 225,000 Btu/h	Maximum capacity[c]	83% AFUE (nonweatherized) or 78% AFUE (weatherized) or 80% $E_t^{b,d}$	DOE 10 CFR 430 Appendix N or Section 42, Combustion, UL 727
Warm-air furnace, oil fired	< 225,000 Btu/h	Maximum capacity[c]	80% E_t before 1/1/2023 82% E_t^d after 1/1/2023	Section 42, Combustion, UL 727
Electric furnaces for applications outside the US	< 225,000 Btu/h	All	96% AFUE	DOE 10 CFR 430 Appendix N
Warm-air duct furnaces, gas fired	All capacities	Maximum capacity[c]	80% E_c^e	Section 2.10, Efficiency, ANSI Z83.8
Warm-air unit heaters, gas fired	All capacities	Maximum capacity[c]	80% $E_c^{e,f}$	Section 2.10, Efficiency, ANSI Z83.8
Warm-air unit heaters, oil fired	All capacities	Maximum capacity[c]	80% $E_c^{e,f}$	Section 40, Combustion, UL 731

For SI: 1 British thermal unit per hour = 0.2931 W.

a. Chapter 6 contains a complete specification of the referenced standards, which include test procedures, including the reference year version of the test procedure.
b. Combination units (i.e., furnaces contained within the same cabinet as an air conditioner) not covered by DOE 10 CFR 430 (i.e., three-phase power or with cooling capacity greater than or equal to 65,000 Btu/h) may comply with either rating. All other units greater than 225,000 Btu/h sold in the US must meet the AFUE standards for consumer products and test using USDOE's AFUE test procedure at DOE 10 CFR 430, Subpart B, Appendix N.
c. Compliance of multiple firing rate units shall be at the maximum firing rate.
d. E_t = thermal efficiency. Units must also include an interrupted or intermittent ignition device (IID), have jacket losses not exceeding 0.75 percent of the input rating, and have either power venting or a flue damper. A vent damper is an acceptable alternative to a flue damper for those furnaces where combustion air is drawn from the conditioned space.
e. E_c = combustion efficiency (100 percent less flue losses). See test procedure for detailed discussion.
f. Units must also include an interrupted or intermittent ignition device (IID) and have either power venting or an automatic flue damper.
g. This table is a replica of ASHRAE 90.1 Table 6.8.1-5 Warm-Air Furnaces and Combination Warm-Air Furnaces/Air-Conditioning Units, Warm-Air Duct Furnaces, and Unit Heaters—Minimum Efficiency Requirements.

COMMERCIAL ENERGY EFFICIENCY

TABLE C403.3.2(6)
GAS- AND OIL-FIRED BOILERS—MINIMUM EFFICIENCY REQUIREMENTS[i]

EQUIPMENT TYPE[b]	SUBCATEGORY OR RATING CONDITION	SIZE CATEGORY (INPUT)	MINIMUM EFFICIENCY	EFFICIENCY AS OF 3/2/2022	TEST PROCEDURE[a]
Boilers, hot water	Gas fired	< 300,000 Btu/h[g, h] for applications outside US	82% AFUE	82% AFUE	DOE 10 CFR 430 Appendix N
		≥ 300,000 Btu/h and ≤ 2,500,000 Btu/h[e]	80% E_t^d	80% E_t^d	DOE 10 CFR 431.86
		> 2,500,000 Btu/h[b]	82% E_c^c	82% E_c^c	
	Oil fired[f]	< 300,000 Btu/h[g, h] for applications outside US	84% AFUE	84% AFUE	DOE 10 CFR 430 Appendix N
		≥ 300,000 Btu/h and ≤ 2,500,000 Btu/h[e]	82% E_t^d	82% E_t^d	DOE 10 CFR 431.86
		> 2,500,000 Btu/h[b]	84% E_c^c	84% E_c^c	
Boilers, steam	Gas fired	< 300,000 Btu/h[g] for applications outside US	80% AFUE	80% AFUE	DOE 10 CFR 430 Appendix N
	Gas fired—all, except natural draft	≥ 300,000 Btu/h and ≤ 2,500,000 Btu/h[e]	79% E_t^d	79% E_t^d	DOE 10 CFR 431.86
		> 2,500,000 Btu/h[b]	79% E_t^d	79% E_t^d	
	Gas fired—natural draft	≥ 300,000 Btu/h and ≤ 2,500,000 Btu/h[e]	77% E_t^d	79% E_t^d	
		> 2,500,000 Btu/h[b]	77% E_t^d	79% E_t^d	
	Oil fired[f]	< 300,000 Btu/h[g] for applications outside US	82% AFUE	82% AFUE	DOE 10 CFR 430 Appendix N
		≥ 300,000 Btu/h and ≤ 2,500,000 Btu/h[e]	81% E_t^d	81% E_t^d	DOE 10 CFR 431.86
		> 2,500,000 Btu/h[b]	81% E_t^d	81% E_t^d	

For SI: 1 British thermal unit per hour = 0.2931 W.

a. Chapter 6 contains a complete specification of the referenced standards, which include test procedures, including the reference year version of the test procedure.
b. These requirements apply to boilers with rated input of 8,000,000 Btu/h or less that are not packaged boilers and to all packaged boilers. Minimum efficiency requirements for boilers cover all capacities of packaged boilers.
c. E_c = Combustion efficiency (100 percent less flue losses).
d. E_t = Thermal efficiency.
e. Maximum capacity—minimum and maximum ratings as provided for and allowed by the unit's controls.
f. Includes oil-fired (residual).
g. Boilers shall not be equipped with a constant burning pilot light.
h. A boiler not equipped with a tankless domestic water-heating coil shall be equipped with an automatic means for adjusting the temperature of the water such that an incremental change in inferred heat load produces a corresponding incremental change in the temperature of the water supplied.
i. This table is a replica of ASHRAE 90.1 Table 6.8.1-6 Gas- and Oil-Fired Boilers—Minimum Efficiency Requirements.

TABLE C403.3.2(7)
PERFORMANCE REQUIREMENTS FOR HEAT REJECTION EQUIPMENT—MINIMUM EFFICIENCY REQUIREMENTS[i]

EQUIPMENT TYPE	TOTAL SYSTEM HEAT-REJECTION CAPACITY AT RATED CONDITIONS	SUBCATEGORY OR RATING CONDITION[h]	PERFORMANCE REQUIRED[b, c, d, f, g]	TEST PROCEDURE[a, e]
Propeller or axial fan open-circuit cooling towers	All	95°F entering water 85°F leaving water 75°F entering wb	≥ 40.2 gpm/hp	CTI ATC-105 and CTI STD-201 RS
Centrifugal fan open-circuit cooling towers	All	95°F entering water 85°F leaving water 75°F entering wb	≥ 20.0 gpm/hp	CTI ATC-105 and CTI STD-201 RS
Propeller or axial fan closed-circuit cooling towers	All	102°F entering water 90°F leaving water 75°F entering wb	≥ 16.1 gpm/hp	CTI ATC-105S and CTI STD-201 RS
Centrifugal fan closed-circuit cooling towers	All	102°F entering water 90°F leaving water 75°F entering wb	≥ 7.0 gpm/hp	CTI ATC-105S and CTI STD-201 RS
Propeller or axial fan dry coolers (air-cooled fluid coolers)	All	115°F entering water 105°F leaving water 95°F entering wb	≥ 4.5 gpm/hp	CTI ATC-105DS
Propeller or axial fan evaporative condensers	All	R-448A test fluid 165°F entering gas temperature 105°F condensing temperature 75°F entering wb	≥ 160,000 Btu/h × hp	CTI ATC-106
Propeller or axial fan evaporative condensers	All	Ammonia test fluid 140°F entering gas temperature 96.3°F condensing temperature 75°F entering wb	≥ 134,000 Btu/h × hp	CTI ATC-106
Centrifugal fan evaporative condensers	All	R-448A test fluid 165°F entering gas temperature 105°F condensing temperature 75°F entering wb	≥ 137,000 Btu/h × hp	CTI ATC-106
Centrifugal fan evaporative condensers	All	Ammonia test fluid 140°F entering gas temperature 96.3°F condensing temperature 75°F entering wb	≥ 110,000 Btu/h × hp	CTI ATC-106
Air-cooled condensers	All	125°F condensing temperature 190°F entering gas temperature 15°F subcooling 95°F entering db	≥ 176,000 Btu/h × hp	AHRI 460

For SI: °C = [(°F) − 32]/1.8, L/s × kW = (gpm/hp)/(11.83), COP = (Btu/h × hp)/(2550.7), db = dry bulb temperature, wb = wet bulb temperature.

a. Chapter 6 contains a complete specification of the referenced standards, which include test procedures, including the reference year version of the test procedure.
b. For purposes of this table, open-circuit cooling tower performance is defined as the water-flow rating of the tower at the thermal rating condition listed in the table divided by the fan motor nameplate power.
c. For purposes of this table, closed-circuit cooling tower performance is defined as the process water-flow rating of the tower at the thermal rating condition listed in the table divided by the sum of the fan motor nameplate power and the integral spray pump motor nameplate power.

(continued)

TABLE C403.3.2(7)—continued
PERFORMANCE REQUIREMENTS FOR HEAT REJECTION EQUIPMENT—MINIMUM EFFICIENCY REQUIREMENTS[i]

d. For purposes of this table, dry-cooler performance is defined as the process water-flow rating of the unit at the thermal rating condition listed in the table divided by the total fan motor nameplate power of the unit, and air-cooled condenser performance is defined as the heat rejected from the refrigerant divided by the total fan motor nameplate power of the unit.

e. The efficiencies and test procedures for both open- and closed-circuit cooling towers are not applicable to hybrid cooling towers that contain a combination of separate wet and dry heat exchange sections. The certification requirements do not apply to field-erected cooling towers.

f. All cooling towers shall comply with the minimum efficiency listed in the table for that specific type of tower with the capacity effect of any project-specific accessories and/or options included in the capacity of the cooling tower.

g. For purposes of this table, evaporative condenser performance is defined as the heat rejected at the specified rating condition in the table, divided by the sum of the fan motor nameplate power and the integral spray pump nameplate power.

h. Requirements for evaporative condensers are listed with ammonia (R-717) and R-448A as test fluids in the table. Evaporative condensers intended for use with halocarbon refrigerants other than R-448A must meet the minimum efficiency requirements listed with R-448A as the test fluid. For ammonia, the condensing temperature is defined as the saturation temperature corresponding to the refrigerant pressure at the condenser entrance. For R-448A, which is a zeotropic refrigerant, the condensing temperature is defined as the arithmetic average of the dew point and the bubble point temperatures corresponding to the refrigerant pressure at the condenser entrance.

i. This table is a replica of ASHRAE 90.1 Table 6.8.1-7 Performance Requirements for Heat Rejection Equipment—Minimum Efficiency Requirements.

TABLE C403.3.2(8)
ELECTRICALLY OPERATED VARIABLE-REFRIGERANT-FLOW AIR CONDITIONERS—MINIMUM EFFICIENCY REQUIREMENTS[b]

EQUIPMENT TYPE	SIZE CATEGORY	HEATING SECTION TYPE	SUBCATEGORY OR RATING CONDITION	MINIMUM EFFICIENCY	TEST PROCEDURE[a]
VRF air conditioners, air cooled	< 65,000 Btu/h	All	VRF multisplit system	13.0 SEER	AHRI 1230
	≥ 65,000 Btu/h and < 135,000 Btu/h	Electric resistance (or none)	VRF multisplit system	11.2 EER 13.1 IEER 15.5 IEER	
	≥ 135,000 Btu/h and < 240,000 Btu/h	Electric resistance (or none)	VRF multisplit system	11.0 EER 12.9 IEER 14.9 IEER	
	≥ 240,000 Btu/h	Electric resistance (or none)	VRF multisplit system	10.0 EER 11.6 IEER 13.9 IEER	

For SI: 1 British thermal unit per hour = 0.2931 W.

a. Chapter 6 contains a complete specification of the referenced standards, which include test procedures, including the reference year version of the test procedure.

b. This table is a replica of ASHRAE 90.1 Table 6.8.1-8 Electrically Operated Variable-Refrigerant-Flow Air Conditioners—Minimum Efficiency Requirements.

TABLE C403.3.2(9)
ELECTRICALLY OPERATED VARIABLE-REFRIGERANT-FLOW AND APPLIED HEAT PUMPS—MINIMUM EFFICIENCY REQUIREMENTS[b]

EQUIPMENT TYPE	SIZE CATEGORY	HEATING SECTION TYPE	SUBCATEGORY OR RATING CONDITION	MINIMUM EFFICIENCY	TEST PROCEDURE[a]
VRF air cooled (cooling mode)	< 65,000 Btu/h	All		13.0 SEER	AHRI 1230
	≥ 65,000 Btu/h and < 135,000 Btu/h	Electric resistance (or none)	VRF multisplit system	11.0 EER 12.9 IEER 14.6 IEER	
			VRF multisplit system with heat recovery	10.8 EER 12.7 IEER 14.4 IEER	
	≥ 135,000 Btu/h and < 240,000 Btu/h		VRF multisplit system	10.6 EER 12.3 IEER 13.9 IEER	
			VRF multisplit system with heat recovery	10.4 EER 12.1 IEER 13.7 IEER	
	≥ 240,000 Btu/h		VRF multisplit system	9.5 EER 11.0 IEER 12.7 IEER	
			VRF multisplit system with heat recovery	9.3 EER 10.8 IEER 12.5 IEER	
VRF water source (cooling mode)	< 65,000 Btu/h	All	VRF multisplit systems 86°F entering water	12.0 EER 16.0 IEER	AHRI 1230
			VRF multisplit systems with heat recovery 86°F entering water	11.8 EER 15.8 IEER	
	≥ 65,000 Btu/h and < 135,000 Btu/h		VRF multisplit system 86°F entering water	12.0 EER 16.0 IEER	
			VRF multisplit system with heat recovery 86°F entering water	11.8 EER 15.8 IEER	
	≥ 135,000 Btu/h and < 240,000 Btu/h		VRF multisplit system 86°F entering water	10.0 EER 14.0 IEER	
			VRF multisplit system with heat recovery 86°F entering water	9.8 EER 13.8 IEER	
	≥ 240,000 Btu/h		VRF multisplit system 86°F entering water	10.0 EER 12.0 IEER	
			VRF multisplit system with heat recovery 86°F entering water	9.8 EER 11.8 IEER	
VRF groundwater source (cooling mode)	< 135,000 Btu/h	All	VRF multisplit system 59°F entering water	16.2 EER	AHRI 1230
			VRF multisplit system with heat recovery 59°F entering water	16.0 EER	
	≥ 135,000 Btu/h		VRF multisplit system 59°F entering water	13.8 EER	
			VRF multisplit system with heat recovery 59°F entering water	13.6 EER	

(continued)

TABLE C403.3.2(9)—continued
ELECTRICALLY OPERATED VARIABLE-REFRIGERANT-FLOW AND
APPLIED HEAT PUMPS—MINIMUM EFFICIENCY REQUIREMENTS[b]

EQUIPMENT TYPE	SIZE CATEGORY	HEATING SECTION TYPE	SUBCATEGORY OR RATING CONDITION	MINIMUM EFFICIENCY	TEST PROCEDURE[a]
VRF ground source (cooling mode)	< 135,000 Btu/h		VRF multisplit system 77°F entering water	13.4 EER	AHRI 1230
	< 135,000 Btu/h		VRF multisplit system with heat recovery 77°F entering water	13.2 EER	
	≥ 135,000 Btu/h		VRF multisplit system 77°F entering water	11.0 EER	
	≥ 135,000 Btu/h		VRF multisplit system with heat recovery 77°F entering water	10.8 EER	
VRF air cooled (heating mode)	< 65,000 Btu/h (cooling capacity)	All	VRF multisplit system	7.7 HSPF	AHRI 1230
	≥ 65,000 Btu/h and < 135,000 Btu/h (cooling capacity)		VRF multisplit system 47°F db/43°F wb outdoor air	3.3 COP_H	
	≥ 65,000 Btu/h and < 135,000 Btu/h (cooling capacity)		17°F db/15°F wb outdoor air	2.25 COP_H	
	≥ 135,000 Btu/h (cooling capacity)		VRF multisplit system 47°F db/43°F wb outdoor air	3.2 COP_H	
	≥ 135,000 Btu/h (cooling capacity)		17°F db/15°F wb outdoor air	2.05 COP_H	
VRF water source (heating mode)	< 65,000 Btu/h (cooling capacity)		VRF multisplit system 68°F entering water	4.2 COP_H / 4.3 COP_H	AHRI 1230
	≥ 65,000 Btu/h and < 135,000 Btu/h (cooling capacity)		VRF multisplit system 68°F entering water	4.2 COP_H / 4.3 COP_H	
	≥ 135,000 Btu/h and < 240,000 Btu/h (cooling capacity)		VRF multisplit system 68°F entering water	3.9 COP_H / 4.0 COP_H	
	≥ 240,000 Btu/h (cooling capacity)		VRF multisplit system 68°F entering water	3.9 COP_H	
VRF groundwater source (heating mode)	< 135,000 Btu/h (cooling capacity)		VRF multisplit system 50°F entering water	3.6 COP_H	AHRI 1230
	≥ 135,000 Btu/h (cooling capacity)		VRF multisplit system 50°F entering water	3.3 COP_H	
VRF ground source (heating mode)	< 135,000 Btu/h (cooling capacity)		VRF multisplit system 32°F entering water	3.1 COP_H	AHRI 1230
	≥ 135,000 Btu/h (cooling capacity)		VRF multisplit system 32°F entering water	2.8 COP_H	

For SI: °C = [(°F) − 32]/1.8, 1 British thermal unit per hour = 0.2931 W, db = dry bulb temperature, wb = wet bulb temperature.

a. Chapter 6 contains a complete specification of the referenced standards, which include test procedures, including the reference year version of the test procedure.
b. This table is a replica of ASHRAE 90.1 Table 6.8.1-9 Electrically Operated Variable-Refrigerant-Flow and Applied Heat Pumps—Minimum Efficiency Requirements.

TABLE C403.3.2(10)
FLOOR-MOUNTED AIR CONDITIONERS AND CONDENSING UNITS SERVING COMPUTER ROOMS—MINIMUM EFFICIENCY REQUIREMENTS[b]

EQUIPMENT TYPE	STANDARD MODEL	NET SENSIBLE COOLING CAPACITY	MINIMUM NET SENSIBLE COP	RATING CONDITIONS RETURN AIR (dry bulb/dew point)	TEST PROCEDURE[a]
Air cooled	Downflow	< 80,000 Btu/h	2.70	85°F/52°F (Class 2)	AHRI 1360
		≥ 80,000 Btu/h and < 295,000 Btu/h	2.58		
		≥ 295,000 Btu/h	2.36		
	Upflow—ducted	< 80,000 Btu/h	2.67		
		≥ 80,000 Btu/h and < 295,000 Btu/h	2.55		
		≥ 295,000 Btu/h	2.33		
	Upflow—nonducted	< 65,000 Btu/h	2.16	75°F/52°F (Class 1)	
		≥ 65,000 Btu/h and < 240,000 Btu/h	2.04		
		≥ 240,000 Btu/h	1.89		
	Horizontal	< 65,000 Btu/h	2.65	95°F/52°F (Class 3)	
		≥ 65,000 Btu/h and < 240,000 Btu/h	2.55		
		≥ 240,000 Btu/h	2.47		
Air cooled with fluid economizer	Downflow	< 80,000 Btu/h	2.70	85°F/52°F (Class 1)	AHRI 1360
		≥ 80,000 Btu/h and < 295,000 Btu/h	2.58		
		≥ 295,000 Btu/h	2.36		
	Upflow—ducted	< 80,000 Btu/h	2.67		
		≥ 80,000 Btu/h and < 295,000 Btu/h	2.55		
		≥ 295,000 Btu/h	2.33		
	Upflow—nonducted	< 65,000 Btu/h	2.09	75°F/52°F (Class 1)	
		≥ 65,000 Btu/h and < 240,000 Btu/h	1.99		
		≥ 240,000 Btu/h	1.81		
	Horizontal	< 65,000 Btu/h	2.65	95°F/52°F (Class 3)	
		≥ 65,000 Btu/h and < 240,000 Btu/h	2.55		
		≥ 240,000 Btu/h	2.47		
Water cooled	Downflow	< 80,000 Btu/h	2.82	85°F/52°F (Class 1)	AHRI 1360
		≥ 80,000 Btu/h and < 295,000 Btu/h	2.73		
		≥ 295,000 Btu/h	2.67		
	Upflow—ducted	< 80,000 Btu/h	2.79		
		≥ 80,000 Btu/h and < 295,000 Btu/h	2.70		
		≥ 295,000 Btu/h	2.64		
	Upflow—nonducted	< 65,000 Btu/h	2.43	75°F/52°F (Class 1)	
		≥ 65,000 Btu/h and < 240,000 Btu/h	2.32		
		≥ 240,000 Btu/h	2.20		
	Horizontal	< 65,000 Btu/h	2.79	95°F/52°F (Class 3)	
		≥ 65,000 Btu/h and < 240,000 Btu/h	2.68		
		≥ 240,000 Btu/h	2.60		

(continued)

TABLE C403.3.2(10)—continued
FLOOR-MOUNTED AIR CONDITIONERS AND CONDENSING UNITS SERVING COMPUTER ROOMS—MINIMUM EFFICIENCY REQUIREMENTS[b]

EQUIPMENT TYPE	STANDARD MODEL	NET SENSIBLE COOLING CAPACITY	MINIMUM NET SENSIBLE COP	RATING CONDITIONS RETURN AIR (dry bulb/dew point)	TEST PROCEDURE[a]
Water cooled with fluid economizer	Downflow	< 80,000 Btu/h	2.77	85°F/52°F (Class 1)	AHRI 1360
		≥ 80,000 Btu/h and < 295,000 Btu/h	2.68		
		≥ 295,000 Btu/h	2.61		
	Upflow—ducted	< 80,000 Btu/h	2.74		
		≥ 80,000 Btu/h and < 295,000 Btu/h	2.65		
		≥ 295,000 Btu/h	2.58		
	Upflow—nonducted	< 65,000 Btu/h	2.35	75°F/52°F (Class 1)	
		≥ 65,000 Btu/h and < 240,000 Btu/h	2.24		
		≥ 240,000 Btu/h	2.12		
	Horizontal	< 65,000 Btu/h	2.71	95°F/52°F (Class 3)	
		≥ 65,000 Btu/h and < 240,000 Btu/h	2.60		
		≥ 240,000 Btu/h	2.54		
Glycol cooled	Downflow	< 80,000 Btu/h	2.56	85°F/52°F (Class 1)	AHRI 1360
		≥ 80,000 Btu/h and < 295,000 Btu/h	2.24		
		≥ 295,000 Btu/h	2.21		
	Upflow—ducted	< 80,000 Btu/h	2.53		
		≥ 80,000 Btu/h and < 295,000 Btu/h	2.21		
		≥ 295,000 Btu/h	2.18		
	Upflow, nonducted	< 65,000 Btu/h	2.08	75°F/52°F (Class 1)	
		≥ 65,000 Btu/h and < 240,000 Btu/h	1.90		
		≥ 240,000 Btu/h	1.81		
	Horizontal	< 65,000 Btu/h	2.48	95°F/52°F (Class 3)	
		≥ 65,000 Btu/h and < 240,000 Btu/h	2.18		
		≥ 240,000 Btu/h	2.18		
Glycol cooled with fluid economizer	Downflow	< 80,000 Btu/h	2.51	85°F/52°F (Class 1)	AHRI 1360
		≥ 80,000 Btu/h and < 295,000 Btu/h	2.19		
		≥ 295,000 Btu/h	2.15		
	Upflow—ducted	< 80,000 Btu/h	2.48		
		≥ 80,000 Btu/h and < 295,000 Btu/h	2.16		
		≥ 295,000 Btu/h	2.12		
	Upflow—nonducted	< 65,000 Btu/h	2.00	75°F/52°F (Class 1)	
		≥ 65,000 Btu/h and < 240,000 Btu/h	1.82		
		≥ 240,000 Btu/h	1.73		
	Horizontal	< 65,000 Btu/h	2.44	95°F/52°F (Class 3)	
		≥ 65,000 Btu/h and < 240,000 Btu/h	2.10		
		≥ 240,000 Btu/h	2.10		

For SI: 1 British thermal unit per hour = 0.2931 W, °C = [(°F) − 32]/1.8, COP = (Btu/h × hp)/(2,550.7).

a. Chapter 6 contains a complete specification of the referenced standards, which include test procedures, including the reference year version of the test procedure.
b. This table is a replica of ASHRAE 90.1 Table 6.8.1-10 Floor-Mounted Air Conditioners and Condensing Units Serving Computer Rooms—Minimum Efficiency Requirements.

COMMERCIAL ENERGY EFFICIENCY

TABLE C403.3.2(11)
VAPOR-COMPRESSION-BASED INDOOR POOL DEHUMIDIFIERS—MINIMUM EFFICIENCY REQUIREMENTS[b]

EQUIPMENT TYPE	SUBCATEGORY OR RATING CONDITION	MINIMUM EFFICIENCY	TEST PROCEDURE[a]
Single package indoor (with or without economizer)	Rating Conditions: A or C	3.5 MRE	AHRI 910
Single package indoor water cooled (with or without economizer)	Rating Conditions: A, B or C	3.5 MRE	
Single package indoor air cooled (with or without economizer)	Rating Conditions: A, B or C	3.5 MRE	
Split system indoor air cooled (with or without economizer)	Rating Conditions: A, B or C	3.5 MRE	

a. Chapter 6 contains a complete specification of the referenced standards, which include test procedures, including the reference year version of the test procedure.
b. This table is a replica of ASHRAE 90.1 Table 6.8.1-12 Vapor-Compression-Based Indoor Pool Dehumidifiers—Minimum Efficiency Requirements.

TABLE C403.3.2(12)
ELECTRICALLY OPERATED DX-DOAS UNITS, SINGLE-PACKAGE AND REMOTE CONDENSER, WITHOUT ENERGY RECOVERY—MINIMUM EFFICIENCY REQUIREMENTS[b]

EQUIPMENT TYPE	SUBCATEGORY OR RATING CONDITION	MINIMUM EFFICIENCY	TEST PROCEDURE[a]
Air cooled (dehumidification mode)	—	4.0 ISMRE	AHRI 920
Air-source heat pumps (dehumidification mode)	—	4.0 ISMRE	AHRI 920
Water cooled (dehumidification mode)	Cooling tower condenser water	4.9 ISMRE	AHRI 920
	Chilled water	6.0 ISMRE	
Air-source heat pump (heating mode)	—	2.7 ISCOP	AHRI 920
Water-source heat pump (dehumidification mode)	Ground source, closed loop	4.8 ISMRE	AHRI 920
	Ground-water source	5.0 ISMRE	
	Water source	4.0 ISMRE	
Water-source heat pump (heating mode)	Ground source, closed loop	2.0 ISCOP	AHRI 920
	Ground-water source	3.2 ISCOP	
	Water source	3.5 ISCOP	

a. Chapter 6 contains a complete specification of the referenced standards, which include test procedures, including the reference year version of the test procedure.
b. This table is a replica of ASHRAE 90.1 Table 6.8.1-13 Electrically Operated DX-DOAS Units, Single-Package and Remote Condenser, without Energy Recovery—Minimum Efficiency Requirements.

TABLE C403.3.2(13)
ELECTRICALLY OPERATED DX-DOAS UNITS, SINGLE-PACKAGE AND REMOTE CONDENSER, WITH ENERGY RECOVERY—MINIMUM EFFICIENCY REQUIREMENTS[b]

EQUIPMENT TYPE	SUBCATEGORY OR RATING CONDITION	MINIMUM EFFICIENCY	TEST PROCEDURE[a]
Air cooled (dehumidification mode)	—	5.2 ISMRE	AHRI 920
Air-source heat pumps (dehumidification mode)	—	5.2 ISMRE	AHRI 920
Water cooled (dehumidification mode)	Cooling tower condenser water	5.3 ISMRE	AHRI 920
	Chilled water	6.6 ISMRE	
Air-source heat pump (heating mode)	—	3.3 ISCOP	AHRI 920
Water-source heat pump (dehumidification mode)	Ground source, closed loop	5.2 ISMRE	AHRI 920
	Ground-water source	5.8 ISMRE	
	Water source	4.8 ISMRE	
Water-source heat pump (heating mode)	Ground source, closed loop	3.8 ISCOP	AHRI 920
	Ground-water source	4.0 ISCOP	
	Water source	4.8 ISCOP	

a. Chapter 6 contains a complete specification of the referenced standards, which include test procedures, including the reference year version of the test procedure.
b. This table is a replica of ASHRAE 90.1 Table 6.8.1-14 Electrically Operated DX-DOAS Units, Single-Package and Remote Condenser, with Energy Recovery—Minimum Efficiency Requirements.

TABLE C403.3.2(14)
ELECTRICALLY OPERATED WATER-SOURCE HEAT PUMPS—MINIMUM EFFICIENCY REQUIREMENTS[c]

EQUIPMENT TYPE	SIZE CATEGORY[b]	HEATING SECTION TYPE	SUBCATEGORY OR RATING CONDITION	MINIMUM EFFICIENCY	TEST PROCEDURE[a]
Water-to-air, water loop (cooling mode)	< 17,000 Btu/h	All	86°F entering water	12.2 EER	ISO 13256-1
	≥ 17,000 Btu/h and < 65,000 Btu/h			13.0 EER	
	≥ 65,000 Btu/h and < 135,000 Btu/h			13.0 EER	
Water-to-air, ground water (cooling mode)	< 135,000 Btu/h	All	59°F entering water	18.0 EER	ISO 13256-1
Brine-to-air, ground loop (cooling mode)	< 135,000 Btu/h	All	77°F entering water	14.1 EER	ISO 13256-1
Water-to-water, water loop (cooling mode)	< 135,000 Btu/h	All	86°F entering water	10.6 EER	ISO 13256-2
Water-to-water, ground water (cooling mode)	< 135,000 Btu/h	All	59°F entering water	16.3 EER	ISO 13256-2
Brine-to-water, ground loop (cooling mode)	< 135,000 Btu/h	All	77°F entering water	12.1 EER	ISO 13256-2
Water-to-water, water loop (heating mode)	< 135,000 Btu/h (cooling capacity)	—	68°F entering water	4.3 COP_H	ISO 13256-1
Water-to-air, ground water (heating mode)	< 135,000 Btu/h (cooling capacity)	—	50°F entering water	3.7 COP_H	ISO 13256-1
Brine-to-air, ground loop (heating mode)	< 135,000 Btu/h (cooling capacity)	—	32°F entering water	3.2 COP_H	ISO 13256-1
Water-to-water, water loop (heating mode)	< 135,000 Btu/h (cooling capacity)	—	68°F entering water	3.7 COP_H	ISO 13256-1
Water-to-water, ground water (heating mode)	< 135,000 Btu/h (cooling capacity)	—	50°F entering water	3.1 COP_H	ISO 13256-2
Brine-to-water, ground loop (heating mode)	< 135,000 Btu/h (cooling capacity)	—	32°F entering water	2.5 COP_H	ISO 13256-2

For SI: 1 British thermal unit per hour = 0.2931 W, °C = [(°F) − 32]/1.8.

a. Chapter 6 contains a complete specification of the referenced standards, which include test procedures, including the reference year version of the test procedure.
b. Single-phase, US air-cooled heat pumps less than 19 kW are regulated as consumer products by DOE 10 CFR 430. SCOPC, SCOP2C, SCOPH and SCOP2H values for single-phase products are set by the USDOE.
c. This table is a replica of ASHRAE 90.1 Table 6.8.1-15 Electrically Operated Water-Source Heat Pumps—Minimum Efficiency Requirements.

COMMERCIAL ENERGY EFFICIENCY

TABLE C403.3.2(15)
HEAT-PUMP AND HEAT RECOVERY CHILLER PACKAGES—MINIMUM EFFICIENCY REQUIREMENTS[g]

EQUIPMENT TYPE	SIZE CATEGORY, ton_R	COOLING-ONLY OPERATION COOLING EFFICIENCY[e] AIR-SOURCE EER (FL/IPLV), Btu/W × h WATER-SOURCE POWER INPUT PER CAPACITY (FL/IPLV), kW/ton_R		HEATING SOURCE CONDITIONS (entering/leaving water) OR OAT (db/wb), °F	HEAT-PUMP HEATING FULL-LOAD EFFICIENCY (COP_H)[b], W/W				HEAT RECOVERY CHILLER FULL-LOAD EFFICIENCY (COP_{HR})[c,d], W/W SIMULTANEOUS COOLING AND HEATING FULL-LOAD EFFICIENCY (COP_{SHC})[c], W/W				Test Procedure[a]
					Leaving Heating Water Temperature				Leaving Heating Water Temperature				
		Path A	Path B		Low 105°F	Medium 120°F	High 140°F	Boost 140°F	Low 105°F	Medium 120°F	High 140°F	Boost 140°F	
Air source	All sizes	≥ 9.595 FL ≥ 13.02 IPLV.IP	≥ 9.215 FL ≥ 15.01 IPLV.IP	47 db 43 wb[e]	≥ 3.290	≥ 2.770	≥ 2.310	NA	NA	NA	NA	NA	AHRI 550/590
		≥ 9.595 FL ≥ 13.30 IPLV.IP	≥ 9.215 FL ≥ 15.30 IPLV.IP	17 db 15 wb[e]	≥ 2.230	≥ 1.950	≥ 1.630	NA	NA	NA	NA	NA	
Water-source electrically operated positive displacement	< 75	≤ 0.7885 FL ≤ 0.6316 IPLV.IP	≤ 0.7875 FL ≤ 0.5145 IPLV.IP	54/44[f]	≥ 4.640	≥ 3.680	≥ 2.680	NA	≥ 8.330	≥ 6.410	≥ 4.420	NA	AHRI 550/590
				75/65[f]	NA	NA	NA	≥ 3.550	NA	NA	NA	6.150	
	≥ 75 and < 150	≤ 0.7579 FL ≤ 0.5895 IPLV.IP	≤ 0.7140 FL ≤ 0.4620 IPLV.IP	54/44[f]	≥ 4.640	≥ 3.680	≥ 2.680	NA	≥ 8.330	≥ 6.410	≥ 4.420	NA	
				75/65[f]	NA	NA	NA	≥ 3.550	NA	NA	NA	6.150	
	≥ 150 and < 300	≤ 0.6947 FL ≤ 0.5684 IPLV.IP	≤ 0.7140 FL ≤ 0.4620 IPLV.IP	54/44[f]	≥ 4.640	≥ 3.680	≥ 2.680	NA	≥ 8.330	≥ 6.410	≥ 4.420	NA	
				75/65[f]	NA	NA	NA	≥ 3.550	NA	NA	NA	6.150	
	≥ 300 and < 600	≤ 0.6421 FL ≤ 0.5474 IPLV.IP	≤ 0.6563 FL ≤ 0.4305 IPLV.IP	54/44[f]	≥ 4.930	≥ 3.960	≥ 2.970	NA	≥ 8.900	≥ 6.980	≥ 5.000	NA	
				75/65[f]	NA	NA	NA	≥ 3.900	NA	NA	NA	6.850	
	≥ 600	≤ 0.5895 FL ≤ 0.5263 IPLV.IP	≤ 0.6143 FL ≤ 0.3990 IPLV.IP	54/44[f]	≥ 4.930	≥ 3.960	≥ 2.970	NA	≥ 8.900	≥ 6.980	≥ 5.000	NA	
				75/65[f]	NA	NA	NA	≥ 3.900	NA	NA	NA	6.850	
Water-source electrically operated centrifugal	< 75	≤ 0.6421 FL ≤ 0.5789 IPLV.IP	≤ 0.7316 FL ≤ 0.4632 IPLV.IP	54/44[f]	≥ 4.640	≥ 3.680	≥ 2.680	NA	≥ 8.330	≥ 6.410	≥ 4.420	NA	AHRI 550/590
				75/65[f]	NA	NA	NA	≥ 3.550	NA	NA	NA	≥ 6.150	
	≥ 75 and < 150	≤ 0.5895 FL ≤ 0.5474 IPLV.IP	≤ 0.6684 FL ≤ 0.4211 IPLV.IP	54/44[f]	≥ 4.640	≥ 3.680	≥ 2.680	NA	≥ 8.330	≥ 6.410	≥ 4.420	NA	
				75/65[f]	NA	NA	NA	≥ 3.550	NA	NA	NA	≥ 6.150	
	≥ 150 and < 300	≤ 0.5895 FL ≤ 0.5263 IPLV.IP	≤ 0.6263 FL ≤ 0.4105 IPLV.IP	54/44[f]	≥ 4.640	≥ 3.680	≥ 2.680	NA	≥ 8.330	≥ 6.410	≥ 4.420	NA	
				75/65[f]	NA	NA	NA	≥ 3.550	NA	NA	NA	≥ 6.150	
	≥ 300 and < 600	≤ 0.5895 FL ≤ 0.5263 IPLV.IP	≤ 0.6158 FL ≤ 0.4000 IPLV.IP	54/44[f]	≥ 4.930	≥ 3.960	≥ 2.970	NA	≥ 8.900	≥ 6.980	≥ 5.000	NA	
				75/65[f]	NA	NA	NA	≥ 3.900	NA	NA	NA	≥ 6.850	
	≥ 600	≤ 0.5895 FL ≤ 0.5263 IPLV.IP	≤ 0.6158 FL ≤ 0.4000 IPLV.IP	54/44[f]	≥ 4.930	≥ 3.960	≥ 2.970	NA	≥ 8.900	≥ 6.980	≥ 5.000	NA	
				75/65[f]	NA	NA	NA	≥ 3.900	NA	NA	NA	≥ 6.850	

For SI: °C = [(°F) − 32]/1.8.

a. Chapter 6 contains a complete specification of the referenced standards, which include test procedures, including the reference year version of the test procedure.
b. Cooling-only rating conditions are standard rating conditions defined in AHRI 550/590, Table 1.
c. Heating full-load rating conditions are at rating conditions defined in AHRI 550/590, Table 1.
d. For water-cooled heat recovery chillers that have capabilities for heat rejection to a heat recovery condenser and a tower condenser, the COP_{HR} applies to operation at full load with 100 percent heat recovery (no tower rejection). Units that only have capabilities for partial heat recovery shall meet the requirements of Table C403.3.2(3).
e. Outdoor air entering dry-bulb (db) and wet-bulb (wb) temperature.
f. Source-water entering and leaving water temperature.
g. This table is a replica of ASHRAE 90.1 Table 6.8.1-16 Heat-Pump and Heat Recovery Chiller Packages—Minimum Efficiency Requirements.

TABLE C403.3.2(16)
CEILING-MOUNTED COMPUTER-ROOM AIR CONDITIONERS—MINIMUM EFFICIENCY REQUIREMENTS[b]

EQUIPMENT TYPE	STANDARD MODEL	NET SENSIBLE COOLING CAPACITY	MINIMUM NET SENSIBLE COP	RATING CONDITIONS RETURN AIR (dry bulb/dew point)	TEST PROCEDURE[a]
Air cooled with free air discharge condenser	Ducted	< 29,000 Btu/h	2.05	75°F/52°F (Class 1)	AHRI 1360
		≥ 29,000 Btu/h and < 65,000 Btu/h	2.02		
		≥ 65,000 Btu/h	1.92		
	Nonducted	< 29,000 Btu/h	2.08		
		≥ 29,000 Btu/h and < 65,000 Btu/h	2.05		
		≥ 65,000 Btu/h	1.94		
Air cooled with free air discharge condenser with fluid economizer	Ducted	< 29,000 Btu/h	2.01	75°F/52°F (Class 1)	AHRI 1360
		≥ 29,000 Btu/h and < 65,000 Btu/h	1.97		
		≥ 65,000 Btu/h	1.87		
	Nonducted	< 29,000 Btu/h	2.04		
		≥ 29,000 Btu/h and < 65,000 Btu/h	2.00		
		≥ 65,000 Btu/h	1.89		
Air cooled with ducted condenser	Ducted	< 29,000 Btu/h	1.86	75°F/52°F (Class 1)	AHRI 1360
		≥ 29,000 Btu/h and < 65,000 Btu/h	1.83		
		≥ 65,000 Btu/h	1.73		
	Nonducted	< 29,000 Btu/h	1.89		
		≥ 29,000 Btu/h and < 65,000 Btu/h	1.86		
		≥ 65,000 Btu/h	1.75		
Air cooled with fluid economizer and ducted condenser	Ducted	< 29,000 Btu/h	1.82	75°F/52°F (Class 1)	AHRI 1360
		≥ 29,000 Btu/h and < 65,000 Btu/h	1.78		
		≥ 65,000 Btu/h	1.68		
	Nonducted	< 29,000 Btu/h	1.85		
		≥ 29,000 Btu/h and < 65,000 Btu/h	1.81		
		≥ 65,000 Btu/h	1.70		
Water cooled	Ducted	< 29,000 Btu/h	2.38	75°F/52°F (Class 1)	AHRI 1360
		≥ 29,000 Btu/h and < 65,000 Btu/h	2.28		
		≥ 65,000 Btu/h	2.18		
	Nonducted	< 29,000 Btu/h	2.41		
		≥ 29,000 Btu/h and < 65,000 Btu/h	2.31		
		≥ 65,000 Btu/h	2.20		

(continued)

TABLE C403.3.2(16)—continued
CEILING-MOUNTED COMPUTER-ROOM AIR CONDITIONERS—MINIMUM EFFICIENCY REQUIREMENTS[b]

EQUIPMENT TYPE	STANDARD MODEL	NET SENSIBLE COOLING CAPACITY	MINIMUM NET SENSIBLE COP	RATING CONDITIONS RETURN AIR (dry bulb/dew point)	TEST PROCEDURE[a]
Water cooled with fluid economizer	Ducted	< 29,000 Btu/h	2.33	75°F/52°F (Class 1)	AHRI 1360
		≥ 29,000 Btu/h and < 65,000 Btu/h	2.23		
		≥ 65,000 Btu/h	2.13		
	Nonducted	< 29,000 Btu/h	2.36		
		≥ 29,000 Btu/h and < 65,000 Btu/h	2.26		
		≥ 65,000 Btu/h	2.16		
Glycol cooled	Ducted	< 29,000 Btu/h	1.97	75°F/52°F (Class 1)	AHRI 1360
		≥ 29,000 Btu/h and < 65,000 Btu/h	1.93		
		≥ 65,000 Btu/h	1.78		
	Nonducted	< 29,000 Btu/h	2.00		
		≥ 29,000 Btu/h and < 65,000 Btu/h	1.98		
		≥ 65,000 Btu/h	1.81		
Glycol cooled with fluid economizer	Ducted	< 29,000 Btu/h	1.92	75°F/52°F (Class 1)	AHRI 1360
		≥ 29,000 Btu/h and < 65,000 Btu/h	1.88		
		≥ 65,000 Btu/h	1.73		
	Nonducted	< 29,000 Btu/h	1.95		
		≥ 29,000 Btu/h and < 65,000 Btu/h	1.93		
		≥ 65,000 Btu/h	1.76		

For SI: 1 British thermal unit per hour = 0.2931 W, °C = [(°F) − 32]/1.8, COP = (Btu/h × hp)/(2,550.7).

a. Chapter 6 contains a complete specification of the referenced standards, which include test procedures, including the reference year version of the test procedure.
b. This is a replica of ASHRAE 90.1 Table 6.8.1-17 Ceiling-Mounted Computer-Room Air Conditioners—Minimum Efficiency Requirements.

TABLE C403.3.3
MAXIMUM HOT GAS BYPASS CAPACITY

RATED CAPACITY	MAXIMUM HOT GAS BYPASS CAPACITY (% of total capacity)
≤ 240,000 Btu/h	50
> 240,000 Btu/h	25

For SI: 1 British thermal unit per hour = 0.2931 W.

TABLE C403.3.4
BOILER TURNDOWN

BOILER SYSTEM DESIGN INPUT (Btu/h)	MINIMUM TURNDOWN RATIO
≥ 1,000,000 and ≤ 5,000,000	3 to 1
> 5,000,000 and ≤ 10,000,000	4 to 1
> 10,000,000	5 to 1

For SI: 1 British thermal unit per hour = 0.2931 W.

C403.3.4 Boiler turndown. *Boiler systems* with design input of greater than 1,000,000 Btu/h (293 kW) shall comply with the turndown ratio specified in Table C403.3.4.

The system turndown requirement shall be met through the use of multiple single-input boilers, one or more *modulating boilers* or a combination of single-input and *modulating boilers*.

C403.4 Heating and cooling system controls. Each heating and cooling system shall be provided with controls in accordance with Sections C403.4.1 through C403.4.5.

C403.4.1 Thermostatic controls. The supply of heating and cooling energy to each *zone* shall be controlled by individual thermostatic controls capable of responding to temperature within the *zone*. Where humidification or dehumidification or both is provided, not fewer than one humidity control device shall be provided for each humidity control system.

Exception: Independent perimeter systems that are designed to offset only building envelope heat losses, gains or both serving one or more perimeter *zones* also served by an interior system provided that both of the following conditions are met:

1. The perimeter system includes not fewer than one thermostatic control *zone* for each building exposure having exterior walls facing only one orientation (within ±45 degrees) (0.8 rad) for more than 50 contiguous feet (15 240 mm).
2. The perimeter system heating and cooling supply is controlled by thermostats located within the *zones* served by the system.

C403.4.1.1 Heat pump supplementary heat. Heat pumps having supplementary electric resistance heat shall have controls that limit supplemental heat operation to only those times when one of the following applies:

1. The vapor compression cycle cannot provide the necessary heating energy to satisfy the thermostat setting.
2. The heat pump is operating in defrost mode.
3. The vapor compression cycle malfunctions.
4. The thermostat malfunctions.

C403.4.1.2 Deadband. Where used to control both heating and cooling, *zone* thermostatic controls shall be configured to provide a temperature range or deadband of not less than 5°F (2.8°C) within which the supply of heating and cooling energy to the *zone* is shut off or reduced to a minimum.

Exceptions:

1. Thermostats requiring manual changeover between heating and cooling modes.
2. Occupancies or applications requiring precision in indoor temperature control as *approved* by the *code official*.

C403.4.1.3 Setpoint overlap restriction. Where a *zone* has a separate heating and a separate cooling thermostatic control located within the *zone*, a limit switch, mechanical stop or direct digital control system with software programming shall be configured to prevent the heating setpoint from exceeding the cooling setpoint and to maintain a deadband in accordance with Section C403.4.1.2.

C403.4.1.4 Heated or cooled vestibules. The heating system for heated vestibules and air curtains with integral heating shall be provided with controls configured to shut off the source of heating when the outdoor air temperature is greater than 45°F (7°C). Vestibule heating and cooling systems shall be controlled by a thermostat located in the vestibule configured to limit heating to a temperature not greater than 60°F (16°C) and cooling to a temperature not less than 85°F (29°C).

Exception: Control of heating or cooling provided by site-recovered energy or transfer air that would otherwise be exhausted.

C403.4.1.5 Hot water boiler outdoor temperature setback control. Hot water boilers that supply heat to the building through one- or two-pipe heating systems shall have an outdoor setback control that lowers the boiler water temperature based on the outdoor temperature.

C403.4.2 Off-hour controls. Each *zone* shall be provided with thermostatic setback controls that are controlled by either an automatic time clock or programmable control system.

Exceptions:

1. *Zones* that will be operated continuously.
2. *Zones* with a full HVAC load demand not exceeding 6,800 Btu/h (2 kW) and having a manual shutoff switch located with *ready access*.

C403.4.2.1 Thermostatic setback. Thermostatic setback controls shall be configured to set back or temporarily operate the system to maintain zone temperatures down to 55°F (13°C) or up to 85°F (29°C).

C403.4.2.2 Automatic setback and shutdown. Automatic time clock or programmable controls shall be capable of starting and stopping the system for seven different daily schedules per week and retaining their programming and time setting during a loss of power for not fewer than 10 hours. Additionally, the controls shall have a manual override that allows temporary operation of the system for up to 2 hours; a manually operated timer configured to operate the system for up to 2 hours; or an occupancy sensor.

C403.4.2.3 Automatic start and stop. Automatic start and stop controls shall be provided for each HVAC system. The automatic start controls shall be configured to automatically adjust the daily start time of the HVAC system in order to bring each space to the desired occupied temperature immediately prior to scheduled occupancy. Automatic stop controls shall be provided for each HVAC system with direct digital control of individual *zones*. The automatic stop controls shall be configured to reduce the HVAC system's heating temperature setpoint and increase the cooling temperature setpoint by not less than 2°F (1.11°C) before scheduled unoccupied periods based on the thermal lag and acceptable drift in space temperature that is within comfort limits.

C403.4.3 Hydronic systems controls. The heating of fluids that have been previously mechanically cooled and the cooling of fluids that have been previously mechanically heated shall be limited in accordance with Sections C403.4.3.1 through C403.4.3.3. Hydronic heating systems comprised of multiple-packaged boilers and designed to deliver conditioned water or steam into a common distribution system shall include automatic controls configured to sequence operation of the boilers. Hydronic heating systems composed of a single boiler and greater than 500,000 Btu/h (146.5 kW) input design capacity shall include either a multistaged or modulating burner.

C403.4.3.1 Three-pipe system. Hydronic systems that use a common return system for both hot water and chilled water are prohibited.

C403.4.3.2 Two-pipe changeover system. Systems that use a common distribution system to supply both heated and chilled water shall be designed to allow a deadband between changeover from one mode to the other of not less than 15°F (8.3°C) outside air temperatures; be designed to and provided with controls that will allow operation in one mode for not less than 4 hours before changing over to the other mode; and be provided with controls that allow heating and cooling supply temperatures at the changeover point to be not more than 30°F (16.7°C) apart.

C403.4.3.3 Hydronic (water loop) heat pump systems. Hydronic heat pump systems shall comply with Sections C403.4.3.3.1 through C403.4.3.3.3.

C403.4.3.3.1 Temperature deadband. Hydronic heat pumps connected to a common heat pump water loop with central devices for heat rejection and heat addition shall have controls that are configured to provide a heat pump water supply temperature deadband of not less than 20°F (11°C) between initiation of heat rejection and heat addition by the central devices.

Exception: Where a system loop temperature optimization controller is installed and can determine the most efficient operating temperature based on real-time conditions of demand and capacity, deadbands of less than 20°F (11°C) shall be permitted.

C403.4.3.3.2 Heat rejection. The following shall apply to hydronic water loop heat pump systems in Climate Zones 3 through 8:

1. Where a closed-circuit cooling tower is used directly in the heat pump loop, either an automatic valve shall be installed to bypass the flow of water around the closed-circuit cooling tower, except for any flow necessary for freeze protection, or low-leakage positive-closure dampers shall be provided.

2. Where an open-circuit cooling tower is used directly in the heat pump loop, an automatic valve shall be installed to bypass all heat pump water flow around the open-circuit cooling tower.

3. Where an open-circuit or closed-circuit cooling tower is used in conjunction with a separate heat exchanger to isolate the open-circuit cooling tower from the heat pump loop, heat loss shall be controlled by shutting down the circulation pump on the cooling tower loop.

Exception: Where it can be demonstrated that a heat pump system will be required to reject heat throughout the year.

C403.4.3.3.3 Two-position valve. Each hydronic heat pump on the hydronic system having a total pump system power exceeding 10 hp (7.5 kW) shall have a two-position automatic valve interlocked to shut off the water flow when the compressor is off.

C403.4.4 Part-load controls. Hydronic systems greater than or equal to 300,000 Btu/h (87.9 kW) in design output capacity supplying heated or chilled water to comfort conditioning systems shall include controls that are configured to do all of the following:

1. Automatically reset the supply-water temperatures in response to varying building heating and cooling demand using coil valve position, zone-return water temperature, building-return water temperature or outside air temperature. The temperature shall be reset by not less than 25 percent of the design supply-to-return water temperature difference.

2. Automatically vary fluid flow for hydronic systems with a combined pump motor capacity of 2 hp (1.5 kW) or larger with three or more control valves or other devices by reducing the system design flow rate by not less than 50 percent or the maximum reduction allowed by the equipment manufacturer for proper operation of equipment by valves that modulate or step open and close, or pumps that modulate or turn on and off as a function of load.

3. Automatically vary pump flow on heating-water systems, chilled-water systems and heat rejection loops serving water-cooled unitary air conditioners as follows:

    3.1. Where pumps operate continuously or operate based on a time schedule, pumps with nominal output motor power of 2 hp or more shall have a variable speed drive.

    3.2. Where pumps have automatic direct digital control configured to operate pumps only when zone heating or cooling is required, a variable speed drive shall be provided for pumps with motors having the same or greater nominal output power indicated in Table C403.4.4 based on the climate zone and system served.

4. Where a variable speed drive is required by Item 3 of this section, pump motor power input shall be not more than 30 percent of design wattage at 50

percent of the design water flow. Pump flow shall be controlled to maintain one control valve nearly wide open or to satisfy the minimum differential pressure.

Exceptions:

1. Supply-water temperature reset is not required for chilled-water systems supplied by off-site district chilled water or chilled water from ice storage systems.
2. Variable pump flow is not required on dedicated coil circulation pumps where needed for freeze protection.
3. Variable pump flow is not required on dedicated equipment circulation pumps where configured in primary/secondary design to provide the minimum flow requirements of the equipment manufacturer for proper operation of equipment.
4. Variable speed drives are not required on heating water pumps where more than 50 percent of annual heat is generated by an electric boiler.

TABLE C403.4.4
VARIABLE SPEED DRIVE (VSD) REQUIREMENTS FOR DEMAND-CONTROLLED PUMPS

CHILLED WATER AND HEAT REJECTION LOOP PUMPS IN THESE CLIMATE ZONES	HEATING WATER PUMPS IN THESE CLIMATE ZONES	VSD REQUIRED FOR MOTORS WITH RATED OUTPUT OF:
0A, 0B, 1A, 1B, 2B	—	≥ 2 hp
2A, 3B	—	≥ 3 hp
3A, 3C, 4A, 4B	7, 8	≥ 5 hp
4C, 5A, 5B, 5C, 6A, 6B	3C, 5A, 5C, 6A, 6B	≥ 7.5 hp
—	4A, 4C, 5B	≥ 10 hp
7, 8	4B	≥ 15 hp
—	2A, 2B, 3A, 3B	≥ 25 hp
—	0B, 1B	≥ 100 hp
—	0A, 1A	≥ 200 hp

For SI: 1 hp = 0.746 kW.

C403.4.5 Pump isolation. Chilled water plants including more than one chiller shall be capable of and configured to reduce flow automatically through the chiller plant when a chiller is shut down. Chillers piped in series for the purpose of increased temperature differential shall be considered as one chiller.

Boiler systems including more than one boiler shall be capable of and configured to reduce flow automatically through the boiler system when a boiler is shut down.

C403.5 Economizers. Economizers shall comply with Sections C403.5.1 through C403.5.5.

An air or water economizer shall be provided for the following cooling systems:

1. Chilled water systems with a total cooling capacity, less cooling capacity provided with air economizers, as specified in Table C403.5(1).
2. Individual fan systems with cooling capacity greater than or equal to 54,000 Btu/h (15.8 kW) in buildings having other than a *Group R* occupancy,

   The total supply capacity of all fan cooling units not provided with economizers shall not exceed 20 percent of the total supply capacity of all fan cooling units in the building or 300,000 Btu/h (88 kW), whichever is greater.

3. Individual fan systems with cooling capacity greater than or equal to 270,000 Btu/h (79.1 kW) in buildings having a *Group R* occupancy.

   The total supply capacity of all fan cooling units not provided with economizers shall not exceed 20 percent of the total supply capacity of all fan cooling units in the building or 1,500,000 Btu/h (440 kW), whichever is greater.

Exceptions: Economizers are not required for the following systems.

1. Individual fan systems not served by chilled water for buildings located in *Climate Zones* 0A, 0B, 1A and 1B.
2. Where more than 25 percent of the air designed to be supplied by the system is to spaces that are designed to be humidified above 35°F (1.7°C) dew-point temperature to satisfy process needs.
3. Systems expected to operate less than 20 hours per week.
4. Systems serving supermarket areas with open refrigerated casework.
5. Where the cooling efficiency is greater than or equal to the efficiency requirements in Table C403.5(2).
6. Systems that include a heat recovery system in accordance with Section C403.10.5.
7. VRF systems installed with a dedicated outdoor air system.

TABLE C403.5(1)
MINIMUM CHILLED-WATER SYSTEM COOLING CAPACITY FOR DETERMINING ECONOMIZER COOLING REQUIREMENTS

CLIMATE ZONES (COOLING)	TOTAL CHILLED-WATER SYSTEM CAPACITY LESS CAPACITY OF COOLING UNITS WITH AIR ECONOMIZERS	
	Local water-cooled chilled-water systems	Air-cooled chilled-water systems or district chilled-water systems
0A, 1A	Economizer not required	Economizer not required
0B, 1B, 2A, 2B	960,000 Btu/h	1,250,000 Btu/h
3A, 3B, 3C, 4A, 4B, 4C	720,000 Btu/h	940,000 Btu/h
5A, 5B, 5C, 6A, 6B, 7, 8	1,320,000 Btu/h	1,720,000 Btu/h

For SI: 1 British thermal unit per hour = 0.2931 W.

TABLE C403.5(2)
EQUIPMENT EFFICIENCY PERFORMANCE EXCEPTION FOR ECONOMIZERS

CLIMATE ZONES	COOLING EQUIPMENT PERFORMANCE IMPROVEMENT (EER OR IPLV)
2A, 2B	10% efficiency improvement
3A, 3B	15% efficiency improvement
4A, 4B	20% efficiency improvement

C403.5.1 Integrated economizer control. Economizer systems shall be integrated with the mechanical cooling system and be configured to provide partial cooling even where additional mechanical cooling is required to provide the remainder of the cooling load. Controls shall not be capable of creating a false load in the mechanical cooling systems by limiting or disabling the economizer or any other means, such as hot gas bypass, except at the lowest stage of mechanical cooling.

Units that include an air economizer shall comply with the following:

1. Unit controls shall have the mechanical cooling capacity control interlocked with the air economizer controls such that the outdoor air damper is at the 100-percent open position when mechanical cooling is on and the outdoor air damper does not begin to close to prevent coil freezing due to minimum compressor run time until the leaving air temperature is less than 45°F (7°C).

2. Direct expansion (DX) units that control 75,000 Btu/h (22 kW) or greater of rated capacity of the capacity of the mechanical cooling directly based on occupied space temperature shall have not fewer than two stages of mechanical cooling capacity.

3. Other DX units, including those that control space temperature by modulating the airflow to the space, shall be in accordance with Table C403.5.1.

TABLE C403.5.1
DX COOLING STAGE REQUIREMENTS FOR MODULATING AIRFLOW UNITS

RATING CAPACITY	MINIMUM NUMBER OF MECHANICAL COOLING STAGES	MINIMUM COMPRESSOR DISPLACEMENT[a]
≥ 65,000 Btu/h and < 240,000 Btu/h	3 stages	≤ 35% of full load
≥ 240,000 Btu/h	4 stages	≤ 25% full load

For SI: 1 British thermal unit per hour = 0.2931 W.

a. For mechanical cooling stage control that does not use variable compressor displacement, the percent displacement shall be equivalent to the mechanical cooling capacity reduction evaluated at the full load rating conditions for the compressor.

C403.5.2 Economizer heating system impact. HVAC system design and economizer controls shall be such that economizer operation does not increase building heating energy use during normal operation.

Exception: Economizers on variable air volume (VAV) systems that cause zone level heating to increase because of a reduction in supply air temperature.

C403.5.3 Air economizers. Where economizers are required by Section C403.5, air economizers shall comply with Sections C403.5.3.1 through C403.5.3.5.

C403.5.3.1 Design capacity. Air economizer systems shall be configured to modulate *outdoor air* and return air dampers to provide up to 100 percent of the design supply air quantity as *outdoor air* for cooling.

C403.5.3.2 Control signal. Economizer controls and dampers shall be configured to sequence the dampers with the mechanical cooling equipment and shall not be controlled by only mixed-air temperature.

Exception: The use of mixed-air temperature limit control shall be permitted for systems controlled from space temperature (such as single-*zone* systems).

C403.5.3.3 High-limit shutoff. Air economizers shall be configured to automatically reduce *outdoor air* intake to the design minimum *outdoor air* quantity when *outdoor air* intake will not reduce cooling energy usage. High-limit shutoff control types for specific climates shall be chosen from Table C403.5.3.3. High-limit shutoff control settings for these control types shall be those specified in Table C403.5.3.3.

C403.5.3.4 Relief of excess outdoor air. Systems shall be capable of relieving excess *outdoor air* during air economizer operation to prevent overpressurizing the building. The relief air outlet shall be located to avoid recirculation into the building.

C403.5.3.5 Economizer dampers. Return, exhaust/relief and outdoor air dampers used in economizers shall comply with Section C403.7.7.

C403.5.4 Water-side economizers. Where economizers are required by Section C403.5, water-side economizers shall comply with Sections C403.5.4.1 and C403.5.4.2.

C403.5.4.1 Design capacity. Water economizer systems shall be configured to cool supply air by indirect evaporation and providing up to 100 percent of the expected system cooling load at *outdoor air* temperatures of not greater than 50°F (10°C) dry bulb/45°F (7°C) wet bulb.

Exceptions:

1. Systems primarily serving computer rooms in which 100 percent of the expected system cooling load at 40°F (4°C) dry bulb/35°F (1.7°C) wet bulb is met with evaporative water economizers.

2. Systems primarily serving computer rooms with dry cooler water economizers that satisfy 100 percent of the expected system cooling load at 35°F (1.7°C) dry bulb.

3. Systems where dehumidification requirements cannot be met using outdoor air temperatures of 50°F (10°C) dry bulb/45°F

TABLE C403.5.3.3
HIGH-LIMIT SHUTOFF CONTROL SETTING FOR AIR ECONOMIZERS[b]

DEVICE TYPE	CLIMATE ZONE	REQUIRED HIGH LIMIT (ECONOMIZER OFF WHEN):	
		Equation	Description
Fixed dry bulb	0B, 1B, 2B, 3B, 3C, 4B, 4C, 5B, 5C, 6B, 7, 8	$T_{OA} > 75°F$	Outdoor air temperature exceeds 75°F
	5A, 6A	$T_{OA} > 70°F$	Outdoor air temperature exceeds 70°F
	0A, 1A, 2A, 3A, 4A	$T_{OA} > 65°F$	Outdoor air temperature exceeds 65°F
Differential dry bulb	0B, 1B, 2B, 3B, 3C, 4B, 4C, 5A, 5B, 5C, 6A, 6B, 7, 8	$T_{OA} > T_{RA}$	Outdoor air temperature exceeds return air temperature
Fixed enthalpy with fixed dry-bulb temperatures	All	$h_{OA} > 28$ Btu/lb[a] or $T_{OA} > 75°F$	Outdoor air enthalpy exceeds 28 Btu/lb of dry air[a] or Outdoor air temperature exceeds 75°F
Differential enthalpy with fixed dry-bulb temperature	All	$h_{OA} > h_{RA}$ or $T_{OA} > 75°F$	Outdoor air enthalpy exceeds return air enthalpy or Outdoor air temperature exceeds 75°F

For SI: °C = (°F – 32)/1.8, 1 Btu/lb = 2.33 kJ/kg.

a. At altitudes substantially different than sea level, the fixed enthalpy limit shall be set to the enthalpy value at 75°F and 50-percent relative humidity. As an example, at approximately 6,000 feet elevation, the fixed enthalpy limit is approximately 30.7 Btu/lb.
b. Devices with selectable setpoints shall be capable of being set to within 2°F and 2 Btu/lb of the setpoint listed.

(7°C) wet bulb and where 100 percent of the expected system cooling load at 45°F (7°C) dry bulb/40°F (4°C) wet bulb is met with evaporative water economizers.

C403.5.4.2 Maximum pressure drop. Precooling coils and water-to-water heat exchangers used as part of a water economizer system shall either have a water-side pressure drop of less than 15 feet (45 kPa) of water or a secondary loop shall be created so that the coil or heat exchanger pressure drop is not seen by the circulating pumps when the system is in the normal cooling (noneconomizer) mode.

C403.5.5 Economizer fault detection and diagnostics. Air-cooled unitary direct-expansion units listed in the tables in Section C403.3.2 and variable refrigerant flow (VRF) units that are equipped with an economizer in accordance with Sections C403.5 through C403.5.4 shall include a fault detection and diagnostics system complying with the following:

1. The following temperature sensors shall be permanently installed to monitor system operation:
   1.1. Outside air.
   1.2. Supply air.
   1.3. Return air.
2. Temperature sensors shall have an accuracy of ±2°F (1.1°C) over the range of 40°F to 80°F (4°C to 26.7°C).
3. Refrigerant pressure sensors, where used, shall have an accuracy of ±3 percent of full scale.
4. The unit controller shall be configured to provide system status by indicating the following:
   4.1. Free cooling available.
   4.2. Economizer enabled.
   4.3. Compressor enabled.
   4.4. Heating enabled.
   4.5. Mixed air low limit cycle active.
   4.6. The current value of each sensor.
5. The unit controller shall be capable of manually initiating each operating mode so that the operation of compressors, economizers, fans and the heating system can be independently tested and verified.
6. The unit shall be configured to report faults to a fault management application available for *access* by day-to-day operating or service personnel, or annunciated locally on zone thermostats.
7. The fault detection and diagnostics system shall be configured to detect the following faults:
   7.1. Air temperature sensor failure/fault.
   7.2. Not economizing when the unit should be economizing.
   7.3. Economizing when the unit should not be economizing.
   7.4. Damper not modulating.
   7.5. Excess outdoor air.

C403.6 Requirements for mechanical systems serving multiple zones. Sections C403.6.1 through C403.6.9 shall apply to mechanical systems serving multiple zones.

C403.6.1 Variable air volume and multiple-zone systems. Supply air systems serving multiple zones shall be variable air volume (VAV) systems that have zone controls configured to reduce the volume of air that is reheated, recooled or mixed in each zone to one of the following:

1. Twenty percent of the zone design peak supply for systems with *direct digital control* (DDC) and 30 percent for other systems.
2. Systems with DDC where all of the following apply:
    2.1. The airflow rate in the deadband between heating and cooling does not exceed 20 percent of the zone design peak supply rate or higher allowed rates under Items 3, 4 and 5 of this section.
    2.2. The first stage of heating modulates the zone supply air temperature setpoint up to a maximum setpoint while the airflow is maintained at the deadband flow rate.
    2.3. The second stage of heating modulates the airflow rate from the deadband flow rate up to the heating maximum flow rate that is less than 50 percent of the zone design peak supply rate.
3. The outdoor airflow rate required to meet the minimum ventilation requirements of Chapter 4 of the *International Mechanical Code*.
4. Any higher rate that can be demonstrated to reduce overall system annual energy use by offsetting reheat/recool energy losses through a reduction in outdoor air intake for the system as approved by the code official.
5. The airflow rate required to comply with applicable codes or accreditation standards such as pressure relationships or minimum air change rates.

Exception: The following individual zones or entire air distribution systems are exempted from the requirement for VAV control:

1. *Zones* or supply air systems where not less than 75 percent of the energy for reheating or for providing warm air in mixing systems is provided from a site-recovered, including condenser heat, or site-solar energy source.
2. Systems that prevent reheating, recooling, mixing or simultaneous supply of air that has been previously cooled, either mechanically or through the use of economizer systems, and air that has been previously mechanically heated.

C403.6.2 Single-duct VAV systems, terminal devices. Single-duct VAV systems shall use terminal devices capable of and configured to reduce the supply of primary supply air before reheating or recooling takes place.

C403.6.3 Dual-duct and mixing VAV systems, terminal devices. Systems that have one warm air duct and one cool air duct shall use terminal devices that are configured to reduce the flow from one duct to a minimum before mixing of air from the other duct takes place.

C403.6.4 Single-fan dual-duct and mixing VAV systems, economizers. Individual dual-duct or mixing heating and cooling systems with a single fan and with total capacities greater than 90,000 Btu/h [(26.4 kW) 7.5 tons] shall not be equipped with air economizers.

C403.6.5 Supply-air temperature reset controls. Multiple-zone HVAC systems shall include controls that are capable of and configured to automatically reset the supply-air temperature in response to representative building loads, or to outdoor air temperature. The controls shall be configured to reset the supply air temperature not less than 25 percent of the difference between the design supply-air temperature and the design room air temperature. Controls that adjust the reset based on zone humidity are allowed in Climate Zones 0B, 1B, 2B, 3B, 3C and 4 through 8. HVAC zones that are expected to experience relatively constant loads shall have maximum airflow designed to accommodate the fully reset supply-air temperature.

Exceptions:

1. Systems that prevent reheating, recooling or mixing of heated and cooled supply air.
2. Seventy-five percent of the energy for reheating is from site-recovered or site-solar energy sources.
3. Systems in Climate Zones 0A, 1A and 3A with less than 3,000 cfm (1500 L/s) of design outside air.
4. Systems in Climate Zone 2A with less than 10,000 cfm (5000 L/s) of design outside air.
5. Systems in Climate Zones 0A, 1A, 2A and 3A with not less than 80 percent outside air and employing exhaust air energy recovery complying with Section C403.7.4.

C403.6.5.1 Dehumidification control interaction. In Climate Zones 0A, 1A, 2A and 3A, the system design shall allow supply-air temperature reset while dehumidification is provided. When dehumidification control is active, air economizers shall be locked out.

C403.6.6 Multiple-zone VAV system ventilation optimization control. Multiple-zone VAV systems with direct digital control of individual zone boxes reporting to a central control panel shall have automatic controls configured to reduce outdoor air intake flow below design rates in response to changes in system *ventilation* efficiency (E_v) as defined by the *International Mechanical Code*.

Exceptions:

1. VAV systems with zonal transfer fans that recirculate air from other zones without directly mixing it with outdoor air, dual-duct dual-fan VAV systems, and VAV systems with fan-powered terminal units.

2. Systems where total design exhaust airflow is more than 70 percent of total design outdoor air intake flow requirements.

C403.6.7 Parallel-flow fan-powered VAV air terminal control. Parallel-flow fan-powered VAV air terminals shall have automatic controls configured to:

1. Turn off the terminal fan except when space heating is required or where required for ventilation.
2. Turn on the terminal fan as the first stage of heating before the heating coil is activated.
3. During heating for warmup or setback temperature control, either:
   3.1. Operate the terminal fan and heating coil without primary air.
   3.2. Reverse the terminal damper logic and provide heating from the central air handler by primary air.

C403.6.8 Setpoints for direct digital control. For systems with direct digital control of individual zones reporting to the central control panel, the static pressure setpoint shall be reset based on the *zone* requiring the most pressure. In such case, the setpoint is reset lower until one *zone* damper is nearly wide open. The direct digital controls shall be capable of monitoring zone damper positions or shall have an alternative method of indicating the need for static pressure that is configured to provide all of the following:

1. Automatic detection of any *zone* that excessively drives the reset logic.
2. Generation of an alarm to the system operational location.
3. Allowance for an operator to readily remove one or more *zones* from the reset algorithm.

C403.6.9 Static pressure sensor location. Static pressure sensors used to control VAV fans shall be located such that the controller setpoint is not greater than 1.2 inches w.c. (299 Pa). Where this results in one or more sensors being located downstream of major duct splits, not less than one sensor shall be located on each major branch to ensure that static pressure can be maintained in each branch.

C403.7 Ventilation and exhaust systems. In addition to other requirements of Section C403 applicable to the provision of ventilation air or the exhaust of air, ventilation and exhaust systems shall be in accordance with Sections C403.7.1 through C403.7.7.

C403.7.1 Demand control ventilation. Demand control ventilation (DCV) shall be provided for all single-zone systems required to comply with Sections C403.5 through C403.5.3 and spaces larger than 500 square feet (46.5 m^2) and with an average occupant load of 15 people or greater per 1,000 square feet (93 m^2) of floor area, as established in Table 403.3.1.1 of the *International Mechanical Code*, and served by systems with one or more of the following:

1. An air-side economizer.
2. Automatic modulating control of the outdoor air damper.
3. A design outdoor airflow greater than 3,000 cfm (1416 L/s).

Exceptions:

1. Systems with energy recovery complying with Section C403.7.4.2.
2. Multiple-zone systems without direct digital control of individual zones communicating with a central control panel.
3. Multiple-zone systems with a design outdoor airflow less than 750 cfm (354 L/s).
4. Spaces where more than 75 percent of the space design outdoor airflow is required for makeup air that is exhausted from the space or transfer air that is required for makeup air that is exhausted from other spaces.
5. Spaces with one of the following occupancy classifications as defined in Table 403.3.1.1 of the *International Mechanical Code*: correctional cells, education laboratories, barber, beauty and nail salons, and bowling alley seating areas.

C403.7.2 Enclosed parking garage ventilation controls. Enclosed parking garages used for storing or handling automobiles operating under their own power shall employ carbon monoxide detectors applied in conjunction with nitrogen dioxide detectors and automatic controls configured to stage fans or modulate fan average airflow rates to 50 percent or less of design capacity, or intermittently operate fans less than 20 percent of the occupied time or as required to maintain acceptable contaminant levels in accordance with *International Mechanical Code* provisions. Failure of contamination-sensing devices shall cause the exhaust fans to operate continuously at design airflow.

Exceptions:

1. Garages with a total exhaust capacity less than 8,000 cfm (3,755 L/s) with ventilation systems that do not utilize heating or mechanical cooling.
2. Garages that have a garage area to ventilation system motor nameplate power ratio that exceeds 1,125 cfm/hp (710 L/s/kW) and do not utilize heating or mechanical cooling.

C403.7.3 Ventilation air heating control. Units that provide ventilation air to multiple zones and operate in conjunction with zone heating and cooling systems shall not use heating or heat recovery to warm supply air to a temperature greater than 60°F (16°C) when representative building loads or outdoor air temperatures indicate that the majority of zones require cooling.

C403.7.4 Energy recovery systems. Energy recovery ventilation systems shall be provided as specified in either Section C403.7.4.1 or C403.7.4.2, as applicable.

C403.7.4.1 Nontransient dwelling units. Nontransient dwelling units shall be provided with outdoor air energy recovery ventilation systems with an *enthalpy recovery ratio* of not less than 50 percent at cooling design condition and not less than 60 percent at heating design condition.

Exceptions:

1. Nontransient dwelling units in Climate Zone 3C.
2. Nontransient dwelling units with not more than 500 square feet (46 m^2) of *conditioned floor area* in Climate Zones 0, 1, 2, 3, 4C and 5C.
3. *Enthalpy recovery ratio* requirements at heating design condition in Climate Zones 0, 1 and 2.
4. *Enthalpy recovery ratio* requirements at cooling design condition in Climate Zones 4, 5, 6, 7 and 8.

C403.7.4.2 Spaces other than nontransient dwelling units. Where the supply airflow rate of a fan system serving a space other than a nontransient dwelling unit exceeds the values specified in Tables C403.7.4.2(1) and C403.7.4.2(2), the system shall include an energy recovery system. The energy recovery system shall provide an *enthalpy recovery ratio* of not less than 50 percent at design conditions. Where an air economizer is required, the energy recovery system shall include a bypass or controls that permit operation of the economizer as required by Section C403.5.

Exception: An energy recovery ventilation system shall not be required in any of the following conditions:

1. Where energy recovery systems are prohibited by the *International Mechanical Code*.
2. Laboratory fume hood systems that include not fewer than one of the following features:
    2.1. Variable-air-volume hood exhaust and room supply systems configured to reduce exhaust and makeup air volume to 50 percent or less of design values.
    2.2. Direct makeup (auxiliary) air supply equal to or greater than 75 percent of the exhaust rate, heated not warmer than 2°F (1.1°C) above room setpoint, cooled to not cooler than 3°F (1.7°C) below room setpoint, with no humidification added, and no simultaneous heating and cooling used for dehumidification control.
3. Systems serving spaces that are heated to less than 60°F (15.5°C) and that are not cooled.
4. Where more than 60 percent of the outdoor heating energy is provided from site-recovered or site-solar energy.
5. *Enthalpy recovery ratio* requirements at heating design condition in *Climate Zones* 1 and 2.
6. *Enthalpy recovery ratio* requirements at cooling design condition in *Climate Zones* 0, 3C, 4C, 5B, 5C, 6B, 7 and 8.
7. Systems requiring dehumidification that employ energy recovery in series with the cooling coil.
8. Where the largest source of air exhausted at a single location at the building exterior is less than 75 percent of the design *outdoor air* flow rate.
9. Systems expected to operate less than 20 hours per week at the *outdoor air* percentage covered by Table C403.7.4.2(1).
10. Systems exhausting toxic, flammable, paint or corrosive fumes or dust.
11. Commercial kitchen hoods used for collecting and removing grease vapors and smoke.

C403.7.5 Kitchen exhaust systems. Replacement air introduced directly into the exhaust hood cavity shall not be greater than 10 percent of the hood exhaust airflow rate. Conditioned supply air delivered to any space shall not exceed the greater of the following:

1. The ventilation rate required to meet the space heating or cooling load.
2. The hood exhaust flow minus the available transfer air from adjacent space where available transfer air is considered to be that portion of outdoor ventilation air not required to satisfy other exhaust needs, such as restrooms, and not required to maintain pressurization of adjacent spaces.

Where total kitchen hood exhaust airflow rate is greater than 5,000 cfm (2360 L/s), each hood shall be a factory-built commercial exhaust hood listed by a nationally recognized testing laboratory in compliance with UL 710. Each hood shall have a maximum exhaust rate as specified in Table C403.7.5 and shall comply with one of the following:

1. Not less than 50 percent of all replacement air shall be transfer air that would otherwise be exhausted.
2. Demand ventilation systems on not less than 75 percent of the exhaust air that are configured to provide not less than a 50-percent reduction in

exhaust and replacement air system airflow rates, including controls necessary to modulate airflow in response to appliance operation and to maintain full capture and containment of smoke, effluent and combustion products during cooking and idle.

3. Listed energy recovery devices with a sensible heat recovery effectiveness of not less than 40 percent on not less than 50 percent of the total exhaust airflow.

Where a single hood, or hood section, is installed over appliances with different duty ratings, the maximum allowable flow rate for the hood or hood section shall be based on the requirements for the highest appliance duty rating under the hood or hood section.

Exception: Where not less than 75 percent of all the replacement air is transfer air that would otherwise be exhausted.

TABLE C403.7.4.2(1)
ENERGY RECOVERY REQUIREMENT (Ventilation systems operating less than 8,000 hours per year)

CLIMATE ZONE	PERCENT (%) OUTDOOR AIR AT FULL DESIGN AIRFLOW RATE							
	≥ 10% and < 20%	≥ 20% and < 30%	≥ 30% and < 40%	≥ 40% and < 50%	≥ 50% and < 60%	≥ 60% and < 70%	≥ 70% and < 80%	≥ 80%
	Design Supply Fan Airflow Rate (cfm)							
3B, 3C, 4B, 4C, 5B	NR	NR	NR	NR	NR	NR	NR	NR
0B, 1B, 2B, 5C	NR	NR	NR	NR	≥ 26,000	≥ 12,000	≥ 5,000	≥ 4,000
6B	≥ 28,000	≥ 26,5000	≥ 11,000	≥ 5,500	≥ 4,500	≥ 3,500	≥ 2,500	≥ 1,500
0A, 1A, 2A, 3A, 4A, 5A, 6A	≥ 26,000	≥ 16,000	≥ 5,500	≥ 4,500	≥ 3,500	≥ 2,000	≥ 1,000	> 120
7, 8	≥ 4,500	≥ 4,000	≥ 2,500	≥ 1,000	> 140	> 120	> 100	> 80

For SI: 1 cfm = 0.4719 L/s.
NR = Not Required.

TABLE C403.7.4.2(2)
ENERGY RECOVERY REQUIREMENT (Ventilation systems operating not less than 8,000 hours per year)

CLIMATE ZONE	PERCENT (%) OUTDOOR AIR AT FULL DESIGN AIRFLOW RATE							
	≥ 10% and < 20%	≥ 20% and < 30%	≥ 30% and < 40%	≥ 40% and < 50%	≥ 50% and < 60%	≥ 60% and < 70%	≥ 70% and < 80%	≥ 80%
	Design Supply Fan Airflow Rate (cfm)							
3C	NR	NR	NR	NR	NR	NR	NR	NR
0B, 1B, 2B, 3B, 4C, 5C	NR	≥ 19,500	≥ 9,000	≥ 5,000	≥ 4,000	≥ 3,000	≥ 1,500	≥ 120
0A, 1A, 2A, 3A, 4B, 5B	≥ 2,500	≥ 2,000	≥ 1,000	≥ 500	≥ 140	≥ 120	≥ 100	≥ 80
4A, 5A, 6A, 6B, 7, 8	≥ 200	≥ 130	≥ 100	≥ 80	≥ 70	≥ 60	≥ 50	≥ 40

For SI: 1 cfm = 0.4719 L/s.
NR = Not Required.

TABLE C403.7.5
MAXIMUM NET EXHAUST FLOW RATE, CFM PER LINEAR FOOT OF HOOD LENGTH

TYPE OF HOOD	LIGHT-DUTY EQUIPMENT	MEDIUM-DUTY EQUIPMENT	HEAVY-DUTY EQUIPMENT	EXTRA-HEAVY-DUTY EQUIPMENT
Wall-mounted canopy	140	210	280	385
Single island	280	350	420	490
Double island (per side)	175	210	280	385
Eyebrow	175	175	NA	NA
Backshelf/Pass-over	210	210	280	NA

For SI: 1 cfm = 0.4719 L/s; 1 foot = 304.8 mm.
NA = Not Allowed.

C403.7.6 Automatic control of HVAC systems serving guestrooms. In Group R-1 buildings containing more than 50 guestrooms, each guestroom shall be provided with controls complying with the provisions of Sections C403.7.6.1 and C403.7.6.2. Card key controls comply with these requirements.

C403.7.6.1 Temperature setpoint controls. Controls shall be provided on each HVAC system that are capable of and configured with three modes of temperature control.

1. When the guestroom is rented but unoccupied, the controls shall automatically raise the cooling setpoint and lower the heating setpoint by not less than 4°F (2°C) from the occupant setpoint within 30 minutes after the occupants have left the guestroom.

2. When the guestroom is unrented and unoccupied, the controls shall automatically raise the cooling setpoint to not lower than 80°F (27°C) and lower the heating setpoint to not higher than 60°F (16°C). Unrented and unoccupied guestroom mode shall be initiated within 16 hours of the guestroom being continuously occupied or where a *networked guestroom control system* indicates that the guestroom is unrented and the guestroom is unoccupied for more than 20 minutes. A *networked guestroom control system* that is capable of returning the thermostat setpoints to default occupied setpoints 60 minutes prior to the time a guestroom is scheduled to be occupied is not precluded by this section. Cooling that is capable of limiting relative humidity with a setpoint not lower than 65-percent relative humidity during unoccupied periods is not precluded by this section.

3. When the guestroom is occupied, HVAC setpoints shall return to their occupied setpoints once occupancy is sensed.

C403.7.6.2 Ventilation controls. Controls shall be provided on each HVAC system that are capable of and configured to automatically turn off the ventilation and exhaust fans within 20 minutes of the occupants leaving the guestroom, or *isolation devices* shall be provided to each guestroom that are capable of automatically shutting off the supply of outdoor air to and exhaust air from the guestroom.

Exception: Guestroom ventilation systems are not precluded from having an automatic daily pre-occupancy purge cycle that provides daily outdoor air ventilation during unrented periods at the design ventilation rate for 60 minutes, or at a rate and duration equivalent to one air change.

C403.7.7 Shutoff dampers. Outdoor air intake and exhaust openings and stairway and shaft vents shall be provided with Class I motorized dampers. The dampers shall have an air leakage rate not greater than 4 cfm/ft^2 (20.3 L/s × m^2) of damper surface area at 1.0 inch water gauge (249 Pa) and shall be labeled by an *approved agency* when tested in accordance with AMCA 500D for such purpose.

Outdoor air intake and exhaust dampers shall be installed with automatic controls configured to close when the systems or spaces served are not in use or during unoccupied period warm-up and setback operation, unless the systems served require outdoor or exhaust air in accordance with the *International Mechanical Code* or the dampers are opened to provide intentional economizer cooling.

Stairway and shaft vent dampers shall be installed with automatic controls configured to open upon the activation of any fire alarm initiating device of the building's fire alarm system or the interruption of power to the damper.

Exception: Nonmotorized gravity dampers shall be an alternative to motorized dampers for exhaust and relief openings as follows:

1. In buildings less than three stories in height above grade plane.

2. In buildings of any height located in *Climate Zones* 0, 1, 2 or 3.

3. Where the design exhaust capacity is not greater than 300 cfm (142 L/s).

Nonmotorized gravity dampers shall have an air leakage rate not greater than 20 cfm/ft^2 (101.6 L/s × m^2) where not less than 24 inches (610 mm) in either dimension and 40 cfm/ft^2 (203.2 L/s × m^2) where less than 24 inches (610 mm) in either dimension. The rate of air leakage shall be determined at 1.0 inch water gauge (249 Pa) when tested in accordance with AMCA 500D for such purpose. The dampers shall be labeled by an *approved agency*.

C403.8 Fans and fan controls. Fans in HVAC systems shall comply with Sections C403.8.1 through C403.8.6.1.

C403.8.1 Allowable fan horsepower. Each HVAC system having a total fan system motor nameplate horsepower exceeding 5 hp (3.7 kW) at fan system design conditions shall not exceed the allowable *fan system motor nameplate hp* (Option 1) or *fan system bhp* (Option 2) shown in Table C403.8.1(1). This includes supply fans, exhaust fans, return/relief fans, and fan-powered terminal units associated with systems providing heating or cooling capability. Single-zone variable air volume systems shall comply with the constant volume fan power limitation.

Exceptions:

1. Hospital, vivarium and laboratory systems that utilize flow control devices on exhaust or return to maintain space pressure relationships necessary for occupant health and safety or environmental control shall be permitted to use variable volume fan power limitation.

2. Individual exhaust fans with motor nameplate horsepower of 1 hp (0.746 kW) or less are

exempt from the allowable fan horsepower requirement.

C403.8.2 Motor nameplate horsepower. For each fan, the fan brake horsepower (bhp) shall be indicated on the construction documents and the selected motor shall be not larger than the first available motor size greater than the following:

1. For fans less than 6 bhp (4476 W), 1.5 times the fan brake horsepower.

2. For fans 6 bhp (4476 W) and larger, 1.3 times the fan brake horsepower.

Exceptions:

1. Fans equipped with electronic speed control devices to vary the fan airflow as a function of load.

2. Fans with a fan nameplate electrical input power of less than 0.89 kW.

TABLE C403.8.1(1)
FAN POWER LIMITATION

	LIMIT	CONSTANT VOLUME	VARIABLE VOLUME
Option 1: Fan system motor nameplate hp	Allowable nameplate motor hp	$hp \leq CFM_S \times 0.0011$	$hp \leq CFM_S \times 0.0015$
Option 2: Fan system bhp	Allowable fan system bhp	$bhp \leq CFM_S \times 0.00094 + A$	$bhp \leq CFM_S \times 0.0013 + A$

For SI: 1 bhp = 735.5 W, 1 hp = 745.5 W, 1 cfm = 0.4719 L/s.

where:
CFM_S = The maximum design supply airflow rate to conditioned spaces served by the system in cubic feet per minute.
hp = The maximum combined motor nameplate horsepower.
bhp = The maximum combined fan brake horsepower.
A = Sum of $[PD \times CFM_D / 4131]$.

where:
PD = Each applicable pressure drop adjustment from Table C403.8.1(2) in. w.c.
CFM_D = The design airflow through each applicable device from Table C403.8.1(2) in cubic feet per minute.

TABLE C403.8.1(2)
FAN POWER LIMITATION PRESSURE DROP ADJUSTMENT

DEVICE	ADJUSTMENT
Credits	
Return air or exhaust systems required by code or accreditation standards to be fully ducted, or systems required to maintain air pressure differentials between adjacent rooms	0.5 inch w.c. (2.15 inches w.c. for laboratory and vivarium systems)
Return and exhaust airflow control devices	0.5 inch w.c.
Exhaust filters, scrubbers or other exhaust treatment	The pressure drop of device calculated at fan system design condition
Particulate filtration credit: MERV 9 thru 12	0.5 inch w.c.
Particulate filtration credit: MERV 13 thru 15	0.9 inch w.c.
Particulate filtration credit: MERV 16 and greater and electronically enhanced filters	Pressure drop calculated at 2 times the clean filter pressure drop at fan system design condition.
Carbon and other gas-phase air cleaners	Clean filter pressure drop at fan system design condition.
Biosafety cabinet	Pressure drop of device at fan system design condition.
Energy recovery device, other than coil runaround loop	For each airstream, (2.2 × energy recovery effectiveness - 0.5) inch w.c.
Coil runaround loop	0.6 inch w.c. for each airstream.
Evaporative humidifier/cooler in series with another cooling coil	Pressure drop of device at fan system design conditions.
Sound attenuation section (fans serving spaces with design background noise goals below NC35)	0.15 inch w.c.
Exhaust system serving fume hoods	0.35 inch w.c.
Laboratory and vivarium exhaust systems in high-rise buildings	0.25 inch w.c./100 feet of vertical duct exceeding 75 feet.
Deductions	
Systems without central cooling device	- 0.6 inch w.c.
Systems without central heating device	- 0.3 inch w.c.
Systems with central electric resistance heat	- 0.2 inch w.c.

For SI: 1 inch w.c. = 249 Pa, 1 inch = 25.4 mm, 1 foot = 304.8 mm.
w.c. = Water Column, NC = Noise Criterion.

3. Systems complying with Section C403.8.1 fan system motor nameplate hp (Option 1).
4. Fans with motor nameplate horsepower less than 1 hp (746 W).

C403.8.3 Fan efficiency. Each fan and fan array shall have a fan energy index (FEI) of not less than 1.00 at the design point of operation, as determined in accordance with AMCA 208 by an *approved* independent testing laboratory and labeled by the manufacturer. Each fan and fan array used for a variable-air-volume system shall have an FEI of not less than 0.95 at the design point of operation, as determined in accordance with AMCA 208 by an approved independent testing laboratory and labeled by the manufacturer. The FEI for fan arrays shall be calculated in accordance with AMCA 208 Annex C.

Exceptions: The following fans are not required to have a fan energy index:

1. Fans that are not embedded fans with motor nameplate horsepower of less than 1.0 hp (0.75 kW) or with a nameplate electrical input power of less than 0.89 kW.
2. Embedded fans that have a motor nameplate horsepower of 5 hp (3.7 kW) or less, or with a fan system electrical input power of 4.1 kW or less.
3. Multiple fans operated in series or parallel as the functional equivalent of a single fan that have a combined motor nameplate horsepower of 5 hp (3.7 kW) or less or with a fan system electrical input power of 4.1 kW or less.
4. Fans that are part of equipment covered in Section C403.3.2.
5. Fans included in an equipment package certified by an *approved agency* for air or energy performance.
6. Ceiling fans, which are defined as nonportable devices suspended from a ceiling or overhead structure for circulating air via the rotation of the blades.
7. Fans used for moving gases at temperatures above 482°F (250°C).
8. Fans used for operation in explosive atmospheres.
9. Reversible fans used for tunnel ventilation.
10. Fans that are intended to operate only during emergency conditions.
11. Fans outside the scope of AMCA 208.

C403.8.4 Fractional hp fan motors. Motors for fans that are not less than $1/_{12}$ hp (0.062 kW) and less than 1 hp (0.746 kW) shall be electronically commutated motors or shall have a minimum motor efficiency of 70 percent, rated in accordance with DOE 10 CFR 431. These motors shall have the means to adjust motor speed for either balancing or remote control. The use of belt-driven fans to sheave adjustments for airflow balancing instead of a varying motor speed shall be permitted.

Exceptions: The following motors are not required to comply with this section

1. Motors in the airstream within fan coils and terminal units that only provide heating to the space served.
2. Motors in space-conditioning equipment that comply with Section C403.3.2 or Sections C403.8.1. through C403.8.3.
3. Motors that comply with Section C405.8.

C403.8.5 Low-capacity ventilation fans. Mechanical ventilation system fans with motors less than $1/_{12}$ hp (0.062 kW) in capacity shall meet the efficacy requirements of Table C403.8.5 at one or more rating points.

Exceptions:

1. Where ventilation fans are a component of a listed heating or cooling appliance.
2. Dryer exhaust duct power ventilators, domestic range hoods and domestic range booster fans that operate intermittently.

TABLE C403.8.5
LOW-CAPACITY VENTILATION FAN EFFICACY[a]

FAN LOCATION	AIRFLOW RATE MINIMUM (CFM)	MINIMUM EFFICACY (CFM/WATT)	AIRFLOW RATE MAXIMUM (CFM)
HRV or ERV	Any	1.2 cfm/watt	Any
In-line fan	Any	3.8 cfm/watt	Any
Bathroom, utility room	10	2.8 cfm/watt	< 90
Bathroom, utility room	90	3.5 cfm/watt	Any

For SI: 1 cfm/ft = 47.82 W.

a. Airflow shall be tested in accordance with HVI 916 and listed. Efficacy shall be listed or shall be derived from listed power and airflow. Fan efficacy for fully ducted HRV, ERV, balanced and in-line fans shall be determined at a static pressure not less than 0.2 inch w.c. Fan efficacy for ducted range hoods, bathroom and utility room fans shall be determined at a static pressure not less than 0.1 inch w.c.

C403.8.6 Fan control. Controls shall be provided for fans in accordance with Section C403.8.6.1 and as required for specific systems provided in Section C403.

C403.8.6.1 Fan airflow control. Each cooling system listed in Table C403.8.6.1 shall be designed to vary the indoor fan airflow as a function of load and shall comply with the following requirements:

1. Direct expansion (DX) and chilled water cooling units that control the capacity of the mechanical cooling directly based on space temperature shall have not fewer than two stages of fan control. Low or minimum speed shall not be greater than 66 percent of full speed. At low or minimum speed, the fan system shall draw not more than 40 percent of

the fan power at full fan speed. Low or minimum speed shall be used during periods of low cooling load and ventilation-only operation.

2. Other units including DX cooling units and chilled water units that control the space temperature by modulating the airflow to the space shall have modulating fan control. Minimum speed shall be not greater than 50 percent of full speed. At minimum speed the fan system shall draw not more than 30 percent of the power at full fan speed. Low or minimum speed shall be used during periods of low cooling load and ventilation-only operation.

3. Units that include an air-side economizer in accordance with Section C403.5 shall have not fewer than two speeds of fan control during economizer operation.

Exceptions:

1. Modulating fan control is not required for chilled water and evaporative cooling units with fan motors of less than 1 hp (0.746 kW) where the units are not used to provide *ventilation air* and the indoor fan cycles with the load.

2. Where the volume of outdoor air required to comply with the ventilation requirements of the *International Mechanical Code* at low speed exceeds the air that would be delivered at the speed defined in Section C403.8.6, the minimum speed shall be selected to provide the required *ventilation air*.

TABLE C403.8.6.1
COOLING SYSTEMS

COOLING SYSTEM TYPE	FAN MOTOR SIZE	MECHANICAL COOLING CAPACITY
DX cooling	Any	≥ 65,000 Btu/h
Chilled water and evaporative cooling	≥ 1/4 hp	Any

For SI: 1 British thermal unit per hour = 0.2931 W; 1 hp = 0.746 kW.

C403.9 Large-diameter ceiling fans. Where provided, *large-diameter ceiling fans* shall be tested and labeled in accordance with AMCA 230.

C403.10 Heat rejection equipment. Heat rejection equipment, including air-cooled condensers, dry coolers, open-circuit cooling towers, closed-circuit cooling towers and evaporative condensers, shall comply with this section.

Exception: Heat rejection devices where energy usage is included in the equipment efficiency ratings listed in Tables C403.3.2(6) and C403.3.2(7).

C403.10.1 Fan speed control. Each fan system powered by an individual motor or array of motors with connected power, including the motor service factor, totaling 5 hp (3.7 kW) or more shall have controls and devices configured to automatically modulate the fan speed to control the leaving fluid temperature or condensing temperature and pressure of the heat rejection device. Fan motor power input shall be not more than 30 percent of design wattage at 50 percent of the design airflow.

Exceptions:

1. Fans serving multiple refrigerant or fluid cooling circuits.

2. Condenser fans serving flooded condensers.

C403.10.2 Multiple-cell heat rejection equipment. Multiple-cell heat rejection equipment with variable speed fan drives shall be controlled to operate the maximum number of fans allowed that comply with the manufacturer's requirements for all system components and so that all fans operate at the same fan speed required for the instantaneous cooling duty, as opposed to staged on and off operation. The minimum fan speed shall be the minimum allowable speed of the fan drive system in accordance with the manufacturer's recommendations.

C403.10.3 Limitation on centrifugal fan open-circuit cooling towers. Centrifugal fan open-circuit cooling towers with a combined rated capacity of 1,100 gpm (4164 L/m) or greater at 95°F (35°C) condenser water return, 85°F (29°C) condenser water supply, and 75°F (24°C) outdoor air wet-bulb temperature shall meet the energy efficiency requirement for axial fan open-circuit cooling towers listed in Table C403.3.2(8).

Exception: Centrifugal open-circuit cooling towers that are designed with inlet or discharge ducts or require external sound attenuation.

C403.10.4 Tower flow turndown. Open-circuit cooling towers used on water-cooled chiller systems that are configured with multiple- or variable-speed condenser water pumps shall be designed so that all open-circuit cooling tower cells can be run in parallel with the larger of the flow that is produced by the smallest pump at its minimum expected flow rate or at 50 percent of the design flow for the cell.

C403.10.5 Heat recovery for service water heating. Condenser heat recovery shall be installed for heating or reheating of service hot water provided that the facility operates 24 hours a day, the total installed heat capacity of water-cooled systems exceeds 6,000,000 Btu/hr (1758 kW) of heat rejection, and the design service water heating load exceeds 1,000,000 Btu/h (293 kW).

The required heat recovery system shall have the capacity to provide the smaller of the following:

1. Sixty percent of the peak heat rejection load at design conditions.

2. The preheating required to raise the peak service hot water draw to 85°F (29°C).

Exceptions:

1. Facilities that employ condenser heat recovery for space heating or reheat purposes with a heat recovery design exceeding 30 percent of the peak water-cooled condenser load at design conditions.

2. Facilities that provide 60 percent of their service water heating from site solar or site recovered energy or from other sources.

C403.10.6 Heat recovery for space conditioning in healthcare facilities. Where heating water is used for space heating, a condenser heat recovery system shall be installed provided that all of the following are true:

1. The building is a Group I-2, Condition 2 occupancy.
2. The total design chilled water capacity for the Group I-2, Condition 2 occupancy, either air cooled or water cooled, required at cooling design conditions exceeds 3,600,000 Btu/h (1100 kw) of cooling.
3. Simultaneous heating and cooling occurs above 60°F (16°C) outdoor air temperature.

The required heat recovery system shall have a cooling capacity that is not less than 7 percent of the total design chilled water capacity of the Group I-2, Condition 2 occupancy at peak design conditions.

Exceptions:

1. Buildings that provide 60 percent or more of their reheat energy from on-site renewable energy or site-recovered energy.
2. Buildings in Climate Zones 5C, 6B, 7 and 8.

C403.11 Refrigeration equipment performance. Refrigeration equipment performance shall be determined in accordance with Sections C403.11.1 and C403.11.2 for commercial refrigerators, freezers, refrigerator-freezers, walk-in coolers, walk-in freezers and refrigeration equipment. The energy use shall be verified through certification under an *approved* certification program or, where a certification program does not exist, the energy use shall be supported by data furnished by the equipment manufacturer.

Exception: Walk-in coolers and walk-in freezers regulated under federal law in accordance with Subpart R of DOE 10 CFR 431.

C403.11.1 Commercial refrigerators, refrigerator-freezers and refrigeration. Refrigeration equipment, defined in DOE 10 CFR Part 431.62, shall have an energy use in kWh/day not greater than the values of Table C403.11.1 when tested and rated in accordance with AHRI Standard 1200.

C403.11.2 Walk-in coolers and walk-in freezers. Walk-in cooler and walk-in freezer refrigeration systems, except for walk-in process cooling refrigeration systems as defined in DOE 10 CFR 431.302, shall meet the requirements of Tables C403.11.2.1(1), C403.11.2.1(2) and C403.11.2.1(3).

C403.11.2.1 Performance standards. *Walk-in coolers* and *walk-in freezers* shall meet the requirements of Tables C403.11.2.1(1), C403.11.2.1(2) and C403.11.2.1(3).

TABLE C403.11.2.1(1)
WALK-IN COOLER AND FREEZER DISPLAY DOOR EFFICIENCY REQUIREMENTS[a]

CLASS DESCRIPTOR	CLASS	MAXIMUM ENERGY CONSUMPTION (kWh/day)[a]
Display door, medium temperature	DD, M	$0.04 \times A_{dd} + 0.41$
Display door, low temperature	DD, L	$0.15 \times A_{dd} + 0.29$

a. A_{dd} is the surface area of the display door.

TABLE C403.11.2.1(2)
WALK-IN COOLER AND FREEZER NONDISPLAY DOOR EFFICIENCY REQUIREMENTS[a]

CLASS DESCRIPTOR	CLASS	MAXIMUM ENERGY CONSUMPTION (kWh/day)[a]
Passage door, medium temperature	PD, M	$0.05 \times A_{nd} + 1.7$
Passage door, low temperature	PD, L	$0.14 \times A_{nd} + 4.8$
Freight door, medium temperature	FD, M	$0.04 \times A_{nd} + 1.9$
Freight door, low temperature	FD, L	$0.12 \times A_{nd} + 5.6$

a. A_{nd} is the surface area of the nondisplay door.

C403.11.3 Refrigeration systems. Refrigerated display cases, *walk-in coolers* or *walk-in freezers* that are served by remote compressors and remote condensers not located in a condensing unit, shall comply with Sections C403.11.3.1 and C403.11.3.2.

Exception: Systems where the working fluid in the refrigeration cycle goes through both subcritical and super-critical states (transcritical) or that use ammonia refrigerant are exempt.

C403.11.3.1 Condensers serving refrigeration systems. Fan-powered condensers shall comply with the following:

1. The design *saturated condensing temperatures* for air-cooled condensers shall not exceed the design dry-bulb temperature plus 10°F (5.6°C) for low-*temperature refrigeration systems*, and the design dry-bulb temperature plus 15°F (8°C) for *medium temperature refrigeration systems* where the *saturated condensing temperature* for blend refrigerants shall be determined using the average of liquid and vapor temperatures as converted from the condenser drain pressure.
2. Condenser fan motors that are less than 1 hp (0.75 kW) shall use electronically commutated motors, permanent split-capacitor-type motors or 3-phase motors.
3. Condenser fans for air-cooled condensers, evaporatively cooled condensers, air- or water-cooled fluid coolers or cooling towers shall

reduce fan motor demand to not more than 30 percent of design wattage at 50 percent of design air volume, and incorporate one of the following continuous variable speed fan control approaches:

3.1. Refrigeration system condenser control for air-cooled condensers shall use variable setpoint control logic to reset the condensing temperature setpoint in response to ambient dry-bulb temperature.

3.2. Refrigeration system condenser control for evaporatively cooled condensers shall use variable setpoint control logic to reset the condensing temperature setpoint in response to ambient wet-bulb temperature.

4. Multiple fan condensers shall be controlled in unison.

5. The minimum condensing temperature setpoint shall be not greater than 70°F (21°C).

C403.11.3.2 Compressor systems. Refrigeration compressor systems shall comply with the following:

1. Compressors and multiple-compressor system suction groups shall include control systems that use floating suction pressure control logic to reset the target suction pressure temperature based on the temperature requirements of the attached refrigeration display cases or walk-ins.

   Exception: Controls are not required for the following:

   1. Single-compressor systems that do not have variable capacity capability.

   2. Suction groups that have a design saturated suction temperature of 30°F (-1.1°C) or higher, suction groups that comprise the high stage of a two-stage or cascade system, or suction groups that primarily serve chillers for secondary cooling fluids.

2. Liquid subcooling shall be provided for all low-temperature compressor systems with a design cooling capacity equal to or greater than 100,000 Btu (29.3 kW) with a design-saturated suction temperature of -10°F (-23°C) or lower. The sub-cooled liquid temperature shall be controlled at a maximum temperature setpoint of 50°F (10°C) at the exit of the subcooler using either compressor economizer (interstage) ports or a separate compressor suction group operating at a saturated suction temperature of 18°F (-7.8°C) or higher.

   2.1. Insulation for liquid lines with a fluid operating temperature less than 60°F (15.6°C) shall comply with Table C403.11.3.

3. Compressors that incorporate internal or external crankcase heaters shall provide a means to cycle the heaters off during compressor operation.

C403.12 Construction of HVAC system elements. Ducts, plenums, piping and other elements that are part of an HVAC system shall be constructed and insulated in accordance with Sections C403.12.1 through C403.12.3.1.

TABLE C403.11.2.1(3)
WALK-IN COOLER AND FREEZER REFRIGERATION SYSTEM EFFICIENCY REQUIREMENTS

CLASS DESCRIPTOR	CLASS	MINIMUM ANNUAL WALK-IN ENERGY FACTOR AWEF (Btu/W-h)[a]	TEST PROCEDURE
Dedicated condensing, medium temperature, indoor system	DC.M.I	5.61	AHRI 1250
Dedicated condensing, medium temperature, outdoor system	DC.M.O	7.60	
Dedicated condensing, low temperature, indoor system, net capacity (q_{net}) < 6,500 Btu/h	DC.L.I, < 6,500	$9.091 \times 10^{-5} \times q_{net} + 1.81$	
Dedicated condensing, low temperature, indoor system, net capacity (q_{net}) ≥ 6,500 Btu/h	DC.L.I, ≥ 6,500	2.40	
Dedicated condensing, low temperature, outdoor system, net capacity (q_{net}) < 6,500 Btu/h	DC.L.O, < 6,500	$6.522 \times 10^{-5} \times q_{net} + 2.73$	
Dedicated condensing, low temperature, outdoor system, net capacity (q_{net}) ≥ 6,500 Btu/h	DC.L.O, ≥ 6,500	3.15	
Unit cooler, medium	UC.M	9.00	
Unit cooler, low temperature, net capacity (q_{net}) < 15,500 Btu/h	UC.L, < 15,500	$1.575 \times 10^{-5} \times q_{net} + 3.91$	
Unit cooler, low temperature, net capacity (q_{net}) ≥ 15,500 Btu/h	UC.L, ≥ 15,500	4.15	

For SI: 1 British thermal unit per hour = 0.2931 W.

a. q_{net} is net capacity (Btu/h) as determined in accordance with AHRI 1250.

TABLE C403.11.1
MINIMUM EFFICIENCY REQUIREMENTS: COMMERCIAL REFRIGERATORS AND FREEZERS AND REFRIGERATION

EQUIPMENT CATEGORY	CONDENSING UNIT CONFIGURATION	EQUIPMENT FAMILY	RATING TEMP., °F	OPERATING TEMP., °F	EQUIPMENT CLASSIFICATION[a, c]	MAXIMUM DAILY ENERGY CONSUMPTION, kWh/day[d, e]	TEST STANDARD
Remote condensing commercial refrigerators and commercial freezers	Remote (RC)	Vertical open (VOP)	38 (M)	≥32	VOP.RC.M	0.64 × TDA + 4.07	AHRI 1200
			0 (L)	<32	VOP.RC.L	2.20 × TDA + 6.85	
		Semivertical open (SVO)	38 (M)	≥32	SVO.RC.M	0.66 × TDA + 3.18	
			0 (L)	<32	SVO.RC.L	2.20 × TDA + 6.85	
		Horizontal open (HZO)	38 (M)	≥32	HZO.RC.M	0.35 × TDA + 2.88	
			0 (L)	<32	HZO.RC.L	0.55 × TDA + 6.88	
		Vertical closed transparent (VCT)	38 (M)	≥32	VCT.RC.M	0.15 × TDA + 1.95	
			0 (L)	<32	VCT.RC.L	0.49 × TDA + 2.61	
		Horizontal closed transparent (HCT)	38 (M)	≥32	HCT.RC.M	0.16 × TDA + 0.13	
			0 (L)	<32	HCT.RC.L	0.34 × TDA + 0.26	
		Vertical closed solid (VCS)	38 (M)	≥32	VCS.RC.M	0.10 × V + 0.26	
			0 (L)	<32	VCS.RC.L	0.21 × V + 0.54	
		Horizontal closed solid (HCS)	38 (M)	≥32	HCS.RC.M	0.10 × V + 0.26	
			0 (L)	<32	HCS.RC.L	0.21 × V + 0.54	
		Service over counter (SOC)	38 (M)	≥32	SOC.RC.M	0.44 × TDA + 0.11	
			0 (L)	<32	SOC.RC.L	0.93 × TDA + 0.22	
Self-contained commercial refrigerators and commercial freezers with and without doors	Self-contained (SC)	Vertical open (VOP)	38 (M)	≥32	VOP.SC.M	1.69 × TDA + 4.71	AHRI 1200
			0 (L)	<32	VOP.SC.L	4.25 × TDA + 11.82	
		Semivertical open (SVO)	38 (M)	≥32	SVO.SC.M	1.70 × TDA + 4.59	
			0 (L)	<32	SVO.SC.L	4.26 × TDA + 11.51	
		Horizontal open (HZO)	38 (M)	≥32	HZO.SC.M	0.72 × TDA + 5.55	
			0 (L)	<32	HZO.RC.L	1.90 × TDA + 7.08	
		Vertical closed transparent (VCT)	38 (M)	≥32	VCT.SC.M	0.10 × V + 0.86	
			0 (L)	<32	VCT.SC.L	0.29 × V + 2.95	
		Vertical closed solid (VCS)	38 (M)	≥32	VCS.SC.M	0.05 × V + 1.36	
			0 (L)	<32	VCS.SC.L	0.22 × V + 1.38	
		Horizontal closed transparent (HCT)	38 (M)	≥32	HCT.SC.M	0.06 × V + 0.37	
			0 (L)	<32	HCT.SC.L	0.08 × V + 1.23	
		Horizontal closed solid (HCS)	38 (M)	≥32	HCS.SC.M	0.05 × V + 0.91	
			0 (L)	<32	HCS.SC.L	0.06 × V + 1.12	
		Service over counter (SOC)	38 (M)	≥32	SOC.SC.M	0.52 × TDA + 1.00	
			0 (L)	<32	SOC.SC.L	1.10 × TDA + 2.10	

(continued)

COMMERCIAL ENERGY EFFICIENCY

TABLE C403.11.1—continued
MINIMUM EFFICIENCY REQUIREMENTS: COMMERCIAL REFRIGERATORS AND FREEZERS AND REFRIGERATION

EQUIPMENT CATEGORY	CONDENSING UNIT CONFIGURATION	EQUIPMENT FAMILY	RATING TEMP., °F	OPERATING TEMP., °F	EQUIPMENT CLASSIFICATION[a,c]	MAXIMUM DAILY ENERGY CONSUMPTION, kWh/day[d,e]	TEST STANDARD
Self-contained commercial refrigerators with transparent doors for pull-down temperature applications	Self-contained (SC)	Pull-down (PD)	38 (M)	≥ 32	PD.SC.M	$0.11 \times V + 0.81$	AHRI 1200
Commercial ice cream freezers	Remote (RC)	Vertical open (VOP)	-15 (I)	≤ -5[b]	VOP.RC.I	$2.79 \times TDA + 8.70$	
		Semivertical open (SVO)			SVO.RC.I	$2.79 \times TDA + 8.70$	
		Horizontal open (HZO)			HZO.RC.I	$0.70 \times TDA + 8.74$	
		Vertical closed transparent (VCT)			VCT.RC.I	$0.58 \times TDA + 3.05$	AHRI 1200
		Horizontal closed transparent (HCT)			HCT.RC.I	$0.40 \times TDA + 0.31$	
		Vertical closed solid (VCS)			VCS.RC.I	$0.25 \times V + 0.63$	
		Horizontal closed solid (HCS)			HCS.RC.I	$0.25 \times V + 0.63$	
		Service over counter (SOC)			SOC.RC.I	$1.09 \times TDA + 0.26$	
	Self-contained (SC)	Vertical open (VOP)			VOP.SC.I	$5.40 \times TDA + 15.02$	
		Semivertical open (SVO)			SVO.SC.I	$5.41 \times TDA + 14.63$	
		Horizontal open (HZO)			HZO.SC.I	$2.42 \times TDA + 9.00$	AHRI 1200
		Vertical closed transparent (VCT)			VCT.SC.I	$0.62 \times TDA + 3.29$	
		Horizontal closed transparent (HCT)			HCT.SC.I	$0.56 \times TDA + 0.43$	
		Vertical closed solid (VCS)			VCS.SC.I	$0.34 \times V + 0.88$	
		Horizontal closed solid (HCS)			HCS.SC.I	$0.34 \times V + 0.88$	
		Service over counter (SOC)			SOC.SC.I	$1.53 \times TDA + 0.36$	

For SI: 1 square foot = 0.0929 m², 1 cubic foot = 0.02832 m³, °C = (°F − 32)/1.8.

a. The meaning of the letters in this column is indicated in the columns to the left.
b. Ice cream freezer is defined in DOE 10 CFR 431.62 as a commercial freezer that is designed to operate at or below -5 °F and that the manufacturer designs, markets or intends for the storing, displaying or dispensing of ice cream.
c. Equipment class designations consist of a combination [in sequential order separated by periods (AAA).(BB).(C)] of the following:
   - (AAA)—An equipment family code (VOP = vertical open, SVO = semivertical open, HZO = horizontal open, VCT = vertical closed transparent doors, VCS = vertical closed solid doors, HCT = horizontal closed transparent doors, HCS = horizontal closed solid doors, and SOC = service over counter);
   - (BB)—An operating mode code (RC = remote condensing and SC = self-contained); and
   - (C)—A rating temperature code [M = medium temperature (38°F), L = low temperature (0°F), or I = ice cream temperature (-15°F)].
   - For example, "VOP.RC.M" refers to the "vertical open, remote condensing, medium temperature" equipment class.
d. V is the volume of the case (ft³) as measured in AHRI 1200, Appendix C.
e. TDA is the total display area of the case (ft²) as measured in AHRI 1200, Appendix D.

C403.12.1 Duct and plenum insulation and sealing. Supply and return air ducts and plenums shall be insulated with not less than R-6 insulation where located in unconditioned spaces and where located outside the building with not less than R-8 insulation in *Climate Zones* 0 through 4 and not less than R-12 insulation in *Climate Zones* 5 through 8. Ducts located underground beneath buildings shall be insulated as required in this section or have an equivalent *thermal distribution efficiency*. Underground ducts utilizing the *thermal distribution efficiency* method shall be *listed* and *labeled* to indicate the *R*-value equivalency. Where located within a building envelope assembly, the duct or plenum shall be separated from the building exterior or unconditioned or exempt spaces by not less than R-8 insulation in *Climate Zones* 0 through 4 and not less than R-12 insulation in *Climate Zones* 5 through 8.

Exceptions:

1. Where located within equipment.
2. Where the design temperature difference between the interior and exterior of the duct or plenum is not greater than 15°F (8°C).

Ducts, air handlers and filter boxes shall be sealed. Joints and seams shall comply with Section 603.9 of the *International Mechanical Code*.

C403.12.2 Duct construction. Ductwork shall be constructed and erected in accordance with the *International Mechanical Code*.

C403.12.2.1 Low-pressure duct systems. Longitudinal and transverse joints, seams and connections of supply and return ducts operating at a static pressure less than or equal to 2 inches water gauge (w.g.) (498 Pa) shall be securely fastened and sealed with welds, gaskets, mastics (adhesives), mastic-plus-embedded-fabric systems or tapes installed in accordance with the manufacturer's instructions. Pressure classifications specific to the duct system shall be clearly indicated on the construction documents in accordance with the *International Mechanical Code*.

Exception: Locking-type longitudinal joints and seams, other than the snap-lock and button-lock types, need not be sealed as specified in this section.

C403.12.2.2 Medium-pressure duct systems. Ducts and plenums designed to operate at a static pressure greater than 2 inches water gauge (w.g.) (498 Pa) but less than 3 inches w.g. (747 Pa) shall be insulated and sealed in accordance with Section C403.12.1. Pressure classifications specific to the duct system shall be clearly indicated on the construction documents in accordance with the *International Mechanical Code*.

C403.12.2.3 High-pressure duct systems. Ducts and plenums designed to operate at static pressures equal to or greater than 3 inches water gauge (747 Pa) shall be insulated and sealed in accordance with Section C403.12.1. In addition, ducts and plenums shall be leak tested in accordance with the SMACNA HVAC Air Duct Leakage Test Manual and shown to have a rate of air leakage (CL) less than or equal to 4.0 as determined in accordance with Equation 4-8.

$$CL = F/P^{0.65} \qquad \text{(Equation 4-8)}$$

where:

F = The measured leakage rate in cfm per 100 square feet (9.3 m^2) of duct surface.

P = The static pressure of the test.

Documentation shall be furnished demonstrating that representative sections totaling not less than 25 percent of the duct area have been tested and that all tested sections comply with the requirements of this section.

C403.12.3 Piping insulation. Piping serving as part of a heating or cooling system shall be thermally insulated in accordance with Table C403.12.3.

Exceptions:

1. Factory-installed piping within HVAC equipment tested and rated in accordance with a test procedure referenced by this code.
2. Factory-installed piping within room fan-coils and unit ventilators tested and rated according to AHRI 440 (except that the sampling and variation provisions of Section 6.5 shall not apply) and AHRI 840, respectively.
3. Piping that conveys fluids that have a design operating temperature range between 60°F (15°C) and 105°F (41°C).
4. Piping that conveys fluids that have not been heated or cooled through the use of fossil fuels or electric power.
5. Strainers, control valves, and balancing valves associated with piping 1 inch (25 mm) or less in diameter.
6. Direct buried piping that conveys fluids at or below 60°F (15°C).
7. In radiant heating systems, sections of piping intended by design to radiate heat.

C403.12.3.1 Protection of piping insulation. Piping insulation exposed to the weather shall be protected from damage, including that caused by sunlight, moisture, equipment maintenance and wind, and shall provide shielding from solar radiation that can cause degradation of the material. Adhesive tape shall not be permitted.

C403.13 Mechanical systems located outside of the building thermal envelope. Mechanical systems providing heat outside of the thermal envelope of a building shall comply with Sections C403.13.1 through C403.13.3.

C403.13.1 Heating outside a building. Systems installed to provide heat outside a building shall be radiant systems.

Such heating systems shall be controlled by an occupancy sensing device or a timer switch, so that the system is automatically de-energized when occupants are not present.

C403.13.2 Snow- and ice-melt system controls. Snow- and ice-melting systems shall include automatic controls configured to shut off the system when the pavement temperature is above 50°F (10°C) and precipitation is not falling, and an automatic or manual control that is configured to shut off when the outdoor temperature is above 40°F (4°C).

C403.13.3 Freeze protection system controls. Freeze protection systems, such as heat tracing of outdoor piping and heat exchangers, including self-regulating heat tracing, shall include automatic controls configured to shut off the systems when outdoor air temperatures are above 40°F (4°C) or when the conditions of the protected fluid will prevent freezing.

C403.14 Operable opening interlocking controls. The heating and cooling systems shall have controls that will interlock these mechanical systems to the set temperatures of 90°F (32°C) for cooling and 55°F (12.7°C) for heating when the conditions of Section C402.5.8 exist. The controls shall configure to shut off the systems entirely when the outdoor temperatures are below 90°F (32°C) or above 55°F (12.7°C).

SECTION C404
SERVICE WATER HEATING

C404.1 General. This section covers the minimum efficiency of, and controls for, service water-heating equipment and insulation of service hot water piping.

C404.2 Service water-heating equipment performance efficiency. Water-heating equipment and hot water storage tanks shall meet the requirements of Table C404.2. The efficiency shall be verified through data furnished by the manufacturer of the equipment or through certification under an *approved* certification program. Water-heating equipment intended to be used to provide space heating shall meet the applicable provisions of Table C404.2.

C404.2.1 High input service water-heating systems. Gas-fired water-heating equipment installed in new buildings shall be in compliance with this section. Where a singular piece of water-heating equipment serves the entire building and the input rating of the equipment is 1,000,000 Btu/h (293 kW) or greater, such equipment shall have a thermal efficiency, E_t, of not less than 92 percent. Where multiple pieces of water-heating equipment serve the building and the combined input rating of the water-heating equipment is 1,000,000 Btu/h (293 kW) or greater, the combined input-capacity-weighted-average thermal efficiency, E_t, shall be not less than 90 percent.

Exceptions:

1. Where not less than 25 percent of the annual service water-heating requirement is provided by *on-site renewable energy* or site-recovered energy, the minimum thermal efficiency requirements of this section shall not apply.

2. The input rating of water heaters installed in individual dwelling units shall not be required to be included in the total input rating of service water-heating equipment for a building.

3. The input rating of water heaters with an input rating of not greater than 100,000 Btu/h (29.3 kW) shall not be required to be included in the total input rating of service water-heating equipment for a building.

TABLE C403.12.3
MINIMUM PIPE INSULATION THICKNESS (in inches)[a, c]

FLUID OPERATING TEMPERATURE RANGE AND USAGE (°F)	INSULATION CONDUCTIVITY		NOMINAL PIPE OR TUBE SIZE (inches)				
	Conductivity Btu × in./(h × ft² × °F)[b]	Mean Rating Temperature, °F	< 1	1 to < 1½	1½ to < 4	4 to < 8	> 8
> 350	0.32–0.34	250	4.5	5.0	5.0	5.0	5.0
251–350	0.29–0.32	200	3.0	4.0	4.5	4.5	4.5
201–250	0.27–0.30	150	2.5	2.5	2.5	3.0	3.0
141–200	0.25–0.29	125	1.5	1.5	2.0	2.0	2.0
105–140	0.21–0.28	100	1.0	1.0	1.5	1.5	1.5
40–60	0.21–0.27	75	0.5	0.5	1.0	1.0	1.0
< 40	0.20–0.26	50	0.5	1.0	1.0	1.0	1.5

For SI: 1 inch = 25.4 mm, °C = [(°F) – 32]/1.8.

a. For piping smaller than 1½ inches and located in partitions within conditioned spaces, reduction of these thicknesses by 1 inch shall be permitted (before thickness adjustment required in Note b) but not to a thickness less than 1 inch.

b. For insulation outside the stated conductivity range, the minimum thickness (T) shall be determined as follows:

$T = r[(1 + t/r)^{K/k} - 1]$

where:

T = Minimum insulation thickness.
r = Actual outside radius of pipe.
t = Insulation thickness listed in the table for applicable fluid temperature and pipe size.
K = Conductivity of alternate material at mean rating temperature indicated for the applicable fluid temperature (Btu × in/h × ft² × °F).
k = The upper value of the conductivity range listed in the table for the applicable fluid temperature.

c. For direct-buried heating and hot water system piping, reduction of these thicknesses by 1½ inches (38 mm) shall be permitted (before thickness adjustment required in Note b but not to thicknesses less than 1 inch.

TABLE C404.2
MINIMUM PERFORMANCE OF WATER-HEATING EQUIPMENT

EQUIPMENT TYPE	SIZE CATEGORY (input)	SUBCATEGORY OR RATING CONDITION	PERFORMANCE REQUIRED[a, b]	TEST PROCEDURE
Water heaters, electric	≤ 12 kW[d]	Tabletop[e], ≥ 20 gallons and ≤ 120 gallons	$0.93 - 0.00132V$, EF	DOE 10 CFR Part 430
		Resistance ≥ 20 gallons and ≤ 55 gallons	$0.960 - 0.0003V$, EF	
		Grid-enabled[f] > 75 gallons and ≤ 120 gallons	$1.061 - 0.00168V$, EF	
	> 12 kW	Resistance	$(0.3 + 27/V_m)$, %/h	ANSI Z21.10.3
	≤ 24 amps and ≤ 250 volts	Heat pump > 55 gallons and ≤ 120 gallons	$2.057 - 0.00113V$, EF	DOE 10 CFR Part 430
Storage water heaters, gas	≤ 75,000 Btu/h	≥ 20 gallons and > 55 gallons	$0.675 - 0.0015V$, EF	DOE 10 CFR Part 430
		> 55 gallons and ≤ 100 gallons	$0.8012 - 0.00078V$, EF	
	> 75,000 Btu/h and ≤ 155,000 Btu/h	< 4,000 Btu/h/gal	80% E_t $(Q/800 + 110\sqrt{V})$SL, Btu/h	ANSI Z21.10.3
	> 155,000 Btu/h	< 4,000 Btu/h/gal	80% E_t $(Q/800 + 110\sqrt{V})$SL, Btu/h	
Instantaneous water heaters, gas	> 50,000 Btu/h and < 200,000 Btu/h[c]	≥ 4,000 Btu/h/gal and < 2 gal	$0.82 - 0.0019V$, EF	DOE 10 CFR Part 430
	≥ 200,000 Btu/h	≥ 4,000 Btu/h/gal and < 10 gal	80% E_t	ANSI Z21.10.3
	≥ 200,000 Btu/h	≥ 4,000 Btu/h/gal and ≥ 10 gal	80% E_t $(Q/800 + 110\sqrt{V})$SL, Btu/h	
Storage water heaters, oil	≤ 105,000 Btu/h	≥ 20 gal and ≤ 50 gallons	$0.68 - 0.0019V$, EF	DOE 10 CFR Part 430
	≥ 105,000 Btu/h	< 4,000 Btu/h/gal	80% E_t $(Q/800 + 110\sqrt{V})$SL, Btu/h	ANSI Z21.10.3
Instantaneous water heaters, oil	≤ 210,000 Btu/h	≥ 4,000 Btu/h/gal and < 2 gal	$0.59 - 0.0019V$, EF	DOE 10 CFR Part 430
	> 210,000 Btu/h	≥ 4,000 Btu/h/gal and < 10 gal	80% E_t	ANSI Z21.10.3
	> 210,000 Btu/h	≥ 4,000 Btu/h/gal and ≥ 10 gal	78% E_t $(Q/800 + 110\sqrt{V})$SL, Btu/h	
Hot water supply boilers, gas and oil	≥ 300,000 Btu/h and < 12,500,000 Btu/h	≥ 4,000 Btu/h/gal and < 10 gal	80% E_t	ANSI Z21.10.3
Hot water supply boilers, gas	≥ 300,000 Btu/h and < 12,500,000 Btu/h	≥ 4,000 Btu/h/gal and ≥ 10 gal	80% E_t $(Q/800 + 110\sqrt{V})$SL, Btu/h	
Hot water supply boilers, oil	> 300,000 Btu/h and < 12,500,000 Btu/h	> 4,000 Btu/h/gal and > 10 gal	78% E_t $(Q/800 + 110\sqrt{V})$SL, Btu/h	
Pool heaters, gas and oil	All	—	82% E_t	ASHRAE 146
Heat pump pool heaters	All	—	4.0 COP	AHRI 1160
Unfired storage tanks	All	—	Minimum insulation requirement R-12.5 (h × ft² × °F)/Btu	(none)

(continued)

TABLE C404.2—continued
MINIMUM PERFORMANCE OF WATER-HEATING EQUIPMENT

For SI: 1 foot = 304.8 mm, 1 square foot = 0.0929 m², °C = [(°F) − 32]/1.8, 1 British thermal unit per hour = 0.2931 W, 1 gallon = 3.785 L, 1 British thermal unit per hour per gallon = 0.078 W/L.

a. Energy factor (EF) and thermal efficiency (E_t) are minimum requirements. In the EF equation, V is the rated volume in gallons.
b. Standby loss (SL) is the maximum Btu/h based on a nominal 70°F temperature difference between stored water and ambient requirements. In the SL equation, Q is the nameplate input rate in Btu/h. In the equations for electric water heaters, V is the rated volume in gallons and V_m is the measured volume in gallons. In the SL equation for oil and gas water heaters and boilers, V is the rated volume in gallons.
c. Instantaneous water heaters with input rates below 200,000 Btu/h shall comply with these requirements where the water heater is designed to heat water to temperatures 180°F or higher.
d. Electric water heaters with an input rating of 12 kW (40,950 Btu/h) or less that are designed to heat water to temperatures of 180°F or greater shall comply with the requirements for electric water heaters that have an input rating greater than 12 kW (40,950 Btu/h).
e. A tabletop water heater is a water heater that is enclosed in a rectangular cabinet with a flat top surface not more than 3 feet in height.
f. A grid-enabled water heater is an electric-resistance water heater that meets all of the following:
  1. Has a rated storage tank volume of more than 75 gallons.
  2. Was manufactured on or after April 16, 2015.
  3. Is equipped at the point of manufacture with an activation lock.
  4. Bears a permanent label applied by the manufacturer that complies with all of the following:
      4.1. Is made of material not adversely affected by water.
      4.2. Is attached by means of nonwater-soluble adhesive.
      4.3. Advises purchasers and end users of the intended and appropriate use of the product with the following notice printed in 16.5 point Arial Narrow Bold font: "IMPORTANT INFORMATION: This water heater is intended only for use as part of an electric thermal storage or demand response program. It will not provide adequate hot water unless enrolled in such a program and activated by your utility company or another program operator. Confirm the availability of a program in your local area before purchasing or installing this product."

C404.3 Heat traps for hot water storage tanks. Storage tank-type water heaters and hot water storage tanks that have vertical water pipes connecting to the inlet and outlet of the tank shall be provided with integral heat traps at those inlets and outlets or shall have pipe-configured heat traps in the piping connected to those inlets and outlets. Tank inlets and outlets associated with solar water heating system circulation loops shall not be required to have heat traps.

C404.4 Insulation of piping. Piping from a water heater to the termination of the heated water fixture supply pipe shall be insulated in accordance with Table C403.12.3. On both the inlet and outlet piping of a storage water heater or heated water storage tank, the piping to a heat trap or the first 8 feet (2438 mm) of piping, whichever is less, shall be insulated. Piping that is heat traced shall be insulated in accordance with Table C403.12.3 or the heat trace manufacturer's instructions. Tubular pipe insulation shall be installed in accordance with the insulation manufacturer's instructions. Pipe insulation shall be continuous except where the piping passes through a framing member. The minimum insulation thickness requirements of this section shall not supersede any greater insulation thickness requirements necessary for the protection of piping from freezing temperatures or the protection of personnel against external surface temperatures on the insulation.

> **Exception:** Tubular pipe insulation shall not be required on the following:
> 1. The tubing from the connection at the termination of the fixture supply piping to a plumbing fixture or plumbing appliance.
> 2. Valves, pumps, strainers and threaded unions in piping that is 1 inch (25 mm) or less in nominal diameter.
> 3. Piping from user-controlled shower and bath mixing valves to the water outlets.
> 4. Cold-water piping of a demand recirculation water system.
> 5. Tubing from a hot drinking-water heating unit to the water outlet.
> 6. Piping at locations where a vertical support of the piping is installed.
> 7. Piping surrounded by building insulation with a thermal resistance (R-value) of not less than R-3.

C404.5 Heated water supply piping. Heated water supply piping shall be in accordance with Section C404.5.1 or C404.5.2. The flow rate through $^1/_4$-inch (6.4 mm) piping shall be not greater than 0.5 gpm (1.9 L/m). The flow rate through $^5/_{16}$-inch (7.9 mm) piping shall be not greater than 1 gpm (3.8 L/m). The flow rate through $^3/_8$-inch (9.5 mm) piping shall be not greater than 1.5 gpm (5.7 L/m).

C404.5.1 Maximum allowable pipe length method. The maximum allowable piping length from the nearest source of heated water to the termination of the fixture supply pipe shall be in accordance with the following. Where the piping contains more than one size of pipe, the largest size of pipe within the piping shall be used for determining the maximum allowable length of the piping in Table C404.5.1.

1. For a public lavatory faucet, use the "Public lavatory faucets" column in Table C404.5.1.
2. For all other plumbing fixtures and plumbing appliances, use the "Other fixtures and appliances" column in Table C404.5.1.

TABLE C404.5.1
PIPING VOLUME AND MAXIMUM PIPING LENGTHS

NOMINAL PIPE SIZE (inches)	VOLUME (liquid ounces per foot length)	MAXIMUM PIPING LENGTH (feet)	
		Public lavatory faucets	Other fixtures and appliances
1/4	0.33	6	50
5/16	0.5	4	50
3/8	0.75	3	50
1/2	1.5	2	43
5/8	2	1	32
3/4	3	0.5	21
7/8	4	0.5	16
1	5	0.5	13
1 1/4	8	0.5	8
1 1/2	11	0.5	6
2 or larger	18	0.5	4

For SI: 1 inch = 25.4 mm, 1 foot = 304.8 mm, 1 liquid ounce = 0.030 L, 1 gallon = 128 ounces.

C404.5.2 Maximum allowable pipe volume method. The water volume in the piping shall be calculated in accordance with Section C404.5.2.1. Water heaters, circulating water systems and heat trace temperature maintenance systems shall be considered to be sources of heated water.

The volume from the nearest source of heated water to the termination of the fixture supply pipe shall be as follows:

1. For a public lavatory faucet: not more than 2 ounces (0.06 L).
2. For other plumbing fixtures or plumbing appliances; not more than 0.5 gallon (1.89 L).

C404.5.2.1 Water volume determination. The volume shall be the sum of the internal volumes of pipe, fittings, valves, meters and manifolds between the nearest source of heated water and the termination of the fixture supply pipe. The volume in the piping shall be determined from the "Volume" column in Table C404.5.1 or from Table C404.5.2.1. The volume contained within fixture shutoff valves, within flexible water supply connectors to a fixture fitting and within a fixture fitting shall not be included in the water volume determination. Where heated water is supplied by a recirculating system or heat-traced piping, the volume shall include the portion of the fitting on the branch pipe that supplies water to the fixture.

C404.6 Heated-water circulating and temperature maintenance systems. Heated-water circulation systems shall be in accordance with Section C404.6.1. Heat trace temperature maintenance systems shall be in accordance with Section C404.6.2. Controls for hot water storage shall be in accordance with Section C404.6.3. Automatic controls, temperature sensors and pumps shall be in a location with *access*. Manual controls shall be in a location with *ready access*.

C404.6.1 Circulation systems. Heated-water circulation systems shall be provided with a circulation pump. The system return pipe shall be a dedicated return pipe or a cold water supply pipe. Gravity and thermo-syphon circulation systems shall be prohibited. Controls for circulating hot water system pumps shall automatically turn off the pump when the water in the circulation loop is at the desired temperature and when there is not a demand for hot water. The controls shall limit the temperature of the water entering the cold water piping to not greater than 104°F (40°C).

C404.6.1.1 Demand recirculation controls. Demand recirculation water systems shall have controls that start the pump upon receiving a signal from the action of a user of a fixture or appliance, sensing the presence of a user of a fixture, or sensing the flow of hot or tempered water to a fixture fitting or appliance.

C404.6.2 Heat trace systems. Electric heat trace systems shall comply with IEEE 515.1. Controls for such systems shall be able to automatically adjust the energy input to the heat tracing to maintain the desired water temperature in the piping in accordance with the times when heated water is used in the occupancy. Heat trace shall be arranged to be turned off automatically when there is not a demand for hot water.

C404.6.3 Controls for hot water storage. The controls on pumps that circulate water between a water heater and a heated-water storage tank shall limit operation of the pump from heating cycle startup to not greater than 5 minutes after the end of the cycle.

TABLE C404.5.2.1
INTERNAL VOLUME OF VARIOUS WATER DISTRIBUTION TUBING

	OUNCES OF WATER PER FOOT OF TUBE								
Nominal Size (inches)	Copper Type M	Copper Type L	Copper Type K	CPVC CTS SDR 11	CPVC SCH 40	CPVC SCH 80	PE-RT SDR 9	Composite ASTM F1281	PEX CTS SDR 9
3/8	1.06	0.97	0.84	N/A	1.17	—	0.64	0.63	0.64
1/2	1.69	1.55	1.45	1.25	1.89	1.46	1.18	1.31	1.18
3/4	3.43	3.22	2.90	2.67	3.38	2.74	2.35	3.39	2.35
1	5.81	5.49	5.17	4.43	5.53	4.57	3.91	5.56	3.91
1 1/4	8.70	8.36	8.09	6.61	9.66	8.24	5.81	8.49	5.81
1 1/2	12.18	11.83	11.45	9.22	13.20	11.38	8.09	13.88	8.09
2	21.08	20.58	20.04	15.79	21.88	19.11	13.86	21.48	13.86

For SI: 1 foot = 304.8 mm, 1 inch = 25.4 mm, 1 liquid ounce = 0.030 L, 1 oz/ft^2 = 305.15 g/m^2.
N/A = Not Available.

* **C404.7 Drain water heat recovery units.** Drain water heat recovery units shall comply with CSA B55.2. Potable waterside pressure loss shall be less than 10 psi (69 kPa) at maximum design flow. For *Group R* occupancies, the efficiency of drain water heat recovery unit efficiency shall be in accordance with CSA B55.1.

C404.8 Energy consumption of pools and permanent spas. The energy consumption of pools and permanent spas shall be controlled by the requirements in Sections C404.8.1 through C404.8.3.

C404.8.1 Heaters. The electric power to all heaters shall be controlled by an on-off switch that is an integral part of the heater, mounted on the exterior of the heater, or external to and within 3 feet (914 mm) of the heater in a location with *ready access*. Operation of such switch shall not change the setting of the heater thermostat. Such switches shall be in addition to a circuit breaker for the power to the heater. Gas-fired heaters shall not be equipped with continuously burning ignition pilots.

C404.8.2 Time switches. Time switches or other control methods that can automatically turn off and on heaters and pump motors according to a preset schedule shall be installed for heaters and pump motors. Heaters and pump motors that have built-in time switches shall be in compliance with this section.

Exceptions:

1. Where public health standards require 24-hour pump operation.
2. Pumps that operate solar- and waste-heat-recovery pool heating systems.

C404.8.3 Covers. Outdoor heated pools and outdoor permanent spas shall be provided with a vapor-retardant cover or other *approved* vapor-retardant means.

Exception: Where more than 75 percent of the energy for heating, computed over an operating season of not fewer than 3 calendar months, is from a heat pump or an on-site renewable energy system, covers or other vapor-retardant means shall not be required.

C404.9 Portable spas. The energy consumption of electric-powered portable spas shall be controlled by the requirements of APSP 14.

SECTION C405
ELECTRICAL POWER AND LIGHTING SYSTEMS

C405.1 General. Lighting system controls, the maximum lighting power for interior and exterior applications, and electrical energy consumption shall comply with this section. *Sleeping units* shall comply with Section C405.2.5 and with either Section C405.1.1 or C405.3. *General lighting* shall consist of all lighting included when calculating the total connected interior lighting power in accordance with Section C405.3.1 and which does not require specific application controls in accordance with Section C405.2.5.

Transformers, uninterruptable power supplies, motors and electrical power processing equipment in data center systems shall comply with Section 8 of ASHRAE 90.4 in addition to this code.

C405.1.1 Lighting for dwelling units. No less than 90 percent of the permanently installed lighting serving dwelling units, excluding kitchen appliance lighting, shall be provided by lamps with an efficacy of not less than 65 lm/W or luminaires with an efficacy of not less than 45 lm/W, or shall comply with Sections C405.2.4 and C405.3.

C405.2 Lighting controls. Lighting systems shall be provided with controls that comply with one of the following.

1. Lighting controls as specified in Sections C405.2.1 through C405.2.8.
2. Luminaire level lighting controls (LLLC) and lighting controls as specified in Sections C405.2.1, C405.2.5 and C405.2.6. The LLLC luminaire shall be independently capable of:

    2.1. Monitoring occupant activity to brighten or dim lighting when occupied or unoccupied, respectively.

    2.2. Monitoring ambient light, both electric light and daylight, and brighten or dim artificial light to maintain desired light level.

    2.3. For each control strategy, configuration and reconfiguration of performance parameters including; bright and dim setpoints, timeouts, dimming fade rates, sensor sensitivity adjustments, and wireless zoning configurations.

Exceptions: Lighting controls are not required for the following:

1. Areas designated as security or emergency areas that are required to be continuously lighted.
2. Interior exit stairways, interior exit ramps and exit passageways.
3. Emergency egress lighting that is normally off.

C405.2.1 Occupant sensor controls. Occupant *sensor controls* shall be installed to control lights in the following space types:

1. Classrooms/lecture/training rooms.
2. Conference/meeting/multipurpose rooms.
3. Copy/print rooms.
4. Lounges/breakrooms.
5. Enclosed offices.
6. Open plan office areas.
7. Restrooms.
8. Storage rooms.
9. Locker rooms.

10. Corridors.
11. Warehouse storage areas.
12. Other spaces 300 square feet (28 m^2) or less that are enclosed by floor-to-ceiling height partitions.

Exception: Luminaires that are required to have specific application controls in accordance with Section C405.2.5.

C405.2.1.1 Occupant sensor control function. Occupant sensor controls in warehouses shall comply with Section C405.2.1.2. Occupant sensor controls in open plan office areas shall comply with Section C405.2.1.3. Occupant sensor controls in corridors shall comply with Section C405.2.1.4. Occupant sensor controls for all other spaces specified in Section C405.2.1 shall comply with the following:

1. They shall automatically turn off lights within 20 minutes after all occupants have left the space.
2. They shall be manual on or controlled to automatically turn on the lighting to not more than 50-percent power.
3. They shall incorporate a manual control to allow occupants to turn off lights.

Exception: Full automatic-on controls with no manual control shall be permitted in corridors, interior parking areas, stairways, restrooms, locker rooms, lobbies, library stacks and areas where manual operation would endanger occupant safety or security.

C405.2.1.2 Occupant sensor control function in warehouse storage areas. Lighting in warehouse storage areas shall be controlled as follows:

1. Lighting in each aisleway shall be controlled independently of lighting in all other aisleways and open areas.
2. Occupant sensors shall automatically reduce lighting power within each controlled area to an unoccupied setpoint of not more than 50 percent of full power within 20 minutes after all occupants have left the controlled area.
3. Lights that are not turned off by occupant sensors shall be turned off by time-switch control complying with Section C405.2.2.1.
4. A manual control shall be provided to allow occupants to turn off lights in the space.

C405.2.1.3 Occupant sensor control function in open plan office areas. Occupant sensor controls in open plan office spaces less than 300 square feet (28 m^2) in area shall comply with Section C405.2.1.1. Occupant sensor controls in all other open plan office spaces shall comply with all of the following:

1. The controls shall be configured so that general lighting can be controlled separately in control zones with floor areas not greater than 600 square feet (55 m^2) within the open plan office space.
2. General lighting in each control zone shall be permitted to automatically turn on upon occupancy within the control zone. General lighting in other unoccupied zones within the open plan office space shall be permitted to turn on to not more than 20 percent of full power or remain unaffected.
3. The controls shall automatically turn off general lighting in all control zones within 20 minutes after all occupants have left the open plan office space.

   Exception: Where general lighting is turned off by time-switch control complying with Section C405.2.2.1.

4. General lighting in each control zone shall turn off or uniformly reduce lighting power to an unoccupied setpoint of not more than 20 percent of full power within 20 minutes after all occupants have left the control zone.

C405.2.1.4 Occupant sensor control function in corridors. Occupant sensor controls in corridors shall uniformly reduce lighting power to an occupied setpoint not more than 50 percent of full power within 20 minutes after all occupants have left the space.

Exception: Corridors provided with less than two footcandles of illumination on the floor at the darkest point with all lights on.

C405.2.2 Time-switch controls. Each area of the building that is not provided with *occupant sensor controls* complying with Section C405.2.1.1 shall be provided with *time-switch controls* complying with Section C405.2.2.1.

Exceptions:

1. Luminaires that are required to have specific application controls in accordance with Section C405.2.4.
2. Spaces where patient care is directly provided.
3. Spaces where an automatic shutoff would endanger occupant safety or security.
4. Lighting intended for continuous operation.
5. Shop and laboratory classrooms.

C405.2.2.1 Time-switch control function. Time-switch *controls* shall comply with all of the following:

1. Automatically turn off lights when the space is scheduled to be unoccupied.
2. Have a minimum 7-day clock.
3. Be capable of being set for seven different day types per week.
4. Incorporate an automatic holiday "shutoff" feature, which turns off all controlled lighting

loads for not fewer than 24 hours and then resumes normally scheduled operations.

5. Have program backup capabilities, which prevent the loss of program and time settings for not fewer than 10 hours, if power is interrupted.
6. Include an override switch that complies with the following:
   6.1. The override switch shall be a manual control.
   6.2. The override switch, when initiated, shall permit the controlled lighting to remain on for not more than 2 hours.
   6.3. Any individual override switch shall control the lighting for an area not larger than 5,000 square feet (465 m²).

Exception: Within mall concourses, auditoriums, sales areas, manufacturing facilities and sports arenas:
1. The time limit shall be permitted to be greater than 2 hours, provided that the switch is a captive key device.
2. The area controlled by the override switch shall not be limited to 5,000 square feet (465 m²) provided that such area is less than 20,000 square feet (1860 m²).

C405.2.3 Light-reduction controls. Where not provided with occupant sensor controls complying with Section C405.2.1.1, general lighting shall be provided with light-reduction controls complying with Section C405.2.3.1.

Exceptions:
1. Luminaires controlled by daylight responsive controls complying with Section C405.2.4.
2. Luminaires controlled by special application controls complying with Section C405.2.5.
3. Where provided with manual control, the following areas are not required to have light-reduction control:
   3.1. Spaces that have only one luminaire with a rated power of less than 60 watts.
   3.2. Spaces that use less than 0.45 watts per square foot (4.9 W/m²).
   3.3. Corridors, lobbies, electrical rooms and/or mechanical rooms.

C405.2.3.1 Light-reduction control function. Spaces required to have light-reduction controls shall have a *manual control* that allows the occupant to reduce the connected lighting load by not less than 50 percent in a reasonably uniform illumination pattern with an intermediate step in addition to full on or off, or with continuous dimming control, using one of the following or another *approved* method:

1. Continuous dimming of all luminaires from full output to less than 20 percent of full power.
2. Switching all luminaires to a reduced output of not less than 30 percent and not more than 70 percent of full power.
3. Switching alternate luminaires or alternate rows of luminaires to achieve a reduced output of not less than 30 percent and not more than 70 percent of full power.

C405.2.4 Daylight-responsive controls. *Daylight-responsive controls* complying with Section C405.2.4.1 shall be provided to control the general lighting within *daylight zones* in the following spaces:

1. Spaces with a total of more than 150 watts of *general lighting* within primary sidelit daylight zones complying with Section C405.2.4.2.
2. Spaces with a total of more than 300 watts of *general lighting* within sidelit daylight zones complying with Section C405.2.4.2.
3. Spaces with a total of more than 150 watts of *general lighting* within toplit daylight zones complying with Section C405.2.4.3.

Exceptions: Daylight responsive controls are not required for the following:

1. Spaces in health care facilities where patient care is directly provided.
2. Sidelit daylight zones on the first floor above grade in Group A-2 and Group M occupancies.
3. New buildings where the total connected lighting power calculated in accordance with Section C405.3.1 is not greater than the adjusted interior lighting power allowance (LPA_{adj}) calculated in accordance with Equation 4-9.

$$LPA_{adj} = [LPA_{norm} \times (1.0 - 0.4 \times UDZFA / TBFA)] \quad \text{(Equation 4-9)}$$

where:

LPA_{adj} = Adjusted building interior lighting power allowance in watts.

LPA_{norm} = Normal building lighting power allowance in watts calculated in accordance with Section C405.3.2 and reduced in accordance with Section C406.3 where Option 2 of Section C406.1 is used to comply with the requirements of Section C406.

$UDZFA$ = Uncontrolled daylight zone floor area is the sum of all sidelit and toplit zones, calculated in accordance with Sections C405.2.4.2 and C405.2.4.3, that do not have daylight responsive controls.

TBFA = Total building floor area is the sum of all floor areas included in the lighting power allowance calculation in Section C405.3.2.

C405.2.4.1 Daylight-responsive control function. Where required, *daylight-responsive controls* shall be provided within each space for control of lights in that space and shall comply with all of the following:

1. Lights in *toplit daylight zones* in accordance with Section C405.2.4.3 shall be controlled independently of lights in sidelit daylight zones in accordance with Section C405.2.4.2.
2. Lights in the primary sidelit daylight zone shall be controlled independently of lights in the secondary sidelit daylight zone.
3. *Daylight responsive controls* within each space shall be configured so that they can be calibrated from within that space by authorized personnel.
4. Calibration mechanisms shall be in a location with *ready access*.
5. *Daylight responsive controls* shall dim lights continuously from full light output to 15 percent of full light output or lower.
6. *Daylight responsive controls* shall be configured to completely shut off all controlled lights.
7. When occupant sensor controls have reduced the lighting power to an unoccupied setpoint in accordance with Sections C405.2.1.2 through C405.2.1.4, daylight responsive controls shall continue to adjust electric light levels in response to available daylight, but shall be configured to not increase the lighting power above the specified unoccupied setpoint.
8. Lights in *sidelit daylight zones* in accordance with Section C405.2.4.2 facing different cardinal orientations [within 45 degrees (0.79 rad) of due north, east, south, west] shall be controlled independently of each other.

Exceptions:

1. Within each space, up to 150 watts of lighting within the primary sidelit daylight zone is permitted to be controlled together with lighting in a primary sidelit daylight zone facing a different cardinal orientation.
2. Within each space, up to 150 watts of lighting within the secondary sidelit daylight zone is permitted to be controlled together with lighting in a secondary sidelit daylight zone facing a different cardinal orientation.

C405.2.4.2 Sidelit daylight zone. The sidelit daylight zone is the floor area adjacent to vertical *fenestration* that complies with all of the following:

1. Where the fenestration is located in a wall, the sidelit daylight zone shall extend laterally to the nearest full-height wall, or up to 1.0 times the height from the floor to the top of the fenestration, and longitudinally from the edge of the fenestration to the nearest full-height wall, or up to 0.5 times the height from the floor to the top of the fenestration, whichever is less, as indicated in Figure C405.2.4.2(1).
2. Where the fenestration is located in a rooftop monitor, the sidelit daylight zone shall extend laterally to the nearest obstruction that is taller than 0.7 times the ceiling height, or up to 1.0 times the height from the floor to the bottom of the fenestration, whichever is less, and longitudinally from the edge of the fenestration to the nearest obstruction that is taller than 0.7 times the ceiling height, or up to 0.25 times the height from the floor to the bottom of the fenestration, whichever is less, as indicated in Figures C405.2.4.2(2) and C405.2.4.2(3).
3. The secondary sidelit daylight zone is directly adjacent to the primary sidelit daylight zone and shall extend laterally to 2.0 times the height from the floor to the top of the fenestration or to the nearest full height wall, whichever is less, and longitudinally from the edge of the fenestration to the nearest full height wall, or up to 0.5 times the height from the floor to the top of the fenestration, whichever is less, as indicated in Figure C405.2.4.2(1). The area of secondary sidelit zones shall not be considered in the calculation of the daylight zones in Section C402.4.1.1.
4. The area of the fenestration is not less than 24 square feet (2.23 m^2).
5. The distance from the fenestration to any building or geological formation that would block *access to* daylight is greater than one-half of the height from the bottom of the fenestration to the top of the building or geologic formation.
6. The visible transmittance of the fenestration is not less than 0.20.
7. The projection factor (determined in accordance with Equation 4-5) for any overhanging projection that is shading the fenestration is not greater than 1.0 for fenestration oriented 45 degrees or less from true north and not greater than 1.5 for all other orientations.

COMMERCIAL ENERGY EFFICIENCY

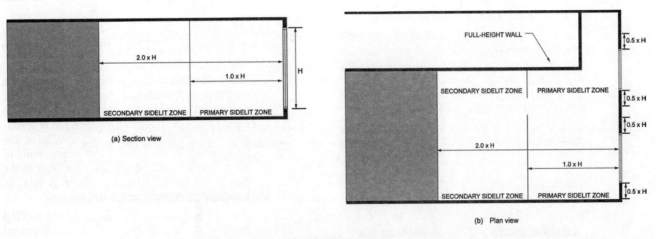

FIGURE C405.2.4.2(1)
PRIMARY AND SECONDARY SIDELIT DAYLIGHT ZONES

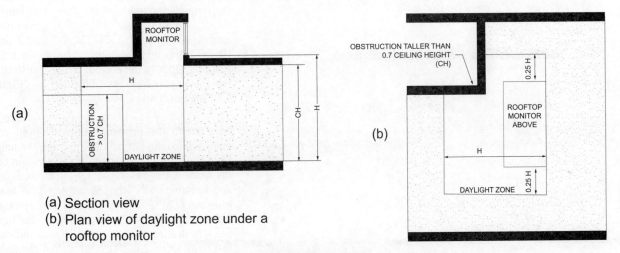

(a) Section view
(b) Plan view of daylight zone under a rooftop monitor

FIGURE C405.2.4.2(2)
DAYLIGHT ZONE UNDER A ROOFTOP MONITOR

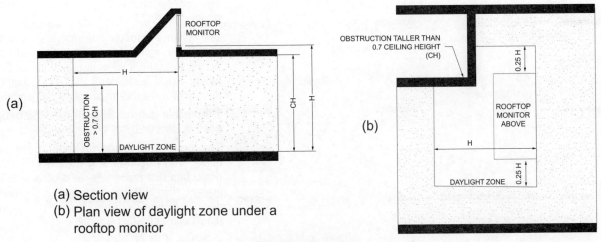

(a) Section view
(b) Plan view of daylight zone under a rooftop monitor

FIGURE C405.2.4.2(3)
DAYLIGHT ZONE UNDER A SLOPED ROOFTOP MONITOR

C405.2.4.3 Toplit daylight zone. The *toplit daylight zone* is the floor area underneath a roof fenestration assembly that complies with all of the following:

1. The toplit daylight zone shall extend laterally and longitudinally beyond the edge of the roof fenestration assembly to the nearest obstruction that is taller than 0.7 times the ceiling height, or up to 0.7 times the ceiling height, whichever is less, as indicated in Figure C405.2.4.3.
2. Direct sunlight is not blocked from hitting the roof fenestration assembly at the peak solar angle on the summer solstice by buildings or geological formations.
3. The product of the visible transmittance of the roof fenestration assembly and the area of the rough opening of the roof fenestration assembly divided by the area of the *toplit* zone is not less than 0.008.

C405.2.4.4 Atriums. Daylight zones at atrium spaces shall be established at the top floor surrounding the atrium and at the floor of the atrium space, and not on intermediate floors, as indicated in Figure C405.2.4.4.

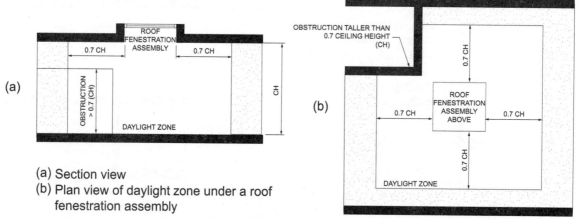

(a) Section view
(b) Plan view of daylight zone under a roof fenestration assembly

**FIGURE C405.2.4.3
TOPLIT DAYLIGHT ZONE**

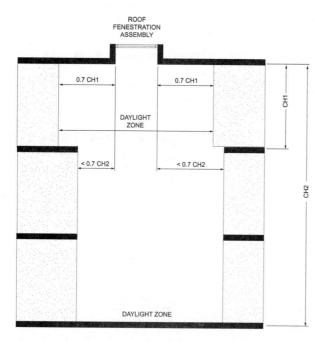

(a) Section view of roof fenestration assembly at atrium

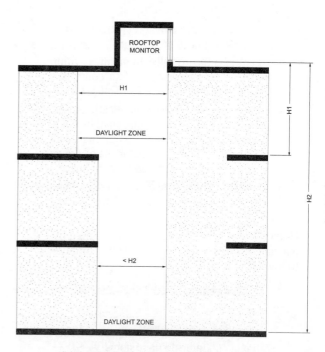

(b) Section view of roof monitor at atrium

**C405.2.4.4
DAYLIGHT ZONES AT A MULTISTORY ATRIUM**

C405.2.5 Specific application controls. Specific application controls shall be provided for the following:

1. The following lighting shall be controlled by an occupant sensor complying with Section C405.2.1.1 or a time-switch control complying with Section C405.2.2.1. In addition, a manual control shall be provided to control such lighting separately from the general lighting in the space:

   1.1. Luminaires for which additional lighting power is claimed in accordance with Section C405.3.2.2.1.
   1.2. Display and accent.
   1.3. Lighting in display cases.
   1.4. Supplemental task lighting, including permanently installed under-shelf or under-cabinet lighting.
   1.5. Lighting equipment that is for sale or demonstration in lighting education.
   1.6. Display lighting for exhibits in galleries, museums and monuments that is in addition to *general lighting*.

2. *Sleeping units* shall have control devices or systems that are configured to automatically switch off all permanently installed luminaires and switched receptacles within 20 minutes after all occupants have left the unit.

   Exceptions:
   1. Lighting and switched receptacles controlled by card key controls.
   2. Spaces where patient care is directly provided.

3. Permanently installed luminaires within *dwelling units* shall be provided with controls complying with Section C405.2.1.1 or C405.2.3.1.

4. Lighting for nonvisual applications, such as plant growth and food warming, shall be controlled by a time switch control complying with Section C405.2.2.1 that is independent of the controls for other lighting within the room or space.

5. Task lighting for medical and dental purposes that is in addition to *general lighting* shall be provided with a *manual control*.

C405.2.6 Manual controls. Where required by this code, manual controls for lights shall comply with the following:

1. They shall be in a location with *ready access* to occupants.
2. They shall be located where the controlled lights are visible, or shall identify the area served by the lights and indicate their status.

C405.2.7 Exterior lighting controls. Exterior lighting systems shall be provided with controls that comply with Sections C405.2.7.1 through C405.2.7.4.

Exceptions:
1. Lighting for covered vehicle entrances and exits from buildings and parking structures where required for eye adaptation.
2. Lighting controlled from within dwelling units.

C405.2.7.1 Daylight shutoff. Lights shall be automatically turned off when daylight is present and satisfies the lighting needs.

C405.2.7.2 Building facade and landscape lighting. Building facade and landscape lighting shall automatically shut off from not later than 1 hour after business closing to not earlier than 1 hour before business opening.

C405.2.7.3 Lighting setback. Lighting that is not controlled in accordance with Section C405.2.7.2 shall comply with the following:

1. Be controlled so that the total wattage of such lighting is automatically reduced by not less than 50 percent by selectively switching off or dimming luminaires at one of the following times:

   1.1. From not later than midnight to not earlier than 6 a.m.
   1.2. From not later than one hour after business closing to not earlier than one hour before business opening.
   1.3. During any time where activity has not been detected for 15 minutes or more.

2. Luminaires serving outdoor parking areas and having a rated input wattage of greater than 78 watts and a mounting height of 24 feet (7315 mm) or less above the ground shall be controlled so that the total wattage of such lighting is automatically reduced by not less than 50 percent during any time where activity has not been detected for 15 minutes or more. Not more than 1,500 watts of lighting power shall be controlled together.

C405.2.7.4 Exterior time-switch control function. Time-switch controls for exterior lighting shall comply with the following:

1. They shall have a clock capable of being programmed for not fewer than 7 days.
2. They shall be capable of being set for seven different day types per week.
3. They shall incorporate an automatic holiday setback feature.

COMMERCIAL ENERGY EFFICIENCY

4. They shall have program backup capabilities that prevent the loss of program and time settings for a period of not less than 10 hours in the event that power is interrupted.

C405.2.8 Parking garage lighting control. Parking garage lighting shall be controlled by an *occupant sensor* complying with Section C405.2.1.1 or a *time-switch control* complying with Section C405.2.2.1. Additional lighting controls shall be provided as follows:

1. Lighting power of each luminaire shall be automatically reduced by not less than 30 percent when there is no activity detected within a lighting zone for 20 minutes. Lighting zones for this requirement shall be not larger than 3,600 square feet (334.5 m^2).

    Exception: Lighting zones provided with less than 1.5 footcandles of illumination on the floor at the darkest point with all lights on are not required to have automatic light-reduction controls.

2. Where lighting for eye adaptation is provided at covered vehicle entrances and exits from buildings and parking structures, such lighting shall be separately controlled by a device that automatically reduces lighting power by at least 50 percent from sunset to sunrise.

3. The power to luminaires within 20 feet (6096 mm) of perimeter wall openings shall automatically reduce in response to daylight by at least 50 percent.

    Exceptions:
    1. Where the opening-to-wall ratio is less than 40 percent as viewed from the interior and encompassing the vertical distance from the driving surface to the lowest structural element.
    2. Where the distance from the opening to any exterior daylight blocking obstruction is less than one-half the height from the bottom of the opening or fenestration to the top of the obstruction.
    3. Where openings are obstructed by permanent screens or architectural elements restricting daylight entering the interior space.

C405.3 Interior lighting power requirements. A building complies with this section where its total connected interior lighting power calculated under Section C405.3.1 is not greater than the interior lighting power allowance calculated under Section C405.3.2.

C405.3.1 Total connected interior lighting power. The total connected interior lighting power shall be determined in accordance with Equation 4-10.

$$TCLP = [LVL + BLL + LED + TRK + \text{Other}]$$

(Equation 4-10)

where:

- $TCLP$ = Total connected lighting power (watts).
- LVL = For luminaires with lamps connected directly to building power, such as line voltage lamps, the rated wattage of the lamp.
- BLL = For luminaires incorporating a ballast or transformer, the rated input wattage of the ballast or transformer when operating that lamp.
- LED = For light-emitting diode luminaires with either integral or remote drivers, the rated wattage of the luminaire.
- TRK = For lighting track, cable conductor, rail conductor, and plug-in busway systems that allow the addition and relocation of luminaires without rewiring, the wattage shall be one of the following:
    1. The specified wattage of the luminaires, but not less than 8 W per linear foot (25 W/lin m).
    2. The wattage limit of the permanent current-limiting devices protecting the system.
    3. The wattage limit of the transformer supplying the system.
- Other = The wattage of all other luminaires and lighting sources not covered previously and associated with interior lighting verified by data supplied by the manufacturer or other *approved* sources.

The connected power associated with the following lighting equipment and applications is not included in calculating total connected lighting power.

1. Television broadcast lighting for playing areas in sports arenas.
2. Emergency lighting automatically off during normal building operation.
3. Lighting in spaces specifically designed for use by occupants with special lighting needs, including those with visual impairment and other medical and age-related issues.
4. Casino gaming areas.
5. Mirror lighting in dressing rooms.
6. Task lighting for medical and dental purposes that is in addition to general lighting.
7. Display lighting for exhibits in galleries, museums and monuments that is in addition to general lighting.
8. Lighting for theatrical purposes, including performance, stage, film production and video production.
9. Lighting for photographic processes.

10. Lighting integral to equipment or instrumentation and installed by the manufacturer.
11. Task lighting for plant growth or maintenance.
12. Advertising signage or directional signage.
13. Lighting for food warming.
14. Lighting equipment that is for sale.
15. Lighting demonstration equipment in lighting education facilities.
16. Lighting approved because of safety considerations.
17. Lighting in retail display windows, provided that the display area is enclosed by ceiling-height partitions.
18. Furniture-mounted supplemental task lighting that is controlled by automatic shutoff.
19. Exit signs.
20. Antimicrobial lighting used for the sole purpose of disinfecting a space.

C405.3.2 Interior lighting power allowance. The total interior lighting power allowance (watts) for an entire building shall be determined according to Table C405.3.2(1) using the Building Area Method or Table C405.3.2(2) using the Space-by-Space Method. The interior lighting power allowance for projects that involve only portions of a building shall be determined according to Table C405.3.2(2) using the Space-by-Space Method. Buildings with unfinished spaces shall use the Space-by-Space Method.

TABLE C405.3.2(1)
INTERIOR LIGHTING POWER ALLOWANCES:
BUILDING AREA METHOD

BUILDING AREA TYPE	LPD (watts/ft²)
Automotive facility	0.75
Convention center	0.64
Courthouse	0.79
Dining: bar lounge/leisure	0.80
Dining: cafeteria/fast food	0.76
Dining: family	0.71
Dormitory[a, b]	0.53
Exercise center	0.72
Fire station[a]	0.56
Gymnasium	0.76
Health care clinic	0.81
Hospital[a]	0.96
Hotel/Motel[a, b]	0.56
Library	0.83
Manufacturing facility	0.82
Motion picture theater	0.44
Multiple-family[c]	0.45
Museum	0.55
Office	0.64

(continued)

TABLE C405.3.2(1)—continued
INTERIOR LIGHTING POWER ALLOWANCES:
BUILDING AREA METHOD

BUILDING AREA TYPE	LPD (watts/ft²)
Parking garage	0.18
Penitentiary	0.69
Performing arts theater	0.84
Police station	0.66
Post office	0.65
Religious building	0.67
Retail	0.84
School/university	0.72
Sports arena	0.76
Town hall	0.69
Transportation	0.50
Warehouse	0.45
Workshop	0.91

For SI: 1 watt per square foot = 10.76 w/m².

a. Where sleeping units are excluded from lighting power calculations by application of Section R404.1, neither the area of the sleeping units nor the wattage of lighting in the sleeping units is counted.
b. Where dwelling units are excluded from lighting power calculations by application of Section R404.1, neither the area of the dwelling units nor the wattage of lighting in the dwelling units is counted.
c. Dwelling units are excluded. Neither the area of the dwelling units nor the wattage of lighting in the dwelling units is counted.

TABLE C405.3.2(2)
INTERIOR LIGHTING POWER ALLOWANCES:
SPACE-BY-SPACE METHOD

COMMON SPACE TYPES[a]	LPD (watts/ft²)
Atrium	
Less than 40 feet in height	0.48
Greater than 40 feet in height	0.60
Audience seating area	
In an auditorium	0.61
In a gymnasium	0.23
In a motion picture theater	0.27
In a penitentiary	0.67
In a performing arts theater	1.16
In a religious building	0.72
In a sports arena	0.33
Otherwise	0.33
Banking activity area	0.61
Breakroom (See Lounge/breakroom)	
Classroom/lecture hall/training room	
In a penitentiary	0.89
Otherwise	0.71
Computer room, data center	0.94
Conference/meeting/multipurpose room	0.97
Copy/print room	0.31

(continued)

TABLE C405.3.2(2)—continued
INTERIOR LIGHTING POWER ALLOWANCES: SPACE-BY-SPACE METHOD

COMMON SPACE TYPES[a]	LPD (watts/ft²)
Corridor	
In a facility for the visually impaired (and not used primarily by the staff)[b]	0.71
In a hospital	0.71
Otherwise	0.41
Courtroom	1.20
Dining area	
In bar/lounge or leisure dining	0.86
In cafeteria or fast food dining	0.40
In a facility for the visually impaired (and not used primarily by the staff)[b]	1.27
In family dining	0.60
In a penitentiary	0.42
Otherwise	0.43
Electrical/mechanical room	0.43
Emergency vehicle garage	0.52
Food preparation area	1.09
Guestroom[c, d]	0.41
Laboratory	
In or as a classroom	1.11
Otherwise	1.33
Laundry/washing area	0.53
Loading dock, interior	0.88
Lobby	
For an elevator	0.65
In a facility for the visually impaired (and not used primarily by the staff)[b]	1.69
In a hotel	0.51
In a motion picture theater	0.23
In a performing arts theater	1.25
Otherwise	0.84
Locker room	0.52
Lounge/breakroom	
In a healthcare facility	0.42
Otherwise	0.59
Office	
Enclosed	0.74
Open plan	0.61
Parking area, interior	0.15
Pharmacy area	1.66
Restroom	
In a facility for the visually impaired (and not used primarily by the staff)[b]	1.26
Otherwise	0.63
Sales area	1.05

COMMON SPACE TYPES[a]	LPD (watts/ft²)
Seating area, general	0.23
Stairwell	0.49
Storage room	0.38
Vehicular maintenance area	0.60
Workshop	1.26
BUILDING TYPE SPECIFIC SPACE TYPES[a]	**LPD (watts/ft²)**
Automotive (see Vehicular maintenance area)	
Convention Center—exhibit space	0.61
Dormitory—living quarters[c, d]	0.50
Facility for the visually impaired[b]	
In a chapel (and not used primarily by the staff)	0.70
In a recreation room (and not used primarily by the staff)	1.77
Fire Station—sleeping quarters[c]	0.23
Gymnasium/fitness center	
In an exercise area	0.90
In a playing area	0.85
Healthcare facility	
In an exam/treatment room	1.40
In an imaging room	0.94
In a medical supply room	0.62
In a nursery	0.92
In a nurse's station	1.17
In an operating room	2.26
In a patient room[c]	0.68
In a physical therapy room	0.91
In a recovery room	1.25
Library	
In a reading area	0.96
In the stacks	1.18
Manufacturing facility	
In a detailed manufacturing area	0.80
In an equipment room	0.76
In an extra-high-bay area (greater than 50 feet floor-to-ceiling height)	1.42
In a high-bay area (25–50 feet floor-to-ceiling height)	1.24
In a low-bay area (less than 25 feet floor-to-ceiling height)	0.86
Museum	
In a general exhibition area	0.31
In a restoration room	1.10
Performing arts theater—dressing room	0.41
Post office—sorting area	0.76

TABLE C405.3.2(2)—continued
INTERIOR LIGHTING POWER ALLOWANCES: SPACE-BY-SPACE METHOD

COMMON SPACE TYPES[a]	LPD (watts/ft²)
Religious buildings	
In a fellowship hall	0.54
In a worship/pulpit/choir area	0.85
Retail facilities	
In a dressing/fitting room	0.51
In a mall concourse	0.82
Sports arena—playing area	
For a Class I facility[e]	2.94
For a Class II facility[f]	2.01
For a Class III facility[g]	1.30
For a Class IV facility[h]	0.86
Transportation facility	
At a terminal ticket counter	0.51
In a baggage/carousel area	0.39
In an airport concourse	0.25
Warehouse—storage area	
For medium to bulky, palletized items	0.33
For smaller, hand-carried items	0.69

For SI: 1 foot = 304.8 mm, 1 watt per square foot = 10.76 w/m².

a. In cases where both a common space type and a building area specific space type are listed, the building area specific space type shall apply.
b. A 'Facility for the Visually Impaired' is a facility that is licensed or will be licensed by local or state authorities for senior long-term care, adult daycare, senior support or people with special visual needs.
c. Where sleeping units are excluded from lighting power calculations by application of Section R404.1, neither the area of the sleeping units nor the wattage of lighting in the sleeping units is counted.
d. Where dwelling units are excluded from lighting power calculations by application of Section R404.1, neither the area of the dwelling units nor the wattage of lighting in the dwelling units is counted.
e. Class I facilities consist of professional facilities; and semiprofessional, collegiate, or club facilities with seating for 5,000 or more spectators.
f. Class II facilities consist of collegiate and semiprofessional facilities with seating for fewer than 5,000 spectators; club facilities with seating for between 2,000 and 5,000 spectators; and amateur league and high school facilities with seating for more than 2,000 spectators.
g. Class III facilities consist of club, amateur league and high school facilities with seating for 2,000 or fewer spectators.
h. Class IV facilities consist of elementary school and recreational facilities; and amateur league and high school facilities without provision for spectators.

C405.3.2.1 Building Area Method. For the Building Area Method, the interior lighting power allowance is calculated as follows:

1. For each building area type inside the building, determine the applicable building area type and the allowed lighting power density for that type from Table C405.3.2(1). For building area types not listed, select the building area type that most closely represents the use of that area. For the purposes of this method, an "area" shall be defined as all contiguous spaces that accommodate or are associated with a single building area type.

2. Determine the floor area for each building area type listed in Table C405.3.2(1) and multiply this area by the applicable value from Table C405.3.2(1) to determine the lighting power (watts) for each building area type.

3. The total interior lighting power allowance (watts) for the entire building is the sum of the lighting power from each building area type.

C405.3.2.2 Space-by-Space Method. Where a building has unfinished spaces, the lighting power allowance for the unfinished spaces shall be the total connected lighting power for those spaces, or 0.2 watts per square foot (10.76 w/m²), whichever is less. For the Space-by-Space Method, the interior lighting power allowance is calculated as follows:

1. For each space enclosed by partitions that are not less than 80 percent of the ceiling height, determine the applicable space type from Table C405.3.2(2). For space types not listed, select the space type that most closely represents the proposed use of the space. Where a space has multiple functions, that space may be divided into separate spaces.

2. Determine the total floor area of all the spaces of each space type and multiply by the value for the space type in Table C405.3.2(2) to determine the lighting power (watts) for each space type.

3. The total interior lighting power allowance (watts) shall be the sum of the lighting power allowances for all space types.

C405.3.2.2.1 Additional interior lighting power. Where using the Space-by-Space Method, an increase in the interior lighting power allowance is permitted for specific lighting functions. Additional power shall be permitted only where the specified lighting is installed and controlled in accordance with Section C405.2.5. This additional power shall be used only for the specified luminaires and shall not be used for any other purpose. An increase in the interior lighting power allowance is permitted in the following cases:

1. For lighting equipment to be installed in sales areas specifically to highlight merchandise, the additional lighting power shall be determined in accordance with Equation 4-11.

   Additional interior lighting power allowance = 1000 W + (Retail Area 1 × 0.45 W/ft²) + (Retail Area 2 × 0.45 W/ft²) + (Retail Area 3 × 1.05 W/ft²) + (Retail Area 4 × 1.87 W/ft²)

For SI units:

Additional interior lighting power allowance = 1000 W + (Retail Area 1 × 4.8 W/m^2) + (Retail Area 2 × 4.84 W/m^2) + (Retail Area 3 × 11 W/m^2) + (Retail Area 4 × 20 W/m^2)

(Equation 4-11)

where:

Retail Area 1 = The floor area for all products not listed in Retail Area 2, 3 or 4.

Retail Area 2 = The floor area used for the sale of vehicles, sporting goods and small electronics.

Retail Area 3 = The floor area used for the sale of furniture, clothing, cosmetics and artwork.

Retail Area 4 = The floor area used for the sale of jewelry, crystal and china.

Exception: Other merchandise categories are permitted to be included in Retail Areas 2 through 4, provided that justification documenting the need for additional lighting power based on visual inspection, contrast or other critical display is approved by the code official.

2. For spaces in which lighting is specified to be installed in addition to the general lighting for the purpose of decorative appearance or for highlighting art or exhibits, provided that the additional lighting power shall be not more than 0.9 W/ft^2 (9.7 W/m^2) in lobbies and not more than 0.75 W/ft^2 (8.1 W/m^2) in other spaces.

C405.4 Lighting for plant growth and maintenance. Not less than 95 percent of the permanently installed luminaires used for plant growth and maintenance shall have a photon efficiency of not less than 1.6 μmol/J as defined in accordance with ANSI/ASABE S640.

C405.5 Exterior lighting power requirements. The total connected exterior lighting power calculated in accordance with Section C405.5.1 shall be not greater than the exterior lighting power allowance calculated in accordance with Section C405.5.2.

C405.5.1 Total connected exterior building exterior lighting power. The total exterior connected lighting power shall be the total maximum rated wattage of all lighting that is powered through the energy service for the building.

Exception: Lighting used for the following applications shall not be included.

1. Lighting *approved* because of safety considerations.
2. Emergency lighting automatically off during normal business operation.
3. Exit signs.
4. Specialized signal, directional and marker lighting associated with transportation.
5. Advertising signage or directional signage.
6. Integral to equipment or instrumentation and installed by its manufacturer.
7. Theatrical purposes, including performance, stage, film production and video production.
8. Athletic playing areas.
9. Temporary lighting.
10. Industrial production, material handling, transportation sites and associated storage areas.
11. Theme elements in theme/amusement parks.
12. Used to highlight features of art, public monuments and the national flag.
13. Lighting for water features and swimming pools.
14. Lighting controlled from within dwelling units, where the lighting complies with Section R404.1.

C405.5.2 Exterior lighting power allowance. The exterior lighting power allowance (watts) is calculated as follows:

1. Determine the Lighting Zone (LZ) for the building according to Table C405.5.2(1), unless otherwise specified by the code official.
2. For each exterior area that is to be illuminated by lighting that is powered through the energy service for the building, determine the applicable area type from Table C405.5.2(2). For area types not listed, select the area type that most closely represents the proposed use of the area.
3. Determine the total area or length of each area type and multiply by the value for the area type in Table C405.5.2(2) to determine the lighting power (watts) allowed for each area type.
4. The total exterior lighting power allowance (watts) is the sum of the base site allowance determined according to Table C405.5.2(2), plus the watts from each area type.

**TABLE C405.5.2(1)
EXTERIOR LIGHTING ZONES**

LIGHTING ZONE	DESCRIPTION
1	Developed areas of national parks, state parks, forest land, and rural areas
2	Areas predominantly consisting of residential zoning, neighborhood business districts, light industrial with limited nighttime use and residential mixed-use areas
3	All other areas not classified as lighting zone 1, 2 or 4
4	High-activity commercial districts in major metropolitan areas as designated by the local land use planning authority

TABLE C405.5.2(2)
LIGHTING POWER ALLOWANCES FOR BUILDING EXTERIORS

	LIGHTING ZONES			
	Zone 1	Zone 2	Zone 3	Zone 4
Base Site Allowance	350 W	400 W	500 W	900 W
Uncovered Parking Areas				
Parking areas and drives	0.03 W/ft^2	0.04 W/ft^2	0.06 W/ft^2	0.08 W/ft^2
Building Grounds				
Walkways and ramps less than 10 feet wide	0.50 W/linear foot	0.50 W/linear foot	0.60 W/linear foot	0.70 W/linear foot
Walkways and ramps 10 feet wide or greater, plaza areas, special feature areas	0.10 W/ft^2	0.10 W/ft^2	0.11 W/ft^2	0.14 W/ft^2
Dining areas	0.65 W/ft^2	0.65 W/ft^2	0.75 W/ft^2	0.95 W/ft^2
Stairways	0.60 W/ft^2	0.70 W/ft^2	0.70 W/ft^2	0.70 W/ft^2
Pedestrian tunnels	0.12 W/ft^2	0.12 W/ft^2	0.14 W/ft^2	0.21 W/ft^2
Landscaping	0.03 W/ft^2	0.04 W/ft^2	0.04 W/ft^2	0.04 W/ft^2
Building Entrances and Exits				
Pedestrian and vehicular entrances and exits	14 W/linear foot of opening	14 W/linear foot of opening	21 W/linear foot of opening	21 W/linear foot of opening
Entry canopies	0.20 W/ft^2	0.25 W/ft^2	0.40 W/ft^2	0.40 W/ft^2
Loading docks	0.35 W/ft^2	0.35 W/ft^2	0.35 W/ft^2	0.35 W/ft^2
Sales Canopies				
Free-standing and attached	0.40 W/ft^2	0.40 W/ft^2	0.60 W/ft^2	0.70 W/ft^2
Outdoor Sales				
Open areas (including vehicle sales lots)	0.20 W/ft^2	0.20 W/ft^2	0.35 W/ft^2	0.50 W/ft^2
Street frontage for vehicle sales lots in addition to "open area" allowance	No allowance	7 W/linear foot	7 W/linear foot	21 W/linear foot

For SI: 1 foot = 304.8 mm, 1 watt per square foot = W/0.0929 m^2.
W = watts.

TABLE C405.5.2(3)
INDIVIDUAL LIGHTING POWER ALLOWANCES FOR BUILDING EXTERIORS

	LIGHTING ZONES			
	Zone 1	Zone 2	Zone 3	Zone 4
Building facades	No allowance	0.075 W/ft^2 of gross above-grade wall area	0.113 W/ft^2 of gross above-grade wall area	0.15 W/ft^2 of gross above-grade wall area
Automated teller machines (ATM) and night depositories	135 W per location plus 45 W per additional ATM per location			
Uncovered entrances and gate-house inspection stations at guarded facilities	0.50 W/ft^2 of area			
Uncovered loading areas for law enforcement, fire, ambulance and other emergency service vehicles	0.35 W/ft^2 of area			
Drive-up windows and doors	200 W per drive through			
Parking near 24-hour retail entrances.	400 W per main entry			

For SI: For SI: 1 watt per square foot = W/0.0929 m^2.
W = watts.

C405.5.2.1 Additional exterior lighting power. Additional exterior lighting power allowances are available for the specific lighting applications listed in Table C405.5.2(3). These additional power allowances shall be used only for the luminaires serving these specific applications and shall not be used to increase any other lighting power allowance.

C405.5.3 Gas lighting. Gas-fired lighting appliances shall not be equipped with continuously burning pilot ignition systems.

C405.6 Dwelling electrical meter. Each dwelling unit located in a Group R-2 building shall have a separate electrical meter.

C405.7 Electrical transformers. Low-voltage dry-type distribution electric transformers shall meet the minimum efficiency requirements of Table C405.7 as tested and rated in accordance with the test procedure listed in DOE 10 CFR 431. The efficiency shall be verified through certification under an approved certification program or, where a certification program does not exist, the equipment efficiency ratings shall be supported by data furnished by the transformer manufacturer.

Exceptions: The following transformers are exempt:

1. Transformers that meet the *Energy Policy Act of 2005* exclusions based on the DOE 10 CFR 431 definition of special purpose applications.
2. Transformers that meet the *Energy Policy Act of 2005* exclusions that are not to be used in general purpose applications based on information provided in DOE 10 CFR 431.
3. Transformers that meet the *Energy Policy Act of 2005* exclusions with multiple voltage taps where the highest tap is not less than 20 percent more than the lowest tap.
4. Drive transformers.
5. Rectifier transformers.
6. Auto-transformers.
7. Uninterruptible power system transformers.
8. Impedance transformers.
9. Regulating transformers.
10. Sealed and nonventilating transformers.
11. Machine tool transformers.
12. Welding transformers.
13. Grounding transformers.
14. Testing transformers.

TABLE C405.7
MINIMUM NOMINAL EFFICIENCY LEVELS FOR DOE 10 CFR 431 LOW-VOLTAGE DRY-TYPE DISTRIBUTION TRANSFORMERS

SINGLE-PHASE TRANSFORMERS		THREE-PHASE TRANSFORMERS	
kVA[a]	Efficiency (%)[b]	kVA[a]	Efficiency (%)[b]
15	97.70	15	97.89
25	98.00	30	98.23
37.5	98.20	45	98.40
50	98.30	75	98.60
75	98.50	112.5	98.74
100	98.60	150	98.83
167	98.70	225	98.94
250	98.80	300	99.02
333	98.90	500	99.14
—	—	750	99.23
—	—	1000	99.28

a. kiloVolt-Amp rating.
b. Nominal efficiencies shall be established in accordance with the DOE 10 CFR 431 test procedure for low-voltage dry-type transformers.

C405.8 Electric motors. Electric motors shall meet the minimum efficiency requirements of Tables C405.8(1) through C405.8(4) when tested and rated in accordance with the DOE 10 CFR 431. The efficiency shall be verified through certification under an approved certification program or, where a certification program does not exist, the equipment efficiency ratings shall be supported by data furnished by the motor manufacturer.

Exception: The standards in this section shall not apply to the following exempt electric motors:

1. Air-over electric motors.
2. Component sets of an electric motor.
3. Liquid-cooled electric motors.
4. Submersible electric motors.
5. Inverter-only electric motors.

C405.9 Vertical and horizontal transportation systems and equipment. Vertical and horizontal transportation systems and equipment shall comply with this section.

C405.9.1 Elevator cabs. For the luminaires in each elevator cab, not including signals and displays, the sum of the lumens divided by the sum of the watts shall be not less than 35 lumens per watt. Ventilation fans in elevators that do not have their own air-conditioning system shall not consume more than 0.33 watts/cfm at the maximum rated speed of the fan. Controls shall be provided that will de-energize ventilation fans and lighting systems when the elevator is stopped, unoccupied and with its doors closed for over 15 minutes.

C405.9.2 Escalators and moving walks. Escalators and moving walks shall comply with ASME A17.1/CSA B44 and shall have automatic controls that reduce speed as permitted in accordance with ASME A17.1/CSA B44 and applicable local code.

Exception: A variable voltage drive system that reduces operating voltage in response to light loading conditions is an alternative to the reduced speed function.

C405.9.2.1 Energy recovery. Escalators shall be designed to recover electrical energy when resisting overspeed in the down direction.

C405.10 Voltage drop. The total *voltage drop* across the combination of customer-owned service conductors, feeder conductors and branch circuit conductors shall not exceed 5 percent.

C405.11 Automatic receptacle control. The following shall have automatic receptacle control complying with Section C405.11.1:

1. At least 50 percent of all 125V, 15- and 20-amp receptacles installed in enclosed offices, conference rooms, rooms used primarily for copy or print functions, breakrooms, classrooms and individual workstations, including those installed in

COMMERCIAL ENERGY EFFICIENCY

modular partitions and module office workstation systems.

2. At least 25 percent of branch circuit feeders installed for modular furniture not shown on the construction documents.

C405.11.1 Automatic receptacle control function. Automatic receptacle controls shall comply with the following:

1. Either split controlled receptacles shall be provided with the top receptacle controlled, or a controlled receptacle shall be located within 12 inches (304.8 mm) of each uncontrolled receptacle.

TABLE C405.8(1)
MINIMUM NOMINAL FULL-LOAD EFFICIENCY FOR NEMA DESIGN A, NEMA DESIGN B, AND IEC DESIGN N MOTORS (EXCLUDING FIRE PUMP) ELECTRIC MOTORS AT 60 HZ[a, b]

MOTOR HORSEPOWER (STANDARD KILOWATT EQUIVALENT)	NOMINAL FULL-LOAD EFFICIENCY (%) AS OF JUNE 1, 2016							
	2 Pole		4 Pole		6 Pole		8 Pole	
	Enclosed	Open	Enclosed	Open	Enclosed	Open	Enclosed	Open
1 (0.75)	77.0	77.0	85.5	85.5	82.5	82.5	75.5	75.5
1.5 (1.1)	84.0	84.0	86.5	86.5	87.5	86.5	78.5	77.0
2 (1.5)	85.5	85.5	86.5	86.5	88.5	87.5	84.0	86.5
3 (2.2)	86.5	85.5	89.5	89.5	89.5	88.5	85.5	87.5
5 (3.7)	88.5	86.5	89.5	89.5	89.5	89.5	86.5	88.5
7.5 (5.5)	89.5	88.5	91.7	91.0	91.0	90.2	86.5	89.5
10 (7.5)	90.2	89.5	91.7	91.7	91.0	91.7	89.5	90.2
15 (11)	91.0	90.2	92.4	93.0	91.7	91.7	89.5	90.2
20 (15)	91.0	91.0	93.0	93.0	91.7	92.4	90.2	91.0
25 (18.5)	91.7	91.7	93.6	93.6	93.0	93.0	90.2	91.0
30 (22)	91.7	91.7	93.6	94.1	93.0	93.6	91.7	91.7
40 (30)	92.4	92.4	94.1	94.1	94.1	94.1	91.7	91.7
50 (37)	93.0	93.0	94.5	94.5	94.1	94.1	92.4	92.4
60 (45)	93.6	93.6	95.0	95.0	94.5	94.5	92.4	93.0
75 (55)	93.6	93.6	95.4	95.0	94.5	94.5	93.6	94.1
100 (75)	94.1	93.6	95.4	95.4	95.0	95.0	93.6	94.1
125 (90)	95.0	94.1	95.4	95.4	95.0	95.0	94.1	94.1
150 (110)	95.0	94.1	95.8	95.8	95.8	95.4	94.1	94.1
200 (150)	95.4	95.0	96.2	95.8	95.8	95.4	94.5	94.1
250 (186)	95.8	95.0	96.2	95.8	95.8	95.8	95.0	95.0
300 (224)	95.8	95.4	96.2	95.8	95.8	95.8	—	—
350 (261)	95.8	95.4	96.2	95.8	95.8	95.8	—	—
400 (298)	95.8	95.8	96.2	95.8	—	—	—	—
450 (336)	95.8	96.2	96.2	96.2	—	—	—	—
500 (373)	95.8	96.2	96.2	96.2	—	—	—	—

a. Nominal efficiencies shall be established in accordance with DOE 10 CFR 431.

b. For purposes of determining the required minimum nominal full-load efficiency of an electric motor that has a horsepower or kilowatt rating between two horsepower or two kilowatt ratings listed in this table, each such motor shall be deemed to have a listed horsepower or kilowatt rating, determined as follows:

1. A horsepower at or above the midpoint between the two consecutive horsepowers shall be rounded up to the higher of the two horsepowers.
2. A horsepower below the midpoint between the two consecutive horsepowers shall be rounded down to the lower of the two horsepowers.
3. A kilowatt rating shall be directly converted from kilowatts to horsepower using the formula: 1 kilowatt = (1/0.746) horsepower. The conversion should be calculated to three significant decimal places, and the resulting horsepower shall be rounded in accordance with No. 1 or No. 2 above, as applicable.

TABLE C405.8(2)
MINIMUM NOMINAL FULL-LOAD EFFICIENCY FOR NEMA DESIGN C AND IEC DESIGN H MOTORS AT 60 HZ[a, b]

MOTOR HORSEPOWER (STANDARD KILOWATT EQUIVALENT)	NOMINAL FULL-LOAD EFFICIENCY (%) AS OF JUNE 1, 2016					
	4 Pole		6 Pole		8 Pole	
	Enclosed	Open	Enclosed	Open	Enclosed	Open
1 (0.75)	85.5	85.5	82.5	82.5	75.5	75.5
1.5 (1.1)	86.5	86.5	87.5	86.5	78.5	77.0
2 (1.5)	86.5	86.5	88.5	87.5	84.0	86.5
3 (2.2)	89.5	89.5	89.5	88.5	85.5	87.5
5 (3.7)	89.5	89.5	89.5	89.5	86.5	88.5
7.5 (5.5)	91.7	91.0	91.0	90.2	86.5	89.5
10 (7.5)	91.7	91.7	91.0	91.7	89.5	90.2
15 (11)	92.4	93.0	91.7	91.7	89.5	90.2
20 (15)	93.0	93.0	91.7	92.4	90.2	91.0
25 (18.5)	93.6	93.6	93.0	93.0	90.2	91.0
30 (22)	93.6	94.1	93.0	93.6	91.7	91.7
40 (30)	94.1	94.1	94.1	94.1	91.7	91.7
50 (37)	94.5	94.5	94.1	94.1	92.4	92.4
60 (45)	95.0	95.0	94.5	94.5	92.4	93.0
75 (55)	95.4	95.0	94.5	94.5	93.6	94.1
100 (75)	95.4	95.4	95.0	95.0	93.6	94.1
125 (90)	95.4	95.4	95.0	95.0	94.1	94.1
150 (110)	95.8	95.8	95.8	95.4	94.1	94.1
200 (150)	96.2	95.8	95.8	95.4	94.5	94.1

a. Nominal efficiencies shall be established in accordance with DOE 10 CFR 431.

b. For purposes of determining the required minimum nominal full-load efficiency of an electric motor that has a horsepower or kilowatt rating between two horsepower or two kilowatt ratings listed in this table, each such motor shall be deemed to have a listed horsepower or kilowatt rating, determined as follows:
   1. A horsepower at or above the midpoint between the two consecutive horsepowers shall be rounded up to the higher of the two horsepowers.
   2. A horsepower below the midpoint between the two consecutive horsepowers shall be rounded down to the lower of the two horsepowers.
   3. A kilowatt rating shall be directly converted from kilowatts to horsepower using the formula: 1 kilowatt = (1/0.746) horsepower. The conversion should be calculated to three significant decimal places, and the resulting horsepower shall be rounded in accordance with No. 1 or No. 2 above, as applicable.

TABLE C405.8(3)
MINIMUM AVERAGE FULL-LOAD EFFICIENCY POLYPHASE SMALL ELECTRIC MOTORS[a]

MOTOR HORSEPOWER	OPEN MOTORS			
	Number of Poles	2	4	6
	Synchronous Speed (RPM)	3600	1800	1200
0.25	—	65.6	69.5	67.5
0.33	—	69.5	73.4	71.4
0.50	—	73.4	78.2	75.3
0.75	—	76.8	81.1	81.7
1	—	77.0	83.5	82.5
1.5	—	84.0	86.5	83.8
2	—	85.5	86.5	N/A
3	—	85.5	86.9	N/A

N/A = Not Applicable.

a. Average full-load efficiencies shall be established in accordance with DOE 10 CFR 431.

TABLE C405.8(4)
MINIMUM AVERAGE FULL-LOAD EFFICIENCY FOR CAPACITOR-START CAPACITOR-RUN AND CAPACITOR-START INDUCTION-RUN SMALL ELECTRIC MOTORS[a]

MOTOR HORSEPOWER	OPEN MOTORS			
	Number of Poles	2	4	6
	Synchronous Speed (RPM)	3600	1800	1200
0.25	—	66.6	68.5	62.2
0.33	—	70.5	72.4	66.6
0.50	—	72.4	76.2	76.2
0.75	—	76.2	81.8	80.2
1	—	80.4	82.6	81.1
1.5	—	81.5	83.8	N/A
2	—	82.9	84.5	N/A
3	—	84.1	N/A	N/A

N/A = Not Applicable.
a. Average full-load efficiencies shall be established in accordance with DOE 10 CFR 431.

2. One of the following methods shall be used to provide control:

    2.1. A scheduled basis using a time-of-day operated control device that turns receptacle power off at specific programmed times and can be programmed separately for each day of the week. The control device shall be configured to provide an independent schedule for each portion of the building of not more than 5,000 square feet (464.5 m^2) and not more than one floor. The occupant shall be able to manually override an area for not more than 2 hours. Any individual override switch shall control the receptacles of not more than 5,000 feet (1524 m).

    2.2. An occupant sensor control that shall turn off receptacles within 20 minutes of all occupants leaving a space.

    2.3. An automated signal from another control or alarm system that shall turn off receptacles within 20 minutes after determining that the area is unoccupied.

3. All controlled receptacles shall be permanently marked in accordance with NFPA 70 and be uniformly distributed throughout the space.

4. Plug-in devices shall not comply.

Exceptions: Automatic receptacle controls are not required for the following:

1. Receptacles specifically designated for equipment requiring continuous operation (24 hours per day, 365 days per year).

2. Spaces where an automatic control would endanger the safety or security of the room or building occupants.

3. Within a single modular office workstation, noncontrolled receptacles are permitted to be located more than 12 inches (304.8 mm), but not more than 72 inches (1828 mm) from the controlled receptacles serving that workstation.

C405.12 Energy monitoring. New buildings with a gross *conditioned floor area* of 25,000 square feet (2322 m^2) or larger shall be equipped to measure, monitor, record and report energy consumption data in compliance with Sections C405.12.1 through C405.12.5.

Exception: R-2 occupancies and individual tenant spaces are not required to comply with this section provided that the space has its own utility services and meters and has less than 5,000 square feet (464.5 m^2) of *conditioned floor area*.

C405.12.1 Electrical energy metering. For all electrical energy supplied to the building and its associated site, including but not limited to site lighting, parking, recreational facilities and other areas that serve the building and its occupants, meters or other measurement devices shall be provided to collect energy consumption data for each end-use category required by Section C405.12.2.

C405.12.2 End-use metering categories. Meters or other *approved* measurement devices shall be provided to collect energy use data for each end-use category indicated in Table C405.12.2. Where multiple meters are used to measure any end-use category, the data acquisition system shall total all of the energy used by that category. Not more than 5 percent of the measured load for each of the end-use categories indicated in Table C405.12.2 shall be permitted to be from a load that is not within that category.

Exceptions:

1. HVAC and water heating equipment serving only an individual dwelling unit shall not require end-use metering.

2. End-use metering shall not be required for fire pumps, stairwell pressurization fans or any system that operates only during testing or emergency.

3. End-use metering shall not be required for an individual tenant space having a floor area not greater than 2,500 square feet (232 m^2) where a dedicated source meter complying with Section C405.12.3 is provided.

TABLE C405.12.2
ENERGY USE CATEGORIES

LOAD CATEGORY	DESCRIPTION OF ENERGY USE
Total HVAC system	Heating, cooling and ventilation, including but not limited to fans, pumps, boilers, chillers and water heating. Energy used by 120-volt equipment, or by 208/120-volt equipment that is located in a building where the main service is 480/277-volt power, is permitted to be excluded from total HVAC system energy use.
Interior lighting	Lighting systems located within the building.
Exterior lighting	Lighting systems located on the building site but not within the building.
Plug loads	Devices, appliances and equipment connected to convenience receptacle outlets.
Process load	Any single load that is not included in an HVAC, lighting or plug load category and that exceeds 5 percent of the peak connected load of the whole building, including but not limited to data centers, manufacturing equipment and commercial kitchens.
Building operations and other miscellaneous loads	The remaining loads not included elsewhere in this table, including but not limited to vertical transportation systems, automatic doors, motorized shading systems, ornamental fountains, ornamental fireplaces, swimming pools, in-ground spas and snow-melt systems.

C405.12.3 Meters. Meters or other measurement devices required by this section shall be configured to automatically communicate energy consumption data to the data acquisition system required by Section C405.12.4. Source meters shall be allowed to be any digital-type meter. Lighting, HVAC or other building systems that can monitor their energy consumption shall be permitted instead of meters. Current sensors shall be permitted, provided that they have a tested accuracy of ±2 percent. Required metering systems and equipment shall have the capability to provide at least hourly data that is fully integrated into the data acquisition system and graphical energy report in accordance with Sections C405.12.4 and C405.12.5.

C405.12.4 Data acquisition system. A data acquisition system shall have the capability to store the data from the required meters and other sensing devices for a minimum of 36 months. The data acquisition system shall have the capability to store real-time energy consumption data and provide hourly, daily, monthly and yearly logged data for each end-use category required by Section C405.12.2.

C405.12.5 Graphical energy report. A permanent and readily accessible reporting mechanism shall be provided in the building that is accessible by building operation and management personnel. The reporting mechanism shall have the capability to graphically provide the energy consumption for each end-use category required by Section C405.12.2 at least every hour, day, month and year for the previous 36 months.

SECTION C406
ADDITIONAL EFFICIENCY REQUIREMENTS

C406.1 Additional energy efficiency credit requirements. New buildings shall achieve a total of 10 credits from Tables C406.1(1) through C406.1(5) where the table is selected based on the use group of the building and from credit calculations as specified in relevant subsections of Section C406. Where a building contains multiple-use groups, credits from each use group shall be weighted by floor area of each group to determine the weighted average building credit. Credits from the tables or calculation shall be achieved where a building complies with one or more of the following:

1. More efficient HVAC performance in accordance with Section C406.2.
2. Reduced lighting power in accordance with Section C406.3.
3. Enhanced lighting controls in accordance with Section C406.4.
4. On-site supply of renewable energy in accordance with Section C406.5.
5. Provision of a dedicated outdoor air system for certain HVAC equipment in accordance with Section C406.6.
6. High-efficiency service water heating in accordance with Section C406.7.
7. Enhanced envelope performance in accordance with Section C406.8.
8. Reduced air infiltration in accordance with Section C406.9
9. Where not required by Section C405.12, include an energy monitoring system in accordance with Section C406.10.
10. Where not required by Section C403.2.3, include a fault detection and diagnostics (FDD) system in accordance with Section C406.11.
11. Efficient kitchen equipment in accordance with Section C406.12.

C406.1.1 Tenant spaces. Tenant spaces shall comply with sufficient options from Tables C406.1(1) through C406.1(5) to achieve a minimum number of 5 credits, where credits are selected from Section C406.2, C406.3, C406.4, C406.6, C406.7 or C406.10. Where the entire building complies using credits from Section C406.5, C406.8 or C406.9, tenant spaces shall be deemed to comply with this section.

> **Exception:** Previously occupied tenant spaces that comply with this code in accordance with Section C501.

C406.2 More efficient HVAC equipment performance. Equipment shall exceed the minimum efficiency requirements listed in the tables in Section C403.3.2. *Variable refrigerant flow systems* listed in the energy efficiency provisions of ANSI/ASHRAE/IES 90.1 in accordance with Section C406.2.1, C406.2.2, C406.2.3 or C406.2.4 shall also meet applicable requirements of Section C403. Energy efficiency credits for heating shall be selected from Section

C406.2.1 or C406.2.3 and energy efficiency credits for cooling shall be selected from Section C406.2.2, C406.2.4 or C406.2.5. Selected credits shall include a heating or cooling energy efficiency credit or both. Equipment not listed in Tables C403.3.2(1) through C403.3.2(9) and *variable refrigerant flow systems* not listed in the energy efficiency provisions of ANSI/ASHRAE/IES 90.1 shall be limited to 10 percent of the total building system capacity for heating equipment where selecting Section C406.2.1 or C406.2.3 and cooling equipment where selecting Section C406.2.2, C406.2.4 or C406.2.5.

C406.2.1 Five-percent heating efficiency improvement. Equipment shall exceed the minimum heating efficiency requirements by 5 percent.

C406.2.2 Five-percent cooling efficiency improvement. Equipment shall exceed the minimum cooling and heat rejection efficiency requirements by 5 percent. Where multiple cooling performance requirements are provided, the equipment shall exceed the annual energy requirement, including IEER, SEER and IPLV.

C406.2.3 Ten-percent heating efficiency improvement. Equipment shall exceed the minimum heating efficiency requirements by 10 percent.

C406.2.4 Ten-percent cooling efficiency improvement. Equipment shall exceed the minimum cooling and heat rejection efficiency requirements by 10 percent. Where multiple cooling performance requirements are provided, the equipment shall exceed the annual energy requirement, including IEER, SEER and IPLV.

C406.2.5 More than 10-percent cooling efficiency improvement. Where equipment exceeds the minimum annual cooling and heat rejection efficiency requirements by more than 10 percent, energy efficiency credits for cooling may be determined using Equation 4-12, rounded to the nearest whole number. Where multiple cooling performance requirements are provided, the equipment shall exceed the annual energy requirement, including IEER, SEER and IPLV.

$$EEC_{HEC} = EEC_{10}\left[1 + ((CEI - 10\text{ percent}) \div 10\text{ percent})\right] \quad \text{(Equation 4-12)}$$

TABLE C406.1(1)
ADDITIONAL ENERGY EFFICIENCY CREDITS FOR GROUP B OCCUPANCIES

SECTION	0A & 1A	0B & 1B	2A	2B	3A	3B	3C	4A	4B	4C	5A	5B	5C	6A	6B	7	8
C406.2.1: 5% heating efficiency improvement	NA	NA	NA	NA	NA	NA	NA	NA	NA	NA	1	NA	NA	1	1	NA	1
C406.2.2: 5% cooling efficiency improvement	6	6	5	5	4	4	3	3	3	2	2	2	1	2	2	2	1
C406.2.3: 10% heating efficiency improvement	NA	NA	NA	NA	NA	NA	NA	NA	NA	NA	2	1	1	2	2	NA	1
C406.2.4: 10% cooling efficiency improvement	11	12	10	9	7	7	6	5	6	4	4	5	3	4	3	3	3
C406.3: Reduced lighting power	9	8	9	9	9	9	10	8	9	9	7	8	8	6	7	7	6
C406.4: Enhanced digital lighting controls	2	2	2	2	2	2	2	2	2	2	2	2	1	2	1	1	1
C406.5: On-site renewable energy	9	9	9	9	9	9	9	9	9	9	9	9	9	9	9	9	9
C406.6: Dedicated outdoor air	4	4	4	4	4	3	2	5	3	2	5	3	2	7	4	5	3
C406.7.2: Recovered or renewable water heating	NA	NA	NA	NA	NA	NA	NA	NA	NA	NA	NA	NA	NA	NA	NA	NA	NA
C406.7.3: Efficient fossil fuel water heater	NA	NA	NA	NA	NA	NA	NA	NA	NA	NA	NA	NA	NA	NA	NA	NA	NA
C406.7.4: Heat pump water heater	NA	NA	NA	NA	NA	NA	NA	NA	NA	NA	NA	NA	NA	NA	NA	NA	NA
C406.8: Enhanced envelope performance	1	4	2	4	4	3	NA	7	4	5	10	7	6	11	10	14	16
C406.9: Reduced air infiltration	2	1	1	2	4	1	NA	8	2	3	11	4	1	15	8	11	6
C406.10: Energy monitoring	4	4	4	4	3	3	3	3	3	3	2	3	2	2	2	2	2
C406.11: Fault detection and diagnostics system	2	2	2	2	1	1	1	1	1	1	1	1	1	1	1	1	1

NA = Not Applicable.

COMMERCIAL ENERGY EFFICIENCY

where:

EEC_{HEC} = Energy efficiency credits for cooling efficiency improvement.

EEC_{10} = Section C406.2.4 credits from Tables C406.1(1) through C406.1(5).

CEI = The lesser of: the improvement above minimum cooling and heat rejection efficiency requirements or 15 percent.

C406.3 Reduced lighting power. Buildings shall comply with Section C406.3.1 or C406.3.2, and dwelling units and sleeping units within the building shall comply with Section C406.3.3.

C406.3.1 Reduced lighting power by more than 10 percent. The total connected interior lighting power calculated in accordance with Section C405.3.1 shall be less than 90 percent of the total lighting power allowance calculated in accordance with Section C405.3.2.

C406.3.2 Reduced lighting power by more than 15 percent. Where the total connected interior lighting power calculated in accordance with Section C405.3.1 is less than 85 percent of the total lighting power allowance calculated in accordance with Section C405.3.2, additional energy efficiency credits shall be determined based on Equation 4-13, rounded to the nearest whole number.

$$AEEC_{LPA} = AEEC_{10} \times 10 \times (LPA - LPD) / LPA$$

(Equation 4-13)

where:

$AEEC_{LPA}$ = Section C406.3.2 additional energy efficiency credits.

$AEEC_{10}$ = Section C406.3.1 credits from Tables C406.1(1) through C406.1(5).

LPA = Total lighting power allowance calculated in accordance with Section C405.3.2.

LPD = Total connected interior lighting power calculated in accordance with Section C405.3.1.

C406.3.3 Lamp efficacy. Not less than 95 percent of the permanently installed lighting, excluding kitchen appli-

TABLE C406.1(2)
ADDITIONAL ENERGY EFFICIENCY CREDITS FOR GROUP R AND I OCCUPANCIES

SECTION	CLIMATE ZONE																
	0A & 1A	0B & 1B	2A	2B	3A	3B	3C	4A	4B	4C	5A	5B	5C	6A	6B	7	8
C406.2.1: 5% heating efficiency improvement	NA	NA	NA	NA	1	NA	NA	1	NA	1	1	1	1	2	1	2	2
C406.2.2: 5% cooling efficiency improvement	3	3	2	2	1	1	1	1	1	NA	1	1	NA	1	1	1	NA
C406.2.3: 10% heating efficiency improvement	NA	NA	NA	NA	1	NA	NA	1	1	1	2	2	1	3	2	3	4
C406.2.4: 10% cooling efficiency improvement	5	5	4	3	2	3	1	2	2	1	1	1	1	1	1	1	1
C406.3: Reduced lighting power	2	2	2	2	2	2	2	2	2	2	2	2	2	2	2	2	2
C406.4: Enhanced digital lighting controls	NA	NA	NA	NA	NA	NA	NA	NA	NA	NA	NA	NA	NA	NA	NA	NA	NA
C406.5: On-site renewable energy	8	8	8	8	7	8	8	7	7	7	7	7	7	7	7	7	7
C406.6: Dedicated outdoor air system	3	4	3	3	4	2	NA	6	3	4	8	5	5	10	7	11	12
C406.7.2: Recovered or renewable water heating	10	9	11	10	13	12	15	14	14	15	14	14	16	14	15	15	15
C406.7.3: Efficient fossil fuel water heater	5	5	6	6	8	7	8	8	8	9	9	9	10	10	9	10	11
C406.7.4: Heat pump water heater	6	5	5	5	5	5	5	5	5	5	5	5	5	5	5	5	5
C406.8: Enhanced envelope performance	3	6	3	5	4	4	1	4	3	3	4	5	3	5	4	6	6
C406.9: Reduced air infiltration	6	5	3	11	6	4	NA	7	3	3	9	5	1	13	6	8	3
C406.10: Energy monitoring	1	1	1	1	1	1	1	1	1	1	1	1	1	1	1	1	1
C406.11: Fault detection and diagnostics system	1	1	1	1	1	NA	1	1	NA	1	1	NA	1	1	1	1	1

NA = Not Applicable.

ance light fixtures, serving dwelling units and sleeping units shall be provided by lamps with an efficacy of not less than 65 lumens per watt or luminaires with an efficacy of not less than 45 lumens per watt.

C406.4 Enhanced digital lighting controls. Interior general lighting in the building shall have the following enhanced lighting controls that shall be located, scheduled and operated in accordance with Sections C405.2.1 through C405.2.3.

1. Luminaires shall be configured for continuous dimming.
2. Luminaires shall be addressed individually. Where individual addressability is not available for the luminaire class type, a controlled group of not more than four luminaries shall be allowed.
3. Not more than eight luminaires shall be controlled together in a *daylight zone*.
4. Fixtures shall be controlled through a digital control system that includes the following function:
   4.1. Control reconfiguration based on digital addressability.
   4.2. Load shedding.
   4.3. Occupancy sensors shall be capable of being reconfigured through the digital control system.
5. Construction documents shall include submittal of a Sequence of Operations, including a specification outlining each of the functions in Item 4.
6. Functional testing of lighting controls shall comply with Section C408.

C406.5 On-site renewable energy. Buildings shall comply with Section C406.5.1 or C406.5.2.

C406.5.1 Basic renewable credit. The total minimum ratings of on-site renewable energy systems, not includ-

TABLE C406.1(3)
ADDITIONAL ENERGY EFFICIENCY CREDITS FOR GROUP E OCCUPANCIES

SECTION	CLIMATE ZONE															
	0A & 1A	0B & 1B	2A	2B	3A	3B	3C	4A	4B	4C	5A	5B	5C	6A	6B	7	8
C406.2.1: 5% heating efficiency improvement	NA	NA	NA	NA	1	1	1	1	1	2	1	2	1	2	2	3	4
C406.2.2: 5% cooling efficiency improvement	4	4	3	3	2	2	2	2	1	1	1	1	NA	1	1	1	NA
C406.2.3: 10% heating efficiency improvement	NA	NA	NA	1	1	1	1	2	3	4	3	4	3	4	3	5	7
C406.2.4: 10% cooling efficiency improvement	7	8	7	6	5	4	3	4	3	1	2	2	1	2	2	2	1
C406.3: Reduced lighting power	8	8	8	9	8	9	9	8	9	9	8	9	8	7	8	7	7
C406.4: Enhanced digital lighting controls	2	2	2	2	2	2	2	2	2	2	2	2	2	2	2	2	1
C406.5: On-site renewable energy	6	6	6	6	6	6	6	6	6	6	6	6	6	6	6	5	5
C406.6: Dedicated outdoor air system	NA	NA	NA	NA	NA	NA	NA	NA	NA	NA	NA	NA	NA	NA	NA	NA	NA
C406.7.2: Recovered or renewable water heating[a]	1	1	1	1	1	1	1	1	1	1	1	1	1	1	1	1	1
C406.7.3: Efficient fossil fuel water heater[a]	NA	1	1	1	1	1	1	2	2	3	2	3	2	3	3	3	5
C406.7.4: Heat pump water heater[a]	NA	NA	NA	NA	NA	NA	NA	1	NA	NA	1	1	NA	1	1	1	1
C406.8: Enhanced envelope performance	3	7	3	4	2	4	1	1	3	1	2	3	NA	4	3	6	9
C406.9: Reduced air infiltration	1	1	1	2	NA	NA	NA	NA	NA	NA	1	NA	NA	4	1	4	3
C406.10: Energy monitoring	3	3	3	3	3	3	3	3	3	2	2	3	2	2	2	2	2
C406.11: Fault detection and diagnostics system	1	2	1	1	1	1	1	1	1	1	1	1	1	1	1	1	2

NA = Not Applicable.

a. For schools with showers or full-service kitchens.

ing systems used for credits under Sections C406.7.2, shall be one of the following:

1. Not less than 0.86 Btu/h per square foot (2.7 W/m²) or 0.25 watts per square foot (2.7 W/m²) of *conditioned floor area*.
2. Not less than 2 percent of the annual energy used within the building for building mechanical and service water-heating equipment and lighting regulated in Section C405.

C406.5.2 Enhanced renewable credit. Where the total minimum ratings of on-site renewable energy systems exceeds the rating in Section C406.5.1, additional energy efficiency credits shall be determined based on Equation 4-14, rounded to the nearest whole number.

$$AEEC_{RRa} = AEEC_{2.5} \times RRa/RR_1 \quad \text{(Equation 4-14)}$$

where:

$AEEC_{RRa}$ = Section C406.5.2 additional energy efficiency credits.

$AEEC_{2.5}$ = Section C406.5 credits from Tables C406.1(1) through C406.1(5).

RRa = Actual total minimum ratings of *on-site renewable energy* systems (in Btu/h, watts per square foot or W/m²).

RR_1 = Minimum ratings of *on-site renewable energy* systems required by Section C406.5.1 (in Btu/h, watts per square foot or W/m²).

C406.6 Dedicated outdoor air system. Buildings containing equipment or systems regulated by Section C403.3.4, C403.4.3, C403.4.4, C403.4.5, C403.6, C403.8.4, C403.8.6, C403.8.6.1, C403.10.1, C403.10.2, C403.10.3 or C403.10.4 shall be equipped with an independent ventilation system designed to provide not less than the minimum 100-percent outdoor air to each individual occupied space, as specified by the *International Mechanical Code*. The ventilation system shall be capable of total energy recovery. The HVAC system shall include supply-air temperature controls that automatically reset the supply-air temperature in response to representative building loads or to outdoor air temperatures. The controls shall reset the supply-air temperature not less

TABLE C406.1(4)
ADDITIONAL ENERGY EFFICIENCY CREDITS FOR GROUP M OCCUPANCIES

SECTION	CLIMATE ZONE															
	0A & 1A	0B & 1B	2A	2B	3A	3B	3C	4A	4B	4C	5A	5B	5C	6A	6B	7	8
C406.2.1: 5% heating efficiency improvement	NA	NA	NA	NA	1	1	NA	1	1	2	2	2	2	3	2	3	4
C406.2.2: 5% cooling efficiency improvement	5	6	4	4	3	3	1	2	2	1	1	2	NA	1	1	1	NA
C406.2.3: 10% heating efficiency improvement	NA	NA	NA	1	1	1	1	2	2	4	3	4	5	5	3	6	8
C406.2.4: 10% cooling efficiency improvement	9	12	9	8	6	6	3	4	4	1	2	3	NA	2	2	2	1
C406.3: Reduced lighting power	13	13	15	14	16	14	17	15	15	14	12	14	14	16	16	14	12
C406.4: Enhanced digital lighting controls	3	3	4	3	4	3	4	4	4	3	3	3	3	4	4	3	3
C406.5: On-site renewable energy	8	8	8	8	8	8	8	8	8	7	7	7	7	7	7	7	6
C406.6: Dedicated outdoor air system	3	4	3	3	3	3	1	3	2	2	2	3	2	4	3	4	4
C406.7.2: Recovered or renewable water heating	NA	NA	NA	NA	NA	NA	NA	NA	NA	NA	NA	NA	NA	NA	NA	NA	NA
C406.7.3: Efficient fossil fuel water heater	NA	NA	NA	NA	NA	NA	NA	NA	NA	NA	NA	NA	NA	NA	NA	NA	NA
C406.7.4: Heat pump water heater	NA	NA	NA	NA	NA	NA	NA	NA	NA	NA	NA	NA	NA	NA	NA	NA	NA
C406.8: Enhanced envelope performance	4	6	3	4	3	3	1	6	4	4	4	5	4	6	5	8	9
C406.9: Reduced air infiltration	1	1	1	2	1	1	NA	3	1	1	3	2	1	7	3	6	3
C406.10: Energy monitoring	4	5	5	5	5	4	4	4	4	3	3	4	3	4	4	4	3
C406.11: Fault detection and diagnostics system	2	2	2	2	1	1	1	1	1	1	1	1	1	1	1	2	2

NA = Not Applicable.

than 25 percent of the difference between the design supply-air temperature and the design room-air temperature.

C406.7 Reduced energy use in service water heating. Buildings shall comply with Section C406.7.1 and Section C406.7.2, C406.7.3 or C406.7.4.

C406.7.1 Building type. To qualify for this credit, the building shall contain one of the following use groups, and the additional energy efficiency credit shall be prorated by conditioned floor area of the portion of the building comprised of the following use groups:

1. Group R-1: Boarding houses, hotels or motels.
2. Group I-2: Hospitals, psychiatric hospitals and nursing homes.
3. Group A-2: Restaurants and banquet halls or buildings containing food preparation areas.
4. Group F: Laundries.
5. Group R-2.
6. Group A-3: Health clubs and spas.
7. Group E: Schools with full-service kitchens or locker rooms with showers.
8. Buildings showing a service hot water load of 10 percent or more of total building energy loads, as shown with an energy analysis as described in Section C407.

C406.7.2 Recovered or renewable water heating. The building service water-heating system shall have one or more of the following that are sized to provide not less than 30 percent of the building's annual hot water requirements, or sized to provide 70 percent of the building's annual hot water requirements if the building is required to comply with Section C403.10.5:

1. Waste heat recovery from service hot water, heat-recovery chillers, building equipment or process equipment.
2. *On-site renewable energy* water-heating systems.

TABLE C406.1(5)
ADDITIONAL ENERGY EFFICIENCY CREDITS FOR OTHER[a] OCCUPANCIES

SECTION	CLIMATE ZONE															
	0A & 1A	0B & 1B	2A	2B	3A	3B	3C	4A	4B	4C	5A	5B	5C	6A	6B	7	8
C406.2.1: 5% heating efficiency improvement	NA	NA	NA	NA	1	1	1	1	1	2	1	2	1	2	2	3	3
C406.2.2: 5% cooling efficiency improvement	5	5	4	4	3	3	2	2	2	1	1	2	1	1	1	1	1
C406.2.3: 10% heating efficiency improvement	NA	NA	NA	1	1	1	1	2	2	3	3	3	3	4	3	5	5
C406.2.4: 10% cooling efficiency improvement	8	9	8	7	5	5	3	4	4	2	2	3	2	2	2	2	2
C406.3: Reduced lighting power	8	8	9	9	9	9	10	8	9	9	7	8	8	8	8	8	7
C406.4: Enhanced digital lighting controls	2	2	2	2	2	2	2	2	2	2	3	2	2	2	2	2	1
C406.5: On-site renewable energy	8	8	8	8	8	8	8	8	7	7	7	7	7	7	7	7	7
C406.6: Dedicated outdoor air system	3	4	3	3	4	3	2	5	3	3	5	4	3	7	5	7	6
C406.7.2: Recovered or renewable water heating[b]	10	9	11	10	13	12	15	14	14	15	14	14	16	14	15	15	15
C406.7.3: Efficient fossil fuel water heater[b]	5	5	6	6	8	7	8	8	8	9	9	9	10	10	9	10	11
C406.7.4: Heat pump water heater[b]	6	5	5	5	5	5	5	5	5	5	5	5	5	5	5	5	5
C406.8: Enhanced envelope performance	3	6	3	4	3	4	1	5	4	3	5	5	4	7	6	9	10
C406.9: Reduced air infiltration	3	2	2	4	4	2	NA	6	2	2	6	4	1	10	5	7	4
C406.10: Energy monitoring	3	3	3	3	3	3	3	3	3	3	2	3	2	2	2	3	2
C406.11: Fault detection and diagnostics system	2	2	2	2	1	1	1	1	1	1	1	1	1	1	1	1	1

NA = Not Applicable.
a. Other occupancy groups include all groups except Groups B, E, I, M and R.
b. For occupancy groups listed in Section C406.7.1.

C406.7.3 Efficient fossil fuel water heater. The combined input-capacity weighted-average equipment rating of all fossil fuel water-heating equipment in the building shall be not less than 95 percent Et or 0.95 EF. This option shall receive only half the listed credits for buildings required to comply with Section C404.2.1.

C406.7.4 Heat pump water heater. Where electric resistance water heaters are allowed, all service hot water system heating requirements shall be met using heat pump technology with a combined input-capacity weighted-average EF of 3.0. Air-source heat pump water heaters shall not draw conditioned air from within the building, except exhaust air that would otherwise be exhausted to the exterior.

C406.8 Enhanced envelope performance. The total UA of the *building thermal envelope* as designed shall be not less than 15 percent below the total UA of the *building thermal envelope* in accordance with Section C402.1.5.

C406.9 Reduced air infiltration. Air infiltration shall be verified by whole-building pressurization testing conducted in accordance with ASTM E779 or ASTM E1827 by an independent third party. The measured air-leakage rate of the building envelope shall not exceed 0.25 cfm/ft^2 (2.0 L/s × m^2) under a pressure differential of 0.3 inches water column (75 Pa), with the calculated surface area being the sum of the above- and below-grade building envelope. A report that includes the tested surface area, floor area, air by volume, stories above grade, and leakage rates shall be submitted to the code official and the building owner.

> **Exception:** For buildings having over 250,000 square feet (25 000 m^2) of *conditioned floor area*, air leakage testing need not be conducted on the whole building where testing is conducted on representative above-grade sections of the building. Tested areas shall total not less than 25 percent of the conditioned floor area and shall be tested in accordance with this section.

C406.10 Energy monitoring. Buildings shall be equipped to measure, monitor, record and report energy consumption data in compliance with Sections C406.10.1 through C406.10.5.

C406.10.1 Electrical energy metering. For all electrical energy supplied to the building and its associated site, including but not limited to site lighting, parking, recreational facilities, and other areas that serve the building and its occupants, meters or other measurement devices shall be provided to collect energy consumption data for each end-use category required by Section C406.10.2.

C406.10.2 End-use metering categories. Meters or other *approved* measurement devices shall be provided to collect energy use data for each end-use category listed in Table 406.10.2. These meters shall have the capability to collect energy consumption data for the whole building or for each separately metered portion of the building. Where multiple meters are used to measure any end-use category, the data acquisition system shall total all of the energy used by that category. Not more than 5 percent of the measured load for each of the end-use categories listed in Table 406.10.2 is permitted to be from a load not within the category.

> **Exceptions:**
> 1. HVAC and water-heating equipment serving only an individual dwelling unit does not require end-use metering.
> 2. End-use metering is not required for fire pumps, stairwell pressurization fans or any system that operates only during testing or emergency.

TABLE C406.10.2
ENERGY USE CATEGORIES

LOAD CATEGORY	DESCRIPTION OF ENERGY USE
Total HVAC system	Heating, cooling and ventilation, including but not limited to fans, pumps, boilers, chillers and water heating. Energy used by 120-volt equipment, or by 208/120-volt equipment that is located in a building where the main service is 480/277-volt power, is permitted to be excluded from total HVAC system energy use.
Interior lighting	Lighting systems located within the building.
Exterior lighting	Lighting systems located on the building site but not within the building.
Plug loads	Devices, appliances and equipment connected to convenience receptacle outlets.
Process loads	Any single load that is not included in an HVAC, lighting or plug load category and that exceeds 5 percent of the peak connected load of the whole building, including but not limited to data centers, manufacturing equipment and commercial kitchens.
Building operations and other miscellaneous loads	The remaining loads not included elsewhere in this table, including but not limited to vertical transportation systems and automatic doors.

C406.10.3 Meters. Meters or other measurement devices required by this section shall be configured to automatically communicate energy consumption data to the data acquisition system required by Section C406.10.4. Source meters shall be allowed to be any digital-type meter. Lighting, HVAC or other building systems that can monitor their energy consumption shall be permitted instead of meters. Current sensors shall be permitted, provided that they have a tested accuracy of ±2 percent. Required metering systems and equipment shall have the capability to provide at least hourly data that is fully integrated into the data acquisition system and graphical energy report in accordance with Sections C406.10.4 and C406.10.5.

C406.10.4 Data acquisition system. A data acquisition system shall have the capability to store the data from the required meters and other sensing devices for a minimum of 36 months. The data acquisition system shall have the capability to store real-time energy consumption data and provide hourly, daily, monthly and yearly logged data for each end-use category required by Section C406.10.2.

C406.10.5 Graphical energy report. A permanent and readily accessible reporting mechanism shall be provided in the building that is accessible by building operation and management personnel. The reporting mechanism shall have the capability to graphically provide the energy consumption for each end-use category required by Section C406.10.2 at least every hour, day, month and year for the previous 36 months.

C406.11 Fault detection and diagnostics system. A fault detection and diagnostics system shall be installed to monitor the HVAC system's performance and automatically identify faults. The system shall do all of the following:

1. Include permanently installed sensors and devices to monitor the HVAC system's performance.
2. Sample the HVAC system's performance at least once every 15 minutes.
3. Automatically identify and report HVAC system faults.
4. Automatically notify authorized personnel of identified HVAC system faults.
5. Automatically provide prioritized recommendations for repair of identified faults based on analysis of data collected from the sampling of the HVAC system performance.
6. Be capable of transmitting the prioritized fault repair recommendations to remotely located authorized personnel.

C406.12 Efficient kitchen equipment. For buildings and spaces designated as Group A-2 or facilities that include a commercial kitchen with at least one gas or electric fryer, all fryers, dishwashers, steam cookers and ovens shall comply with all of the following:

1. Achieve performance levels in accordance with the equipment specifications listed in Tables C406.12(1) through C406.12(4) when rated in accordance with the applicable test procedure.
2. Be installed prior to the issuance of the Certificate of Occupancy.
3. Have associated performance levels listed on the construction documents submitted for permitting.

Energy efficiency credits for efficient kitchen equipment shall be independent of climate zone and determined based on Equation 4-15, rounded to the nearest whole number.

$$AEEC_K = 20 \times Area_K / Area_B \qquad \text{(Equation 4-15)}$$

where:

$AEEC_K$ = Section C406.12 additional energy efficiency credits.

$Area_K$ = Floor area of full-service kitchen (ft² or m²).

$Area_B$ = Gross floor area of building (ft² or m²).

TABLE C406.12(1)
MINIMUM EFFICIENCY REQUIREMENTS: COMMERCIAL FRYERS

FRYER TYPE	HEAVY-LOAD COOKING ENERGY EFFICIENCY	IDLE ENERGY RATE	TEST PROCEDURE
Standard open deep-fat gas fryers	≥ 50%	≤ 9,000 Btu/h	ASTM F1361
Standard open deep-fat electric fryers	≥ 83%	≤ 800 watts	
Large-vat open deep-fat gas fryers	≥ 50%	≤ 12,000 Btu/h	ASTM F2144
Large-vat open deep-fat electric fryers	≥ 80%	≤ 1,100 watts	

For SI: 1 Btu/h = 0.293/W.

TABLE C406.12(2)
MINIMUM EFFICIENCY REQUIREMENTS: COMMERCIAL STEAM COOKERS

FUEL TYPE	PAN CAPACITY	COOKING ENERGY EFFICIENCY[a]	IDLE ENERGY RATE	TEST PROCEDURE
Electric steam	3-pan	50%	400 watts	ASTM F1484
	4-pan	50%	530 watts	
	5-pan	50%	670 watts	
	6-pan and larger	50%	800 watts	
Gas steam	3-pan	38%	6,250 Btu/h	
	4-pan	38%	8,350 Btu/h	
	5-pan	38%	10,400 Btu/h	
	6-pan and larger	38%	12,500 Btu/h	

For SI: Btu/h = 0.293/W.

a. Cooking energy efficiency is based on heavy load (potato) cooking capacity.

TABLE C406.12(3)
MINIMUM EFFICIENCY REQUIREMENTS: COMMERCIAL DISHWASHERS

MACHINE TYPE	HIGH-TEMPERATURE EFFICIENCY REQUIREMENTS		LOW-TEMPERATURE EFFICIENCY REQUIREMENTS		TEST PROCEDURE
	Idle energy rate[a]	Water consumption[b]	Idle energy rate[a]	Water consumption[b]	
Under counter	≤ .50 kW	≤ 0.86 GPR	≤ 0.50 kW	≤ 1.19 GPR	ASTM F1696 ASTM F1920
Stationary single-tank door	≤ .70 kW	≤ 0.89 GPR	≤ 0.60 kW	≤ 1.18 GPR	
Pot, pan and utensil	≤ 1.20 kW	≤ 0.58 GPR	≤ 1.00 kW	≤ 0.58 GPSF	
Single-tank conveyor	≤ 1.50 kW	≤ 0.70 GPR	≤ 1.50 kW	≤ 0.79 GPR	
Multiple-tank conveyor	≤ 2.25 kW	≤ 0.54 GPR	≤ 2.00 kW	≤ 0.54 GPR	
Single-tank flight	Reported	GPH ≤ 2.975x + 55.00	Reported	GPH ≤ 2.975x + 55.00	
Multiple-tank flight	Reported	GPH ≤ 4.96x + 17.00	Reported	GPH ≤ 4.96x + 17.00	

a. Idle results shall be measured with the door closed and represent the total idle energy consumed by the machine, including all tank heaters and controls. Booster heater (internal or external) energy consumption shall not be part of this measurement unless it cannot be separately monitored.
b. GPR = gallons per rack, GPSF = gallons per square foot of rack, GPH = gallons per hour, x = maximum conveyor belt speed (feet/minute) × conveyor belt width (feet).

TABLE C406.12(4)
MINIMUM EFFICIENCY REQUIREMENTS: COMMERCIAL OVENS

FUEL TYPE	CLASSIFICATION	IDLE RATE	COOKING-ENERGY EFFICIENCY, %	TEST PROCEDURE
Convection ovens				
Gas	Full-size	≤ 12,000 Btu/h	≥ 46	ASTM F1496
Electric	Half-size	≤ 1.0 Btu/h	≥ 71	
	Full-size	≤ 1.60 Btu/h		
Combination ovens				
Gas	Steam mode	≤ 200P[a] + 6,511 Btu/h	≥ 41	ASTM F2861
	Convection mode	≤ 150P[a] + 5,425 Btu/h	≥ 56	
Electric	Steam mode	≤ 0.133P[a] + 0.6400 kW	≥ 55	
	Convection mode	≤ 0.080P[a] + 0.4989 kW	≥ 76	
Rack ovens				
Gas	Single	≤ 25,000 Btu/h	≥ 48	ASTM F2093
	Double	≤ 30,000 Btu/h	≥ 52	

For SI: 1 Btu/h = 0.293/W.

a. P = Pan Capacity: the number of steam table pans the combination oven is able to accommodate in accordance with ASTM F1495.

SECTION C407
TOTAL BUILDING PERFORMANCE

C407.1 Scope. This section establishes criteria for compliance using total building performance. The following systems and loads shall be included in determining the total building performance: heating systems, cooling systems, service water heating, fan systems, lighting power, receptacle loads and process loads.

> **Exception:** Energy used to recharge or refuel vehicles that are used for on-road and off-site transportation purposes.

C407.2 Mandatory requirements. Compliance based on total building performance requires that a proposed design meet all of the following:

1. The requirements of the sections indicated within Table C407.2.
2. An annual energy cost that is less than or equal to 80 percent of the annual energy cost of the *standard reference design*. Energy prices shall be taken from a source *approved* by the *code official*, such as the Department of Energy, Energy Information Administration's *State Energy Data System Prices and*

Expenditures reports. *Code officials* shall be permitted to require time-of-use pricing in energy cost calculations. The reduction in energy cost of the proposed design associated with *on-site renewable energy* shall be not more than 5 percent of the total energy cost. The amount of renewable energy purchased from off-site sources shall be the same in the *standard reference design* and the *proposed design*.

Exception: Jurisdictions that require site energy (1 kWh = 3413 Btu) rather than energy cost as the metric of comparison.

TABLE C407.2
REQUIREMENTS FOR TOTAL BUILDING PERFORMANCE

SECTION[a]	TITLE
Envelope	
C402.5	Air leakage—thermal envelope
Mechanical	
C403.1.1	Calculation of heating and cooling loads
C403.1.2	Data centers
C403.2	System design
C403.3	Heating and cooling equipment efficiencies
C403.4, except C403.4.3, C403.4.4 and C403.4.5	Heating and cooling system controls
C403.5.5	Economizer fault detection and diagnostics
C403.7, except C403.7.4.1	Ventilation and exhaust systems
C403.8, except C403.8.6	Fan and fan controls
C403.9	Large-diameter ceiling fans
C403.11, except C403.11.3	Refrigeration equipment performance
C403.12	Construction of HVAC system elements
C403.13	Mechanical systems located outside of the building thermal envelope
C404	Service water heating
C405, except C405.3	Electrical power and lighting systems
C408	Maintenance information and system commisioning

a. Reference to a code section includes all the relative subsections except as indicated in the table.

C407.3 Documentation. Documentation verifying that the methods and accuracy of compliance software tools conform to the provisions of this section shall be provided to the *code official*.

C407.3.1 Compliance report. Permit submittals shall include a report documenting that the proposed design has annual energy costs less than or equal to the annual energy costs of the standard reference design. The compliance documentation shall include the following information:

1. Address of the building.
2. An inspection checklist documenting the building component characteristics of the *proposed design* as specified in Table C407.4.1(1). The inspection checklist shall show the estimated annual energy cost for both the *standard reference design* and the *proposed design*.
3. Name of individual completing the compliance report.
4. Name and version of the compliance software tool.

C407.3.2 Additional documentation. The *code official* shall be permitted to require the following documents:

1. Documentation of the building component characteristics of the *standard reference design*.
2. Thermal zoning diagrams consisting of floor plans showing the thermal zoning scheme for *standard reference design* and *proposed design*.
3. Input and output reports from the energy analysis simulation program containing the complete input and output files, as applicable. The output file shall include energy use totals and energy use by energy source and end-use served, total hours that space conditioning loads are not met and any errors or warning messages generated by the simulation tool as applicable.
4. An explanation of any error or warning messages appearing in the simulation tool output.
5. A certification signed by the builder providing the building component characteristics of the *proposed design* as given in Table C407.4.1(1).
6. Documentation of the reduction in energy use associated with *on-site renewable energy*.

C407.4 Calculation procedure. Except as specified by this section, the *standard reference design* and *proposed design* shall be configured and analyzed using identical methods and techniques.

C407.4.1 Building specifications. The *standard reference design* and *proposed design* shall be configured and analyzed as specified by Table C407.4.1(1). Table C407.4.1(1) shall include by reference all notes contained in Table C402.1.4.

TABLE C407.4.1(1)
SPECIFICATIONS FOR THE STANDARD REFERENCE AND PROPOSED DESIGNS

BUILDING COMPONENT CHARACTERISTICS	STANDARD REFERENCE DESIGN	PROPOSED DESIGN
Space use classification	Same as proposed	The space use classification shall be chosen in accordance with Table C405.3.2(1) or C405.3.2(2) for all areas of the building covered by this permit. Where the space use classification for a building is not known, the building shall be categorized as an office building.
Roofs	Type: insulation entirely above deck	As proposed
	Gross area: same as proposed	As proposed
	U-factor: as specified in Table C402.1.4	As proposed
	Solar absorptance: 0.75	As proposed
	Emittance: 0.90	As proposed
Walls, above-grade	Type: same as proposed	As proposed
	Gross area: same as proposed	As proposed
	U-factor: as specified in Table C402.1.4	As proposed
	Solar absorptance: 0.75	As proposed
	Emittance: 0.90	As proposed
Walls, below-grade	Type: mass wall	As proposed
	Gross area: same as proposed	As proposed
	U-Factor: as specified in Table C402.1.4 with insulation layer on interior side of walls	As proposed
Floors, above-grade	Type: joist/framed floor	As proposed
	Gross area: same as proposed	As proposed
	U-factor: as specified in Table C402.1.4	As proposed
Floors, slab-on-grade	Type: unheated	As proposed
	F-factor: as specified in Table C402.1.4	As proposed
Opaque doors	Type: swinging	As proposed
	Area: Same as proposed	As proposed
	U-factor: as specified in Table C402.1.4	As proposed
Vertical fenestration other than opaque doors	Area 1. The proposed vertical fenestration area; where the proposed vertical fenestration area is less than 40 percent of above-grade wall area. 2. 40 percent of above-grade wall area; where the proposed vertical fenestration area is 40 percent or more of the above-grade wall area.	As proposed
	U-factor: as specified in Table C402.4	As proposed
	SHGC: as specified in Table C402.4 except that for climates with no requirement (NR) SHGC = 0.40 shall be used	As proposed
	External shading and PF: none	As proposed

(continued)

TABLE C407.4.1(1)—continued
SPECIFICATIONS FOR THE STANDARD REFERENCE AND PROPOSED DESIGNS

BUILDING COMPONENT CHARACTERISTICS	STANDARD REFERENCE DESIGN	PROPOSED DESIGN
Skylights	Area 1. The proposed skylight area; where the proposed skylight area is less than that permitted by Section C402.1. 2. The area permitted by Section C402.1; where the proposed skylight area exceeds that permitted by Section C402.1.	As proposed
	U-factor: as specified in Table C402.4	As proposed
	SHGC: as specified in Table C402.4 except that for climates with no requirement (NR) SHGC = 0.40 shall be used.	As proposed
Lighting, interior	The interior lighting power shall be determined in accordance with Section C405.3.2. Where the occupancy of the building is not known, the lighting power density shall be 1.0 watt per square foot based on the categorization of buildings with unknown space classification as offices.	As proposed
Lighting, exterior	The lighting power shall be determined in accordance with Tables C405.5.2(1), C405.5.2(2) and C405.5.2(3). Areas and dimensions of surfaces shall be the same as proposed.	As proposed
Internal gains	Same as proposed	Receptacle, motor and process loads shall be modeled and estimated based on the space use classification. End-use load components within and associated with the building shall be modeled to include, but not be limited to, the following: exhaust fans, parking garage ventilation fans, exterior building lighting, swimming pool heaters and pumps, elevators, escalators, refrigeration equipment and cooking equipment.
Schedules	Same as proposed **Exception:** Thermostat settings and schedules for HVAC systems that utilize radiant heating, radiant cooling and elevated air speed, provided that equivalent levels of occupant thermal comfort are demonstrated by means of equal Standard Effective Temperature as calculated in Normative Appendix B of ASHRAE Standard 55.	Operating schedules shall include hourly profiles for daily operation and shall account for variations between weekdays, weekends, holidays and any seasonal operation. Schedules shall model the time-dependent variations in occupancy, illumination, receptacle loads, thermostat settings, mechanical ventilation, HVAC equipment availability, service hot water usage and any process loads. The schedules shall be typical of the proposed building type as determined by the designer and approved by the jurisdiction.
Mechanical ventilation	Same as proposed	As proposed, in accordance with Section C403.2.2.
Heating systems	Fuel type: same as proposed design	As proposed
	Equipment type[a]: as specified in Tables C407.4.1(2) and C407.4.1(3)	As proposed
	Efficiency: as specified in the tables in Section C403.3.2.	As proposed
	Capacity[b]: sized proportionally to the capacities in the proposed design based on sizing runs, and shall be established such that no smaller number of unmet heating load hours and no larger heating capacity safety factors are provided than in the proposed design.	As proposed

(continued)

TABLE C407.4.1(1)—continued
SPECIFICATIONS FOR THE STANDARD REFERENCE AND PROPOSED DESIGNS

BUILDING COMPONENT CHARACTERISTICS	STANDARD REFERENCE DESIGN	PROPOSED DESIGN
Cooling systems	Fuel type: same as proposed design	As proposed
	Equipment type[c]: as specified in Tables C407.4.1(2) and C407.4.1(3)	As proposed
	Efficiency: as specified in Tables C403.3.2(1), C403.3.2(2) and C403.3.2(3)	As proposed
	Capacity[b]: sized proportionally to the capacities in the proposed design based on sizing runs, and shall be established such that no smaller number of unmet cooling load hours and no larger cooling capacity safety factors are provided than in the proposed design.	As proposed
	Economizer[d]: same as proposed, in accordance with Section C403.5.	As proposed
Service water heating[e]	Fuel type: same as proposed	As proposed
	Efficiency: as specified in Table C404.2	For Group R, as proposed multiplied by SWHF. For other than Group R, as proposed multiplied by efficiency as provided by the manufacturer of the DWHR unit.
	Capacity: same as proposed	As proposed
	Where no service water hot water system exists or is specified in the proposed design, no service hot water heating shall be modeled.	As proposed

For SI: 1 watt per square foot = 10.7 w/m^2.

SWHF = Service Water Heat Recovery factor, DWHR = Drain Water Heat Recovery.

a. Where no heating system exists or has been specified, the heating system shall be modeled as fossil fuel. The system characteristics shall be identical in both the standard reference design and proposed design.
b. The ratio between the capacities used in the annual simulations and the capacities determined by sizing runs shall be the same for both the standard reference design and proposed design.
c. Where no cooling system exists or no cooling system has been specified, the cooling system shall be modeled as an air-cooled single-zone system, one unit per thermal zone. The system characteristics shall be identical in both the standard reference design and proposed design.
d. If an economizer is required in accordance with Table C403.5(1) and where no economizer exists or is specified in the proposed design, then a supply-air economizer shall be provided in the standard reference design in accordance with Section C403.5.
e. The SWHF shall be applied as follows:
   1. Where potable water from the DWHR unit supplies not less than one shower and not greater than two showers, of which the drain water from the same showers flows through the DWHR unit then SWHF = [1 − (DWHR unit efficiency × 0.36)].
   2. Where potable water from the DWHR unit supplies not less than three showers and not greater than four showers, of which the drain water from the same showers flows through the DWHR unit then SWHF = [1 − (DWHR unit efficiency × 0.33)].
   3. Where potable water from the DWHR unit supplies not less than five showers and not greater than six showers, of which the drain water from the same showers flows through the DWHR unit, then SWHF = [1 − (DWHR unit efficiency × 0.26)].
   4. Where Items 1 through 3 are not met, SWHF = 1.0.

C407.4.2 Thermal blocks. The *standard reference design* and *proposed design* shall be analyzed using identical thermal blocks as specified in Section C407.4.2.1, C407.4.2.2 or C407.4.2.3.

C407.4.2.1 HVAC zones designed. Where HVAC *zones* are defined on HVAC design drawings, each HVAC *zone* shall be modeled as a separate thermal block.

Exception: Different HVAC *zones* shall be allowed to be combined to create a single thermal block or identical thermal blocks to which multipliers are applied, provided that:

1. The space use classification is the same throughout the thermal block.
2. All HVAC *zones* in the thermal block that are adjacent to glazed exterior walls face the same orientation or their orientations are within 45 degrees (0.79 rad) of each other.
3. All of the *zones* are served by the same HVAC system or by the same kind of HVAC system.

C407.4.2.2 HVAC zones not designed. Where HVAC *zones* have not yet been designed, thermal blocks shall be defined based on similar internal load densities, occupancy, lighting, thermal and temperature schedules, and in combination with the following guidelines:

1. Separate thermal blocks shall be assumed for interior and perimeter spaces. Interior spaces shall be those located more than 15 feet (4572 mm) from an exterior wall. Perimeter spaces

**TABLE C407.4.1(2)
HVAC SYSTEMS MAP**

CONDENSER COOLING SOURCE[a]	HEATING SYSTEM CLASSIFICATION[b]	STANDARD REFERENCE DESIGN HVC SYSTEM TYPE[c]		
		Single-zone Residential System	Single-zone Nonresidential System	All Other
Water/ground	Electric resistance	System 5	System 5	System 1
	Heat pump	System 6	System 6	System 6
	Fossil fuel	System 7	System 7	System 2
Air/none	Electric resistance	System 8	System 9	System 3
	Heat pump	System 8	System 9	System 3
	Fossil fuel	System 10	System 11	System 4

a. Select "water/ground" where the proposed design system condenser is water or evaporatively cooled; select "air/none" where the condenser is air cooled. Closed-circuit dry coolers shall be considered to be air cooled. Systems utilizing district cooling shall be treated as if the condenser water type were "water." Where mechanical cooling is not specified or the mechanical cooling system in the proposed design does not require heat rejection, the system shall be treated as if the condenser water type were "Air." For proposed designs with ground-source or groundwater-source heat pumps, the standard reference design HVAC system shall be water-source heat pump (System 6).
b. Select the path that corresponds to the proposed design heat source: electric resistance, heat pump (including air source and water source), or fuel fired. Systems utilizing district heating (steam or hot water) and systems without heating capability shall be treated as if the heating system type were "fossil fuel." For systems with mixed fuel heating sources, the system or systems that use the secondary heating source type (the one with the smallest total installed output capacity for the spaces served by the system) shall be modeled identically in the standard reference design and the primary heating source type shall be used to determine standard reference design HVAC system type.
c. Select the standard reference design HVAC system category: The system under "single-zone residential system" shall be selected where the HVAC system in the proposed design is a single-zone system and serves a Group R occupancy. The system under "single-zone nonresidential system" shall be selected where the HVAC system in the proposed design is a single-zone system and serves other than Group R occupancy. The system under "all other" shall be selected for all other cases.

shall be those located closer than 15 feet (4572 mm) from an *exterior wall*.

2. Separate thermal blocks shall be assumed for spaces adjacent to glazed exterior walls: a separate *zone* shall be provided for each orientation, except orientations that differ by not more than 45 degrees (0.79 rad) shall be permitted to be considered to be the same orientation. Each *zone* shall include floor area that is 15 feet (4572 mm) or less from a glazed perimeter wall, except that floor area within 15 feet (4572 mm) of glazed perimeter walls having more than one orientation shall be divided proportionately between *zones*.

3. Separate thermal blocks shall be assumed for spaces having floors that are in contact with the ground or exposed to ambient conditions from *zones* that do not share these features.

4. Separate thermal blocks shall be assumed for spaces having exterior ceiling or roof assemblies from *zones* that do not share these features.

C407.4.2.3 Group R-2 occupancy buildings. Group R-2 occupancy spaces shall be modeled using one thermal block per space except that those facing the same orientations are permitted to be combined into one thermal block. Corner units and units with roof or floor loads shall only be combined with units sharing these features.

C407.5 Calculation software tools. Calculation procedures used to comply with this section shall be software tools capable of calculating the annual energy consumption of all building elements that differ between the *standard reference design* and the *proposed design* and shall include the following capabilities.

1. Building operation for a full calendar year (8,760 hours).

2. Climate data for a full calendar year (8,760 hours) and shall reflect *approved* coincident hourly data for temperature, solar radiation, humidity and wind speed for the building location.

3. Ten or more thermal zones.

4. Thermal mass effects.

5. Hourly variations in occupancy, illumination, receptacle loads, thermostat settings, mechanical ventilation, HVAC equipment availability, service hot water usage and any process loads.

6. Part-load performance curves for mechanical equipment.

7. Capacity and efficiency correction curves for mechanical heating and cooling equipment.

8. Printed *code official* inspection checklist listing each of the *proposed design* component characteristics from Table C407.4.1(1) determined by the analysis to provide compliance, along with their respective performance ratings, including but not limited to *R*-value, *U*-factor, SHGC, HSPF, AFUE, SEER and EF.

TABLE C407.4.1(3)
SPECIFICATIONS FOR THE STANDARD REFERENCE DESIGN HVAC SYSTEM DESCRIPTIONS

SYSTEM NO.	SYSTEM TYPE	FAN CONTROL	COOLING TYPE	HEATING TYPE
1	Variable air volume with parallel fan-powered boxes[a]	VAV[d]	Chilled water[e]	Electric resistance
2	Variable air volume with reheat[b]	VAV[d]	Chilled water[e]	Hot water fossil fuel boiler[f]
3	Packaged variable air volume with parallel fan-powered boxes[a]	VAV[d]	Direct expansion[c]	Electric resistance
4	Packaged variable air volume with reheat[b]	VAV[d]	Direct expansion[c]	Hot water fossil fuel boiler[f]
5	Two-pipe fan coil	Constant volume[i]	Chilled water[e]	Electric resistance
6	Water-source heat pump	Constant volume[i]	Direct expansion[c]	Electric heat pump and boiler[g]
7	Four-pipe fan coil	Constant volume[i]	Chilled water[e]	Hot water fossil fuel boiler[f]
8	Packaged terminal heat pump	Constant volume[i]	Direct expansion[c]	Electric heat pump[h]
9	Packaged rooftop heat pump	Constant volume[i]	Direct expansion[c]	Electric heat pump[h]
10	Packaged terminal air conditioner	Constant volume[i]	Direct expansion	Hot water fossil fuel boiler[f]
11	Packaged rooftop air conditioner	Constant volume[i]	Direct expansion	Fossil fuel furnace

For SI: 1 foot = 304.8 mm, 1 cfm = 0.4719 L/s, 1 Btu/h = 0.293/W, °C = [(°F) − 32]/1.8.

a. **VAV with parallel boxes:** Fans in parallel VAV fan-powered boxes shall be sized for 50 percent of the peak design flow rate and shall be modeled with 0.35 W/cfm fan power. Minimum volume setpoints for fan-powered boxes shall be equal to the minimum rate for the space required for ventilation consistent with Section C403.6.1, Item 3. Supply air temperature setpoint shall be constant at the design condition.

b. **VAV with reheat:** Minimum volume setpoints for VAV reheat boxes shall be 0.4 cfm/ft^2 of floor area. Supply air temperature shall be reset based on zone demand from the design temperature difference to a 10°F temperature difference under minimum load conditions. Design airflow rates shall be sized for the reset supply air temperature; i.e., a 10°F temperature difference.

c. **Direct expansion:** The fuel type for the cooling system shall match that of the cooling system in the proposed design.

d. **VAV:** Where the proposed design system has a supply, return or relief fan motor 25 hp or larger, the corresponding fan in the VAV system of the standard reference design shall be modeled assuming a variable-speed drive. For smaller fans, a forward-curved centrifugal fan with inlet vanes shall be modeled. Where the proposed design's system has a direct digital control system at the zone level, static pressure setpoint reset based on zone requirements in accordance with Section C403.8.6 shall be modeled.

e. **Chilled water:** For systems using purchased chilled water, the chillers are not explicitly modeled and chilled water costs shall be based as determined in Sections C407.2 and C407.4.2. Otherwise, the standard reference design's chiller plant shall be modeled with chillers having the number as indicated in Table C407.4.1(4) as a function of standard reference building chiller plant load and type as indicated in Table C407.4.1(5) as a function of individual chiller load. Where chiller fuel source is mixed, the system in the standard reference design shall have chillers with the same fuel types and with capacities having the same proportional capacity as the proposed design's chillers for each fuel type. Chilled water supply temperature shall be modeled at 44°F design supply temperature and 56°F return temperature. Piping losses shall not be modeled in either building model. Chilled water supply water temperature shall be reset in accordance with Section C403.4.4. Pump system power for each pumping system shall be the same as the proposed design; where the proposed design has no chilled water pumps, the standard reference design pump power shall be 22 W/gpm (equal to a pump operating against a 75-foot head, 65-percent combined impeller and motor efficiency). The chilled water system shall be modeled as primary-only variable flow with flow maintained at the design rate through each chiller using a bypass. Chilled water pumps shall be modeled as riding the pump curve or with variable-speed drives where required in Section C403.4.4. The heat rejection device shall be an axial fan cooling tower with two-speed fans where required in Section C403.10. Condenser water design supply temperature shall be 85°F or 10°F approach to design wet-bulb temperature, whichever is lower, with a design temperature rise of 10°F. The tower shall be controlled to maintain a 70°F leaving water temperature where weather permits, floating up to leaving water temperature at design conditions. Pump system power for each pumping system shall be the same as the proposed design; where the proposed design has no condenser water pumps, the standard reference design pump power shall be 19 W/gpm (equal to a pump operating against a 60-foot head, 60-percent combined impeller and motor efficiency). Each chiller shall be modeled with separate condenser water and chilled water pumps interlocked to operate with the associated chiller.

f. **Fossil fuel boiler:** For systems using purchased hot water or steam, the boilers are not explicitly modeled and hot water or steam costs shall be based on actual utility rates. Otherwise, the boiler plant shall use the same fuel as the proposed design and shall be natural draft. The standard reference design boiler plant shall be modeled with a single boiler where the standard reference design plant load is 600,000 Btu/h and less and with two equally sized boilers for plant capacities exceeding 600,000 Btu/h. Boilers shall be staged as required by the load. Hot water supply temperature shall be modeled at 180°F design supply temperature and 130°F return temperature. Piping losses shall not be modeled in either building model. Hot water supply water temperature shall be reset in accordance with Section C403.4.4. Pump system power for each pumping system shall be the same as the proposed design; where the proposed design has no hot water pumps, the standard reference design pump power shall be 19 W/gpm (equal to a pump operating against a 60-foot head, 60-percent combined impeller and motor efficiency). The hot water system shall be modeled as primary only with continuous variable flow. Hot water pumps shall be modeled as riding the pump curve or with variable speed drives where required by Section C403.4.4.

(continued)

TABLE C407.4.1(3)—continued
SPECIFICATIONS FOR THE STANDARD REFERENCE DESIGN HVAC SYSTEM DESCRIPTIONS

g. **Electric heat pump and boiler:** Water-source heat pumps shall be connected to a common heat pump water loop controlled to maintain temperatures between 60°F and 90°F. Heat rejection from the loop shall be provided by an axial fan closed-circuit evaporative fluid cooler with two-speed fans where required in Section C403.8.6. Heat addition to the loop shall be provided by a boiler that uses the same fuel as the proposed design and shall be natural draft. Where no boilers exist in the proposed design, the standard reference building boilers shall be fossil fuel. The standard reference design boiler plant shall be modeled with a single boiler where the standard reference design plant load is 600,000 Btu/h or less and with two equally sized boilers for plant capacities exceeding 600,000 Btu/h. Boilers shall be staged as required by the load. Piping losses shall not be modeled in either building model. Pump system power shall be the same as the proposed design; where the proposed design has no pumps, the standard reference design pump power shall be 22 W/gpm, which is equal to a pump operating against a 75-foot head, with a 65-percent combined impeller and motor efficiency. Loop flow shall be variable with flow shutoff at each heat pump when its compressor cycles off as required by Section C403.4.4. Loop pumps shall be modeled as riding the pump curve or with variable speed drives where required by Section C403.10.

h. **Electric heat pump:** Electric air-source heat pumps shall be modeled with electric auxiliary heat. The system shall be controlled with a multistage space thermostat and an outdoor air thermostat wired to energize auxiliary heat only on the last thermostat stage and when outdoor air temperature is less than 40°F.

i. **Constant volume:** Fans shall be controlled in the same manner as in the proposed design; i.e., fan operation whenever the space is occupied or fan operation cycled on calls for heating and cooling. Where the fan is modeled as cycling and the fan energy is included in the energy efficiency rating of the equipment, fan energy shall not be modeled explicitly.

**TABLE C407.5.1(4)
NUMBER OF CHILLERS**

TOTAL CHILLER PLANT CAPACITY	NUMBER OF CHILLERS
≤ 300 tons	1
> 300 tons, < 600 tons	2, sized equally
≥ 600 tons	2 minimum, with chillers added so that all are sized equally and none is larger than 800 tons

For SI: 1 ton = 3517 W.

**TABLE C407.5.1(5)
WATER CHILLER TYPES**

INDIVIDUAL CHILLER PLANT CAPACITY	ELECTRIC CHILLER TYPE	FOSSIL FUEL CHILLER TYPE
≤ 100 tons	Reciprocating	Single-effect absorption, direct fired
> 100 tons, < 300 tons	Screw	Double-effect absorption, direct fired
≥ 300 tons	Centrifugal	Double-effect absorption, direct fired

For SI: 1 ton = 3517 W.

C407.5.1 Specific approval. Performance analysis tools complying with the applicable subsections of Section C407 and tested according to ASHRAE Standard 140 shall be permitted to be *approved*. Tools are permitted to be *approved* based on meeting a specified threshold for a jurisdiction. The *code official* shall be permitted to approve tools for a specified application or limited scope.

C407.5.2 Input values. Where calculations require input values not specified by Sections C402, C403, C404 and C405, those input values shall be taken from an *approved* source.

C407.5.3 Exceptional calculation methods. Where the simulation program does not model a design, material or device of the *proposed design*, an exceptional calculation method shall be used where approved by the *code official*. Where there are multiple designs, materials or devices that the simulation program does not model, each shall be calculated separately and exceptional savings determined for each. The total exceptional savings shall not constitute more than half of the difference between the baseline building performance and the proposed building performance. Applications for approval of an exceptional method shall include all of the following:

1. Step-by-step documentation of the exceptional calculation method performed, detailed enough to reproduce the results.
2. Copies of all spreadsheets used to perform the calculations.
3. A sensitivity analysis of energy consumption where each of the input parameters is varied from half to double the value assumed.
4. The calculations shall be performed on a time step basis consistent with the simulation program used.
5. The performance rating calculated with and without the exceptional calculation method.

SECTION C408
MAINTENANCE INFORMATION AND SYSTEM COMMISSIONING

C408.1 General. This section covers the provision of maintenance information and the commissioning of, and the functional testing requirements for, building systems.

C408.1.1 Building operations and maintenance information. The building operations and maintenance documents shall be provided to the owner and shall consist of manufacturers' information, specifications and recommendations; programming procedures and data points; narratives; and other means of illustrating to the owner how the building, equipment and systems are intended to be installed, maintained and operated. Required regular maintenance actions for equipment and systems shall be clearly stated on a readily visible label. The label shall include the title or publication number for the operation and maintenance manual for that particular model and type of product.

C408.2 Mechanical systems and service water-heating systems commissioning and completion requirements. Prior to the final mechanical and plumbing inspections, the *registered design professional or approved agency* shall provide evidence of mechanical systems *commissioning* and completion in accordance with the provisions of this section.

Construction document notes shall clearly indicate provisions for *commissioning* and completion requirements in accordance with this section and are permitted to refer to specifications for further requirements. Copies of all documentation shall be given to the owner or owner's authorized agent and made available to the *code official* upon request in accordance with Sections C408.2.4 and C408.2.5.

Exceptions: The following systems are exempt:

1. Mechanical systems and service water-heating systems in buildings where the total mechanical equipment capacity is less than 480,000 Btu/h (140.7 kW) cooling capacity and 600,000 Btu/h (175.8 kW) combined service water-heating and space-heating capacity.
2. Systems included in Section C403.5 that serve individual *dwelling units* and *sleeping units*.

C408.2.1 Commissioning plan. A *commissioning plan* shall be developed by a *registered design professional* or *approved agency* and shall include the following items:

1. A narrative description of the activities that will be accomplished during each phase of *commissioning*, including the personnel intended to accomplish each of the activities.
2. A listing of the specific equipment, appliances or systems to be tested and a description of the tests to be performed.
3. Functions to be tested including, but not limited to, calibrations and economizer controls.
4. Conditions under which the test will be performed. Testing shall affirm winter and summer design conditions and full outside air conditions.
5. Measurable criteria for performance.

C408.2.2 Systems adjusting and balancing. HVAC systems shall be balanced in accordance with generally accepted engineering standards. Air and water flow rates shall be measured and adjusted to deliver final flow rates within the tolerances provided in the product specifications. Test and balance activities shall include air system and hydronic system balancing.

C408.2.2.1 Air systems balancing. Each supply air outlet and *zone* terminal device shall be equipped with means for air balancing in accordance with the requirements of Chapter 6 of the *International Mechanical Code*. Discharge dampers used for air-system balancing are prohibited on constant-volume fans and variable volume fans with motors 10 hp (18.6 kW) and larger. Air systems shall be balanced in a manner to first minimize throttling losses then, for fans with system power of greater than 1 hp (0.746 kW), fan speed shall be adjusted to meet design flow conditions.

Exception: Fans with fan motors of 1 hp (0.74 kW) or less are not required to be provided with a means for air balancing.

C408.2.2.2 Hydronic systems balancing. Individual hydronic heating and cooling coils shall be equipped with means for balancing and measuring flow. Hydronic systems shall be proportionately balanced in a manner to first minimize throttling losses, then the pump impeller shall be trimmed or pump speed shall be adjusted to meet design flow conditions. Each hydronic system shall have either the capability to measure pressure across the pump, or test ports at each side of each pump.

Exception: The following equipment is not required to be equipped with a means for balancing or measuring flow:

1. Pumps with pump motors of 5 hp (3.7 kW) or less.
2. Where throttling results in not greater than 5 percent of the nameplate horsepower draw above that required if the impeller were trimmed.

C408.2.3 Functional performance testing. Functional performance testing specified in Sections C408.2.3.1 through C408.2.3.3 shall be conducted.

C408.2.3.1 Equipment. Equipment functional performance testing shall demonstrate the installation and operation of components, systems and system-to-system interfacing relationships in accordance with approved plans and specifications such that operation, function and maintenance serviceability for each of the commissioned systems are confirmed. Testing shall include all modes and *sequence of operation*, including under full-load, part-load and the following emergency conditions:

1. All modes as described in the *sequence* of *operation*.
2. Redundant or *automatic* back-up mode.
3. Performance of alarms.
4. Mode of operation upon a loss of power and restoration of power.

Exception: Unitary or packaged HVAC equipment listed in the tables in Section C403.3.2 that do not require supply air economizers.

C408.2.3.2 Controls. HVAC and service water-heating control systems shall be tested to document that control devices, components, equipment and systems are calibrated and adjusted and operate in accordance with approved plans and specifications. Sequences of operation shall be functionally tested to document they operate in accordance with *approved* plans and specifications.

C408.2.3.3 Economizers. Air economizers shall undergo a functional test to determine that they operate in accordance with manufacturer's specifications.

C408.2.4 Preliminary commissioning report. A preliminary report of *commissioning* test procedures and results shall be completed and certified by the *registered design professional* or *approved agency* and provided to the building owner or owner's authorized agent. The report shall be organized with mechanical and service hot water findings in separate sections to allow independent review. The report shall be identified as "Preliminary Commissioning Report," shall include the completed Commissioning Compliance Checklist, Figure C408.2.4, and shall identify:

1. Itemization of deficiencies found during testing required by this section that have not been corrected at the time of report preparation.
2. Deferred tests that cannot be performed at the time of report preparation because of climatic conditions.
3. Climatic conditions required for performance of the deferred tests.
4. Results of functional performance tests.
5. Functional performance test procedures used during the commissioning process, including measurable criteria for test acceptance.

C408.2.4.1 Acceptance of report. Buildings, or portions thereof, shall not be considered as acceptable for a final inspection pursuant to Section C105.2.6 until the *code official* has received the Preliminary Commissioning Report from the building owner or owner's authorized agent.

Project Information: _____ Project Name: _____

Project Address: _____

Commissioning Authority: _____

Commissioning Plan (Section C408.2.1)

☐ Commissioning Plan was used during construction and includes all items required by Section C408.2.1

☐ Systems Adjusting and Balancing has been completed.

☐ HVAC Equipment Functional Testing has been executed. If applicable, deferred and follow-up testing is scheduled to be provided on:_____

☐ HVAC Controls Functional Testing has been executed. If applicable, deferred and follow-up testing is scheduled to be provided on:_____

☐ Economizer Functional Testing has been executed. If applicable, deferred and follow-up testing is scheduled to be provided on:_____

☐ Lighting Controls Functional Testing has been executed. If applicable, deferred and follow-up testing is scheduled to be provided on:_____

☐ Service Water Heating System Functional Testing has been executed. If applicable, deferred and follow-up testing is scheduled to be provided on:_____

☐ Manual, record documents and training have been completed or scheduled

☐ Preliminary Commissioning Report submitted to owner and includes all items required by Section C408.2.4

I hereby certify that the commissioning provider has provided me with evidence of mechanical, service water heating and lighting systems commissioning in accordance with the 2021 IECC.

Signature of Building Owner or Owner's Representative _____ Date_____

FIGURE C408.2.4
COMMISSIONING COMPLIANCE CHECKLIST

C408.2.4.2 Copy of report. The *code official* shall be permitted to require that a copy of the Preliminary Commissioning Report be made available for review by the *code official*.

C408.2.5 Documentation requirements. The *construction documents* shall specify that the documents described in this section be provided to the building owner or owner's authorized agent within 90 days of the date of receipt of the *certificate of occupancy*.

C408.2.5.1 System balancing report. A written report describing the activities and measurements completed in accordance with Section C408.2.2.

C408.2.5.2 Final commissioning report. A report of test procedures and results identified as "Final Commissioning Report" shall be delivered to the building owner or owner's authorized agent. The report shall be organized with mechanical system and service hot water system findings in separate sections to allow independent review. The report shall include the following:

1. Results of functional performance tests.
2. Disposition of deficiencies found during testing, including details of corrective measures used or proposed.
3. Functional performance test procedures used during the commissioning process including measurable criteria for test acceptance, provided herein for repeatability.

Exception: Deferred tests that cannot be performed at the time of report preparation due to climatic conditions.

C408.3 Functional testing of lighting controls. Automatic lighting controls required by this code shall comply with this section.

C408.3.1 Functional testing. Prior to passing final inspection, the *registered design professional* or *approved agency* shall provide evidence that the lighting control systems have been tested to ensure that control hardware and software are calibrated, adjusted, programmed and in proper working condition in accordance with the *construction documents* and manufacturer's instructions. Functional testing shall be in accordance with Sections C408.3.1.1 through C408.3.1.3 for the applicable control type.

C408.3.1.1 Occupant sensor controls. Where *occupant sensor controls* are provided, the following procedures shall be performed:

1. Certify that the *occupant sensor* has been located and aimed in accordance with manufacturer recommendations.
2. For projects with seven or fewer *occupant sensors*, each sensor shall be tested.
3. For projects with more than seven *occupant sensors*, testing shall be done for each unique combination of sensor type and space geometry. Where multiples of each unique combination of sensor type and space geometry are provided, not less than 10 percent and in no case fewer than one, of each combination shall be tested unless the *code official* or design professional requires a higher percentage to be tested. Where 30 percent or more of the tested controls fail, all remaining identical combinations shall be tested.

For *occupant sensor controls* to be tested, verify the following:

   3.1. Where *occupant sensor controls* include status indicators, verify correct operation.

   3.2. The controlled lights turn off or down to the permitted level within the required time.

   3.3. For auto-on *occupant sensor controls*, the lights turn on to the permitted level when an occupant enters the space.

   3.4. For manual-on *occupant sensor controls*, the lights turn on only when manually activated.

   3.5. The lights are not incorrectly turned on by movement in adjacent areas or by HVAC operation.

C408.3.1.2 Time-switch controls. Where *time-switch controls* are provided, the following procedures shall be performed:

1. Confirm that the *time-switch control* is programmed with accurate weekday, weekend and holiday schedules.
2. Provide documentation to the owner of *time-switch controls* programming including weekday, weekend, holiday schedules, and set-up and preference program settings.
3. Verify the correct time and date in the time switch.
4. Verify that any battery back-up is installed and energized.
5. Verify that the override time limit is set to not more than 2 hours.
6. Simulate occupied condition. Verify and document the following:

   6.1. All lights can be turned on and off by their respective area control switch.

   6.2. The switch only operates lighting in the enclosed space in which the switch is located.

7. Simulate unoccupied condition. Verify and document the following:

   7.1. Nonexempt lighting turns off.

   7.2. Manual override switch allows only the lights in the enclosed space where the override switch is located to turn on or remain on until the next scheduled shut-off occurs.

8. Additional testing as specified by the *registered design professional*.

C408.3.1.3 Daylight responsive controls. Where *daylight responsive controls* are provided, the following shall be verified:

1. Control devices have been properly located, field calibrated and set for accurate setpoints and threshold light levels.
2. Daylight controlled lighting loads adjust to light level setpoints in response to available daylight.
3. The calibration adjustment equipment is located for *ready access* only by authorized personnel.

C408.3.2 Documentation requirements. The *construction documents* shall specify that the documents described in this section be provided to the building owner or owner's authorized agent within 90 days of the date of receipt of the *certificate of occupancy*.

C408.3.2.1 Drawings. Construction documents shall include the location and catalogue number of each piece of equipment.

C408.3.2.2 Manuals. An operating and maintenance manual shall be provided and include the following:

1. Name and address of not less than one service agency for installed equipment.
2. A narrative of how each system is intended to operate, including recommended setpoints.
3. Submittal data indicating all selected options for each piece of lighting equipment and lighting controls.
4. Operation and maintenance manuals for each piece of lighting equipment. Required routine maintenance actions, cleaning and recommended relamping shall be clearly identified.
5. A schedule for inspecting and recalibrating all lighting controls.

C408.3.2.3 Report. A report of test results shall be provided and include the following:

1. Results of functional performance tests.
2. Disposition of deficiencies found during testing, including details of corrective measures used or proposed.

CHAPTER 5 [CE]
EXISTING BUILDINGS

User note:

About this chapter: Many buildings are renovated or altered in numerous ways that could affect the energy use of the building as a whole. Chapter 5 requires the application of certain parts of Chapter 4 in order to maintain, if not improve, the conservation of energy by the renovated or altered building.

SECTION C501
GENERAL

C501.1 Scope. The provisions of this chapter shall control the *alteration, repair, addition* and *change of occupancy* of existing buildings and structures.

C501.1.1 Existing buildings. Except as specified in this chapter, this code shall not be used to require the removal, *alteration* or abandonment of, nor prevent the continued use and maintenance of, an existing *building* or *building* system lawfully in existence at the time of adoption of this code.

C501.2 Compliance. *Additions, alterations, repairs,* and changes of occupancy to, or relocation of, existing *buildings* and structures shall comply with Sections C502, C503, C504 and C505 of this code, as applicable, and with the provisions for *alterations, repairs, additions* and changes of occupancy or relocation, respectively, in the *International Building Code, International Existing Building Code, International Fire Code, International Fuel Gas Code, International Mechanical Code, International Plumbing Code, International Property Maintenance Code, International Private Sewage Disposal Code* and NFPA 70. Changes where unconditioned space is changed to conditioned space shall comply with Section C502.

> **Exception:** *Additions, alterations, repairs* or changes of occupancy complying with ANSI/ASHRAE/IESNA 90.1.

C501.3 Maintenance. *Buildings* and structures, and parts thereof, shall be maintained in a safe and sanitary condition. Devices and systems required by this code shall be maintained in conformance to the code edition under which they were installed. The owner or the owner's authorized agent shall be responsible for the maintenance of buildings and structures. The requirements of this chapter shall not provide the basis for removal or abrogation of energy conservation, fire protection and safety systems and devices in existing structures.

C501.4 New and replacement materials. Except as otherwise required or permitted by this code, materials permitted by the applicable code for new construction shall be used. Like materials shall be permitted for *repairs*, provided that hazards to life, health or property are not created. Hazardous materials shall not be used where the code for new construction would not allow use of these materials in buildings of similar occupancy, purpose and location.

C501.5 Historic buildings. Provisions of this code relating to the construction, *repair, alteration,* restoration and movement of structures, and *change of occupancy* shall not be mandatory for *historic buildings* provided that a report has been submitted to the *code official* and signed by a *registered design professional*, or a representative of the State Historic Preservation Office or the historic preservation authority having jurisdiction, demonstrating that compliance with that provision would threaten, degrade or destroy the historic form, fabric or function of the *building*.

SECTION C502
ADDITIONS

C502.1 General. *Additions* to an existing *building, building* system or portion thereof shall conform to the provisions of this code as those provisions relate to new construction without requiring the unaltered portion of the existing *building* or *building* system to comply with this code. *Additions* shall not create an unsafe or hazardous condition or overload existing building systems. An *addition* shall be deemed to comply with this code if the *addition* alone complies or if the existing building and *addition* comply with this code as a single building.

C502.2 Change in space conditioning. Any nonconditioned or low-energy space that is altered to become *conditioned space* shall be required to comply with Section C502.

Exceptions:

1. Where the component performance alternative in Section C402.1.5 is used to comply with this section, the proposed UA shall be not greater than 110 percent of the target UA.

2. Where the total building performance option in Section C407 is used to comply with this section, the annual energy cost of the proposed design shall be not greater than 110 percent of the annual energy cost otherwise permitted by Section C407.2.

C502.3 Compliance. *Additions* shall comply with Sections C502.3.1 through C502.3.6.2.

C502.3.1 Vertical fenestration area. Additions shall comply with the following:

1. Where an addition has a new vertical fenestration area that results in a total building fenestration area less than or equal to that permitted by Section

C402.4.1, the addition shall comply with Section C402.1.5, C402.4.3 or C407.

2. Where an addition with vertical fenestration that results in a total building fenestration area greater than Section C402.4.1 or an addition that exceeds the fenestration area greater than that permitted by Section C402.4.1, the fenestration shall comply with Section C402.4.1.1 for the addition only.

3. Where an addition has vertical fenestration that results in a total building vertical fenestration area exceeding that permitted by Section C402.4.1.1, the addition shall comply with Section C402.1.5 or C407.

C502.3.2 Skylight area. Skylights shall comply with the following:

1. Where an addition has new skylight area that results in a total building fenestration area less than or equal to that permitted by Section C402.4.1, the addition shall comply with Section C402.1.5 or C407.

2. Where an addition has new skylight area that results in a total building skylight area greater than permitted by Section C402.4.1 or where additions have skylight area greater than that permitted by Section C402.4.1, the skylight area shall comply with Section C402.4.1.2 for the addition only.

3. Where an addition has skylight area that results in a total building skylight area exceeding that permitted by Section C402.4.1.2, the addition shall comply with Section C402.1.5 or C407.

C502.3.3 Building mechanical systems. New mechanical systems and equipment that are part of the *addition* and serve the building heating, cooling and ventilation needs shall comply with Sections C403 and C408.

C502.3.4 Service water-heating systems. New service water-heating equipment, controls and service water-heating piping shall comply with Section C404.

C502.3.5 Pools and inground permanently installed spas. New pools and inground permanently installed spas shall comply with Section C404.9.

C502.3.6 Lighting power and systems. New lighting systems that are installed as part of the addition shall comply with Sections C405 and C408.

C502.3.6.1 Interior lighting power. The total interior lighting power for the *addition* shall comply with Section C405.3.2 for the *addition* alone, or the existing building and the *addition* shall comply as a single building.

C502.3.6.2 Exterior lighting power. The total exterior lighting power for the *addition* shall comply with Section C405.5.2 for the *addition* alone, or the existing building and the *addition* shall comply as a single building.

SECTION C503
ALTERATIONS

C503.1 General. *Alterations* to any *building* or structure shall comply with the requirements of Section C503. *Alterations* shall be such that the existing *building* or structure is not less conforming to the provisions of this code than the existing *building* or structure was prior to the *alteration*. *Alterations* to an existing *building*, *building* system or portion thereof shall conform to the provisions of this code as those provisions relate to new construction without requiring the unaltered portions of the existing *building* or *building* system to comply with this code. *Alterations* shall not create an unsafe or hazardous condition or overload existing *building* systems.

Exception: The following *alterations* need not comply with the requirements for new construction, provided that the energy use of the building is not increased:

1. Storm windows installed over existing *fenestration*.

2. Surface-applied window film installed on existing single-pane *fenestration* assemblies reducing solar heat gain, provided that the code does not require the glazing or *fenestration* to be replaced.

3. Existing ceiling, wall or floor cavities exposed during construction, provided that these cavities are filled with insulation.

4. Construction where the existing roof, wall or floor cavity is not exposed.

5. *Roof recover*.

6. *Air barriers* shall not be required for *roof recover* and roof replacement where the *alterations* or renovations to the building do not include *alterations*, renovations or *repairs* to the remainder of the building envelope.

C503.2 Building envelope. New building envelope assemblies that are part of the *alteration* shall comply with Sections C402.1 through C402.5.

Exception: Where the existing building exceeds the fenestration area limitations of Section C402.4.1 prior to alteration, the building is exempt from Section C402.4.1 provided that there is not an increase in fenestration area.

C503.2.1 Roof replacement. *Roof replacements* shall comply with Section C402.1.3, C402.1.4, C402.1.5 or C407 where the existing roof assembly is part of the *building thermal envelope* and contains insulation entirely above the roof deck. In no case shall the *R*-value of the roof insulation be reduced or the *U*-factor of the roof assembly be increased as part of the *roof replacement*.

C503.2.2 Vertical fenestration. The addition of *vertical fenestration* that results in a total building *fenestration* area less than or equal to that specified in Section C402.4.1 shall comply with Section C402.1.5, C402.4.3 or C407. The addition of *vertical fenestration* that results in a total building *fenestration* area greater than Section C402.4.1 shall comply with Section C402.4.1.1 for the space adjacent to the new fenestration only. *Alterations* that result in a total building *vertical fenestration* area exceeding that specified in Section C402.4.1.1 shall comply with Section C402.1.5 or C407. Provided that the vertical fenestration area is not changed, using the same vertical fenestration area in the *standard reference design* as the building prior to alteration shall be an alternative to using the vertical fenestration area specified in Table C407.4.1(1).

C503.2.2.1 Application to replacement fenestration products. Where some or all of an existing *fenestration* unit is replaced with a new *fenestration* product, including sash and glazing, the replacement *fenestration* unit shall meet the applicable requirements for *U*-factor and *SHGC* in Table C402.4.

> **Exception:** An area-weighted average of the *U*-factor of replacement fenestration products being installed in the building for each fenestration product category listed in Table C402.4 shall be permitted to satisfy the *U*-factor requirements for each fenestration product category listed in Table C402.4. Individual fenestration products from different product categories listed in Table C402.4 shall not be combined in calculating the area-weighted average *U*-factor.

C503.2.3 Skylight area. New *skylight* area that results in a total building *skylight* area less than or equal to that specified in Section C402.4.1 shall comply with Section C402.1.5, C402.4 or C407. The addition of *skylight* area that results in a total building skylight area greater than Section C402.4.1 shall comply with Section C402.4.1.2 for the space adjacent to the new skylights. *Alterations* that result in a total building skylight area exceeding that specified in Section C402.4.1.2 shall comply with Section C402.1.5 or C407. Provided that the skylight area is not changed, using the same skylight area in the *standard reference design* as the building prior to alteration shall be an alternative to using the skylight area specified in Table C407.4.1(1).

C503.3 Heating and cooling systems. New heating, cooling and duct systems that are part of the *alteration* shall comply with Sections C403 and C408.

C503.3.1 Economizers. New cooling systems that are part of *alteration* shall comply with Section C403.5.

C503.4 Service hot water systems. New service hot water systems that are part of the *alteration* shall comply with Sections C404 and C408.

C503.5 Lighting systems. New lighting systems that are part of the *alteration* shall comply with Sections C405 and C408.

> **Exception:** *Alterations* that replace less than 10 percent of the luminaires in a space, provided that such *alterations* do not increase the installed interior lighting power.

SECTION C504
REPAIRS

C504.1 General. *Buildings* and structures, and parts thereof, shall be repaired in compliance with Section C501.3 and this section. Work on nondamaged components that is necessary for the required *repair* of damaged components shall be considered to be part of the *repair* and shall not be subject to the requirements for *alterations* in this chapter. Routine maintenance required by Section C501.3, ordinary *repairs* exempt from *permit* and abatement of wear due to normal service conditions shall not be subject to the requirements for *repairs* in this section.

Where a building was constructed to comply with ANSI/ASHRAE/IESNA 90.1, repairs shall comply with the standard and need not comply with Sections C402, C403, C404 and C405.

C504.2 Application. For the purposes of this code, the following shall be considered to be repairs:

1. Glass-only replacements in an existing sash and frame.
2. *Roof repairs*.
3. Air barriers shall not be required for *roof repair* where the repairs to the building do not include *alterations*, renovations or *repairs* to the remainder of the building envelope.
4. Replacement of existing doors that separate conditioned space from the exterior shall not require the installation of a vestibule or revolving door, provided that an existing vestibule that separates a conditioned space from the exterior shall not be removed.
5. *Repairs* where only the bulb, the ballast or both within the existing luminaires in a space are replaced, provided that the replacement does not increase the installed interior lighting power.

SECTION C505
CHANGE OF OCCUPANCY OR USE

C505.1 General. Spaces undergoing a change in occupancy that would result in an increase in demand for either fossil fuel or electrical energy shall comply with this code. Where the use in a space changes from one use in Table C405.3.2(1) or C405.3.2(2) to another use in Table C405.3.2(1) or C405.3.2(2), the installed lighting wattage shall comply with Section C405.3. Where the space undergoing a change in occupancy or use is in a building with a fenestration area that exceeds the limitations of Section C402.4.1, the space is exempt from Section C402.4.1 provided that there is not an increase in fenestration area.

> **Exceptions:**
> 1. Where the component performance alternative in Section C402.1.5 is used to comply with this section, the proposed UA shall not be greater than 110 percent of the target UA.

2. Where the total building performance option in Section C407 is used to comply with this section, the annual energy cost of the proposed design shall not be greater than 110 percent of the annual energy cost otherwise permitted by Section C407.3.

CHAPTER 6 [CE]
REFERENCED STANDARDS

User note:

About this chapter: Chapter 6 lists the full title, edition year and address of the promulgator for all standards that are referenced in the code. The section numbers in which the standards are referenced are also listed.

This chapter lists the standards that are referenced in various sections of this document. The standards are listed herein by the promulgating agency of the standard, the standard identification, the effective date and title, and the section or sections of this document that reference the standard. The application of the referenced standards shall be as specified in Section 108.

AAMA

American Architectural Manufacturers Association
1827 Walden Office Square
Suite 550
Schaumburg, IL 60173-4268

AAMA/WDMA/CSA 101/I.S.2/A440—17: North American Fenestration Standard/Specification for Windows, Doors, and Skylights
  Table C402.5.5

AHAM

Association of Home Appliance Manufacturers
1111 19th Street NW, Suite 402
Washington, DC 20036

ANSI/AHAM RAC-1—2015: Room Air Conditioners
  Table C403.3.2(4)

AHRI

Air-Conditioning, Heating, & Refrigeration Institute
2111 Wilson Blvd, Suite 500
Arlington, VA 22201

210/240—2017 and 2023: Performance Rating of Unitary Air-conditioning and Air-source Heat Pump Equipment
  Table C403.3.2(1), Table C403.3.2(2)

310/380—2017 (CSA-C744-17): Packaged Terminal Air Conditioners and Heat Pumps
  Table C403.3.2(4)

340/360—2019: Performance Rating of Commercial and Industrial Unitary Air-conditioning and Heat Pump Equipment
  Table C403.3.2(1), Table C403.3.2(2)

365(I-P)—2009: Commercial and Industrial Unitary Air-conditioning Condensing Units
  Table C403.3.2(1)

390 (I-P)—2003: Performance Rating of Single Package Vertical Air-conditioners and Heat Pumps
  Table C403.3.2(3)

400 (I-P)—2015: Performance Rating of Liquid to Liquid Heat Exchangers
  C403.3.2

440—2008: Performance Rating of Room Fan Coils—with Addendum 1
  C403.12.3

460—2005: Performance Rating of Remote Mechanical-draft Air-cooled Refrigerant Condensers
  Table C403.3.2(7)

550/590 (I-P)—2018: Performance Rating of Water-chilling and Heat Pump Water-heating Packages Using the Vapor Compression Cycle
  Table C403.3.2(3), Table C403.3.2(15)

560—2018: Absorption Water Chilling and Water Heating Packages
  Table C403.3.2(3)

910—2014: Performance Rating of Indoor Pool Dehumidifiers
  Table C403.3.2(11)

REFERENCED STANDARDS

AHRI—continued

920—2015: Performance Rating of DX-Dedicated Outdoor Air System Units
    Table C403.3.2(12), Table C403.3.2(13)

1160 (I-P)—2014: Performance Rating of Heat Pump Pool Heaters (with Addendum 1)
    Table C404.2

1200 (I-P)—2013: Performance Rating of Commercial Refrigerated Display Merchandisers and Storage Cabinets
    Table C403.11.1

1230—2014: Performance Rating of Variable Refrigerant Flow (VRF) Multi-split Air-Conditioning and Heat Pump Equipment (with Addendum 1)
    Table C403.3.2(9)

1250 (I-P)—2014: Standard for Performance Rating in Walk-in Coolers and Freezers
    Table C403.11.2.1(3)

1360—2017: Performance Rating of Computer and Data Processing Room Air Conditioners
    Table C403.3.2(10), Table C403.3.2(16)

ISO/AHRI/ASHRAE 13256-1 (2012): Water-to-Air and Brine-to-Air Heat Pumps—Testing and Rating for Performance
    Table C403.3.2(14)

AMCA

Air Movement and Control Association International
30 West University Drive
Arlington Heights, IL 60004-1806

208—18: Calculation of the Fan Energy Index
    C403.8.3

220—19: Laboratory Methods of Testing Air Curtain Units for Aerodynamic Performance Rating
    C402.5.9

230—15: Laboratory Methods of Testing Air Circulating Fans for Rating and Certification
    C403.9

500D—18: Laboratory Methods for Testing Dampers for Rating
    C403.7.7

ANSI

American National Standards Institute
25 West 43rd Street, 4th Floor
New York, NY 10036

Z21.10.3/CSA 4.3—17: Gas Water Heaters, Volume III—Storage Water Heaters with Input Ratings Above 75,000 Btu per Hour, Circulating Tank and Instantaneous
    Table C404.2

ANSI Z21.47—2016/CSA 2.3—2016: Gas-Fired Central Furnaces
    Table C403.3.2(4)

ANSI Z83.8—2016/CSA 2.6—2016: Gas Unit Heater, Gas Packaged Heaters, Gas Utility Heaters And Gas-Fired Duct Furnaces
    Table C403.3.2(4)

APSP

Pool & Hot Tub Alliance (formerly the Association of Pool and Spa Professionals
2111 Eisenhower Avenue, Suite 580
Alexandria, VA 22314

14—2019: American National Standard for Portable Electric Spa Energy Efficiency
    C404.7

ASABE

American Society of Agricultural and Biological Engineers
2950 Niles Road
St. Joseph, MI 49085

S640—2017: Quantities and Units of Electromagnetic Radiation for Plants (Photosynthetic Organisms)
    C405.4

ASHRAE

ASHRAE
180 Technology Parkway NW
Peachtree Corners, GA 30092

55—2017: Thermal Environmental Conditions for Human Occupancy
    Table C407.4.1(1)

90.1—2019: Energy Standard for Buildings Except Low-rise Residential Buildings
    C402.1.4, C406.2

90.4—2016: Energy Standard for Data Centers
    C403.1.2 , C405.2.4

140—2014: Standard Method of Test for the Evaluation of Building Energy Analysis Computer Programs
    C407.5.1

146—2011: Testing for Rating Pool Heaters
    Table C404.2

ANSI/ASHRAE/ACCA Standard 183—(RA2017): Peak Cooling and Heating Load Calculations in Buildings, Except Low-rise Residential Buildings
    C403.1.1

ASHRAE—2020: HVAC Systems and Equipment Handbook—2020
    C403.1.1

ISO/AHRI/ASHRAE 13256-1 (2012): Water-to-Air and Brine-to-Air Heat Pumps—Testing and Rating for Performance
    Table C403.3.2(14)

ISO/AHRI/ASHRAE 13256-2 (2012): Water-to-Water and Brine-to-Water Heat Pumps—Testing and Rating for Performance
    Table C403.3.2(14)

ASME

American Society of Mechanical Engineers
Two Park Avenue
New York, NY 10016-5990

ASME A17.1—2019/CSA B44—19: Safety Code for Elevators and Escalators
    C405.9.2

ASTM

ASTM International
100 Barr Harbor Drive, P.O. Box C700
West Conshohocken, PA 19428-2959

C90—2016A: Specification for Load-bearing Concrete Masonry Units
    Table C402.1.3

C1363—11: Standard Test Method for Thermal Performance of Building Materials and Envelope Assemblies by Means of a Hot Box Apparatus
    C303.1.4.1, Table C402.1.4, 402.2.7

C1371—15: Standard Test Method for Determination of Emittance of Materials Near Room Temperature Using Portable Emissometers
    Table C402.3

C1549—2016: Standard Test Method for Determination of Solar Reflectance Near Ambient Temperature Using a Portable Solar Reflectometer
    Table C402.3

D1003—13: Standard Test Method for Haze and Luminous Transmittance of Transparent Plastics
    C402.4.2.2

D8052/D8052M—2017: Standard Test Method for Quantification of Air Leakage in Low-Sloped Membrane Roof Assemblies
    C402.5.2.1.2

E283—2004(2012): Test Method for Determining the Rate of Air Leakage Through Exterior Windows, Curtain Walls and Doors Under Specified Pressure Differences Across the Specimen
    C402.5.1.4, Table C402.5.4, C402.5.10

E408—13: Test Methods for Total Normal Emittance of Surfaces Using Inspection-meter Techniques
    Table C402.3

REFERENCED STANDARDS

ASTM—continued

E779—10(2018): Standard Test Method for Determining Air Leakage Rate by Fan Pressurization
    C402.5.3, C402.5.4, C406.9

E903—2012: Standard Test Method Solar Absorptance, Reflectance and Transmittance of Materials Using Integrating Spheres (Withdrawn 2005)
    Table C402.3

E1677—11: Specification for Air Barrier (AB) Material or Systems for Low-rise Framed Building Walls
    C402.5.2.1.2

E1827—2011(2017): Standard Test Methods for Determining Airtightness of Building Using an Orifice Blower Door
    C402.5.3, C402.5.4, C406.9

E1918—06(2016): Standard Test Method for Measuring Solar Reflectance of Horizontal or Low-sloped Surfaces in the Field
    Table C402.3

E1980—11: Standard Practice for Calculating Solar Reflectance Index of Horizontal and Low-sloped Opaque Surfaces
    Table C402.3

E2178—13: Standard Test Method for Air Permanence of Building Materials
    C402.5.2.1.1

E2357—2018: Standard Test Method for Determining Air Leakage of Air Barriers Assemblies
    C402.5.2.1.2

F1281—2017: Specification for Cross-linked Polyethylene/Aluminum/Cross-linked Polyethylene (PEX-AL_PEX) Pressure Pipe
    Table C404.5.2.1

F1361—2017: Standard Test Method for Performance of Open Deep Fat Fryers
    Table C406.12(1)

F1484—2018: Standard Test Method for Performance of Steam Cookers
    Table C406.12(2)

F1495—2014a: Standard Specification for Combination Oven Electric or Gas Fired
    Table C406.12(4)

F1496—2013: Standard Test Method for Performance of Convection Ovens
    Table C406.12(4)

F1696—2018: Standard Test Method for Energy Performance of Stationary-Rack, Door-Type Commercial Dishwashing Machines
    Table C406.12(3)

F1920—2015: Standard Test Method for Performance of Rack Conveyor Commercial Dishwashing Machines
    Table C406.12(3)

F2093—2018: Standard Test Method for Performance of Rack Ovens
    Table C406.12(4)

F2144—2017: Standard Test Method for Performance of Large Open Vat Fryers
    Table C406.12(1)

F2861—2017: Standard Test Method for Enhanced Performance of Combination Oven in Various Modes
    Table C406.12(4)

CRRC

Cool Roof Rating Council
2435 North Lombard Street
Portland, OR 97217

ANSI/CRRC S100—2020: Standard Test Methods for Determining Radiative Properties of Materials
    Table C402.3, C402.3.1

CSA

CSA Group
8501 East Pleasant Valley Road
Cleveland, OH 44131-5516

AAMA/WDMA/CSA 101/I.S.2/A440—17: North American Fenestration Standard/Specification for Windows, Doors and Unit Skylights
    Table C402.5.5

CSA B55.1—2015: Test Method for Measuring Efficiency and Pressure Loss of Drain Water Heat Recovery Units
    C404.7

CSA B55.2—2015: Drain Water Heat Recovery Units
    C404.7

REFERENCED STANDARDS

CTI

Cooling Technology Institute
P. O. Box 681807
Houston, TX 77268

ATC-105—2019: Acceptance Test Code for Water Cooling Towers
    Table C403.3.2(7)

ATC-105DS—2018: Acceptance Test Code for Dry Fluid Coolers
    Table C403.3.2(7)

ATC-105S—11: Acceptance Test Code for Closed Circuit Cooling Towers
    Table C403.3.2(7), Table C403.3.2(8)

ATC-106—11: Acceptance Test for Mechanical Draft Evaporative Vapor Condensers
    Table C403.3.2(7), Table C403.3.2(8)

CTI STD-201 RS(17): Performance Rating of Evaporative Heat Rejection Equipment
    Table C403.3.2(7), Table C403.3.2(8)

DASMA

Door & Access Systems Manufacturers Association, International
1300 Sumner Avenue
Cleveland, OH 44115-2851

105—2017: Test Method for Thermal Transmittance and Air Infiltration of Garage Doors and Rolling Doors
    C303.1.3, Table C402.5.5

DOE

US Department of Energy
c/o Superintendent of Documents
1000 Independence Avenue SW
Washington, DC 20585

10 CFR, Part 430—2015: Energy Conservation Program for Consumer Products: Test Procedures and Certification and Enforcement Requirement for Plumbing Products; and Certification and Enforcement Requirements for Residential Appliances; Final Rule
    Table C403.3.2(1), Table C403.3.2(2), Table C403.3.2(5), Table C403.3.2(6), Table C403.3.2(14), Table C404.2

10 CFR, Part 431—2015: Energy Efficiency Program for Certain Commercial and Industrial Equipment: Test Procedures and Efficiency Standards; Final Rules
    Table C403.3.2(6), C403.8.4, C403.11, C403.11.1, Table C403.11.1, C403.11.2, C405.7, Table C405.7, C405.8, Table C405.8(1), Table C405.8(2), Table C405.8(3), Table C405.8(4)

ICC

International Code Council, Inc.
500 New Jersey Avenue NW 6th Floor
Washington, DC 20001

IBC—21: International Building Code®
    C201.3, C303.2, C402.5.6, C501.2

ICC 500—2020: Standard for the Design and Construction of Storm Shelters
    C402.4.2

IFC—21: International Fire Code®
    C201.3, C501.2

IFGC—21: International Fuel Gas Code®
    C201.3, C501.2

IMC—21: International Mechanical Code®
    C403.2.2, C403.6, C403.6.6, C403.7.1, C403.7.2, C403.7.4.2, C403.7.5, C403.7.7, C403.12.1, C403.12.2.1, C403.12.2.2, C406.6, C501.2

IPC—21: International Plumbing Code®
    C201.3, C501.2

IPMC—21: International Property Maintenance Code®
    C501.2

IPSDC—21: International Private Sewage Disposal Code®
    C501.2

REFERENCED STANDARDS

IEEE

Institute of Electrical and Electronic Engineers
3 Park Avenue, 17th Floor
New York, NY 10016

IEEE 515.1—2012: IEEE Standard for the Testing, Design, Installation, and Maintenance of Electrical Resistance Trace Heating for Commercial Applications
    C404.6.2

IES

Illuminating Engineering Society
120 Wall Street, 17th Floor
New York, NY 10005-4001

ANSI/ASHRAE/IES 90.1—2019: Energy Standard for Buildings, Except Low-rise Residential Buildings
    C401.2, Table C402.1.3, Table C402.1.4, C406.2, C502.1, C503.1, C504.1

ISO

International Organization for Standardization
Chemin de Blandonnet 8, CP 401, 1214 Vernier
Geneva, Switzerland

ISO/AHRI/ASHRAE 13256-1(: 2017): Water-to-Air and Brine-to-Air Heat Pumps—Testing and Rating for Performance
    Table C403.3.2(14)

ISO/AHRI/ASHRAE 13256-2(2017): Water-to-Water and Brine-to-Water Heat Pumps—Testing and Rating for Performance
    Table C403.3.2(14)

NEMA

National Electrical Manufacturers Association
1300 North 17th Street, Suite 900
Rosslyn, VA 22209

MG1—2016: Motors and Generators
    C202

NFPA

National Fire Protection Association
1 Batterymarch Park
Quincy, MA 02169-7471

70—20: National Electrical Code
    C501.2

NFRC

National Fenestration Rating Council, Inc.
6305 Ivy Lane, Suite 140
Greenbelt, MD 20770

100—2020: Procedure for Determining Fenestration Products U-factors
    C303.1.3, Table 402.1.4, C402.2.1.5, C402.4.1.1

200—2020: Procedure for Determining Fenestration Product Solar Heat Gain Coefficient and Visible Transmittance at Normal Incidence
    C303.1.3, C402.4.1.1

203—2017: Procedure for Determining Visible Transmittance of Tubular Daylighting Devices
    C303.1.3

400—2020: Procedure for Determining Fenestration Product Air Leakage
    Table C402.5.5

SMACNA

Sheet Metal and Air Conditioning Contractors' National Association, Inc.
4021 Lafayette Center Drive
Chantilly, VA 20151-1219

ANSI/SMACNA 016—2012: HVAC Air Duct Leakage Test Manual Second Edition
    C403.12.2.3

REFERENCED STANDARDS

UL

UL LLC
333 Pfingsten Road
Northbrook, IL 60062-2096

710—12: Exhaust Hoods for Commercial Cooking Equipment—with Revisions through November 2013
    C403.7.5

727—18: Oil-fired Central Furnaces
    Table C403.3.2(4), Table C403.3.2(5)

731—18: Oil-fired Unit Heaters
    Table C403.3.2(5)

1784—15: Air Leakage Tests of Door Assemblies—with Revisions through February 2015
    C402.5.6, C402.5.7

US-FTC

United States-Federal Trade Commission
600 Pennsylvania Avenue NW
Washington, DC 20580

CFR Title 16 (2015): *R*-value Rule
    C303.1.4

WDMA

Window and Door Manufacturers Association
2025 M Street NW, Suite 800
Washington, DC 20036-3309

AAMA/WDMA/CSA 101/I.S.2/A440—17: North American Fenestration Standard/Specification for Windows, Doors and Skylights
    Table C402.5.5

APPENDIX CA
BOARD OF APPEALS—COMMERCIAL

The provisions contained in this appendix are not mandatory unless specifically referenced in the adopting ordinance.

User note:
About this appendix: Appendix CA provides criteria for Board of Appeals members. Also provided are procedures by which the Board of Appeals should conduct its business.

SECTION CA101
GENERAL

CA101.1 Scope. A board of appeals shall be established within the jurisdiction for the purpose of hearing applications for modification of the requirements of this code pursuant to the provisions of Section C110. The board shall be established and operated in accordance with this section, and shall be authorized to hear evidence from appellants and the code official pertaining to the application and intent of this code for the purpose of issuing orders pursuant to these provisions.

CA101.2 Application for appeal. Any person shall have the right to appeal a decision of the code official to the board. An application for appeal shall be based on a claim that the intent of this code or the rules legally adopted hereunder have been incorrectly interpreted, the provisions of this code do not fully apply or an equally good or better form of construction is proposed. The application shall be filed on a form obtained from the code official within 20 days after the notice was served.

CA101.2.1 Limitation of authority. The board shall not have authority to waive requirements of this code or interpret the administration of this code.

CA101.2.2 Stays of enforcement. Appeals of notice and orders, other than Imminent Danger notices, shall stay the enforcement of the notice and order until the appeal is heard by the board.

CA101.3 Membership of board. The board shall consist of five voting members appointed by the chief appointing authority of the jurisdiction. Each member shall serve for [INSERT NUMBER OF YEARS] years or until a successor has been appointed. The board member's terms shall be staggered at intervals, so as to provide continuity. The code official shall be an ex officio member of said board but shall not vote on any matter before the board.

CA101.3.1 Qualifications. The board shall consist of five individuals, who are qualified by experience and training to pass on matters pertaining to building construction and are not employees of the jurisdiction.

CA101.3.2 Alternate members. The chief appointing authority is authorized to appoint two alternate members who shall be called by the board chairperson to hear appeals during the absence or disqualification of a member. Alternate members shall possess the qualifications required for board membership, and shall be appointed for the same term or until a successor has been appointed.

CA101.3.3 Vacancies. Vacancies shall be filled for an unexpired term in the same manner in which original appointments are required to be made.

CA101.3.4 Chairperson. The board shall annually select one of its members to serve as chairperson.

CA101.3.5 Secretary. The chief appointing authority shall designate a qualified clerk to serve as secretary to the board. The secretary shall file a detailed record of all proceedings which shall set forth the reasons for the board's decision, the vote of each member, the absence of a member and any failure of a member to vote.

CA101.3.6 Conflict of interest. A member with any personal, professional or financial interest in a matter before the board shall declare such interest and refrain from participating in discussions, deliberations and voting on such matters.

CA101.3.7 Compensation of members. Compensation of members shall be determined by law.

CA101.3.8 Removal from the board. A member shall be removed from the board prior to the end of their terms only for cause. Any member with continued absence from regular meeting of the board may be removed at the discretion of the chief appointing authority.

CA101.4 Rules and procedures. The board shall establish policies and procedures necessary to carry out its duties consistent with the provisions of this code and applicable state law. The procedures shall not require compliance with strict rules of evidence, but shall mandate that only relevant information be presented.

CA101.5 Notice of meeting. The board shall meet upon notice from the chairperson, within 10 days of the filing of an appeal or at stated periodic intervals.

CA101.5.1 Open hearing. All hearings before the board shall be open to the public. The appellant, the appellant's representative, the code official and any person whose interests are affected shall be given an opportunity to be heard.

CA101.5.2 Quorum. Three members of the board shall constitute a quorum.

CA101.5.3 Postponed hearing. When five members are not present to hear an appeal, either the appellant or the appellant's representative shall have the right to request a postponement of the hearing.

CA101.6 Legal counsel. The jurisdiction shall furnish legal counsel to the board to provide members with general legal advice concerning matters before them for consideration. Members shall be represented by legal counsel at the jurisdiction's expense in all matters arising from service within the scope of their duties.

CA101.7 Board decision. The board shall only modify or reverse the decision of the code official by a concurring vote of three or more members.

CA101.7.1 Resolution. The decision of the board shall be by resolution. Every decision shall be promptly filed in writing in the office of the code official within three days and shall be open to the public for inspection. A certified copy shall be furnished to the appellant or the appellant's representative and to the code official.

CA101.7.2 Administration. The code official shall take immediate action in accordance with the decision of the board.

CA101.8 Court review. Any person, whether or not a previous party of the appeal, shall have the right to apply to the appropriate court for a writ of certiorari to correct errors of law. Application for review shall be made in the manner and time required by law following the filing of the decision in the office of the chief administrative officer.

APPENDIX CB

SOLAR-READY ZONE—COMMERCIAL

The provisions contained in this appendix are not mandatory unless specifically referenced in the adopting ordinance.

User note:

About this appendix: Appendix CB is intended to encourage the installation of renewable energy systems by preparing buildings for the future installation of solar energy equipment, piping and wiring.

SECTION CB101
SCOPE

CB101.1 General. These provisions shall be applicable for new construction where solar-ready provisions are required.

SECTION CB102
GENERAL DEFINITION

SOLAR-READY ZONE. A section or sections of the roof or building overhang designated and reserved for the future installation of a solar photovoltaic or solar thermal system.

SECTION CB103
SOLAR-READY ZONE

CB103.1 General. A solar-ready zone shall be located on the roof of buildings that are five stories or less in height above grade plane, and are oriented between 110 degrees and 270 degrees of true north or have low-slope roofs. Solar-ready zones shall comply with Sections CB103.2 through CB103.9.

Exceptions:

1. A building with a permanently installed, on-site renewable energy system.
2. A building with a solar-ready zone that is shaded for more than 70 percent of daylight hours annually.
3. A building where the licensed design professional certifies that the incident solar radiation available to the building is not suitable for a solar-ready zone.
4. A building where the licensed design professional certifies that the solar zone area required by Section CB103.3 cannot be met because of extensive rooftop equipment, skylights, vegetative roof areas or other obstructions.

CB103.2 Construction document requirements for a solar-ready zone. Construction documents shall indicate the solar-ready zone.

CB103.3 Solar-ready zone area. The total solar-ready zone area shall be not less than 40 percent of the roof area calculated as the horizontally projected gross roof area less the area covered by skylights, occupied roof decks, vegetative roof areas and mandatory *access* or set back areas as required by the *International Fire Code*. The solar-ready zone shall be a single area or smaller, separated sub-zone areas. Each sub-zone shall be not less than 5 feet (1524 mm) in width in the narrowest dimension.

CB103.4 Obstructions. Solar ready zones shall be free from obstructions, including pipes, vents, ducts, HVAC equipment, skylights and roof-mounted equipment.

CB103.5 Roof loads and documentation. A collateral dead load of not less than 5 pounds per square foot (5 psf) (24.41 kg/m^2) shall be included in the gravity and lateral design calculations for the solar-ready zone. The structural design loads for roof dead load and roof live load shall be indicated on the construction documents.

CB103.6 Interconnection pathway. Construction documents shall indicate pathways for routing of conduit or piping from the solar-ready zone to the electrical service panel and electrical energy storage system area or service hot water system.

CB103.7 Electrical energy storage system-ready area. The floor area of the electrical energy storage system-ready area shall be not less than 2 feet (610 mm) in one dimension and 4 feet (1219 mm) in another dimension, and located in accordance with Section 1207 of the *International Fire Code*. The location and layout diagram of the electrical energy storage system-ready area shall be indicated on the construction documents.

CB103.8 Electrical service reserved space. The main electrical service panel shall have a reserved space to allow installation of a dual-pole circuit breaker for future solar electric and a dual-pole circuit breaker for future electrical energy storage system installation. These spaces shall be labeled "For Future Solar Electric and Storage." The reserved spaces shall be positioned at the end of the panel that is opposite from the panel supply conductor connection.

CB103.9 Construction documentation certificate. A permanent certificate, indicating the solar-ready zone and other requirements of this section, shall be posted near the electrical distribution panel, water heater or other conspicuous location by the builder or registered design professional.

APPENDIX CC
ZERO ENERGY COMMERCIAL BUILDING PROVISIONS

The provisions contained in this appendix are not mandatory unless specifically referenced in the adopting ordinance.

User note:
About this chapter: Appendix CC provides a model for applying new renewable energy generation when new buildings add electric load to the grid. This renewable energy will avoid the additional emissions that would otherwise occur from conventional power generation.

SECTION CC101
GENERAL

CC101.1 Purpose. The purpose of this appendix is to supplement the *International Energy Conservation Code* and require renewable energy systems of adequate capacity to achieve net zero carbon.

CC101.2 Scope. This appendix applies to new buildings that are addressed by the *International Energy Conservation Code*.

Exceptions:

1. Detached one- and two-family dwellings and townhouses as well as Group R-2 buildings three stories or less in height above grade plane, manufactured homes (mobile dwellings), and manufactured houses (modular dwellings).
2. Buildings that use neither electricity nor fossil fuel.

SECTION CC102
DEFINITIONS

CC102.1 Definitions. The definitions contained in this section supplement or modify the definitions in the *International Energy Conservation Code*.

ADJUSTED OFF-SITE RENEWABLE ENERGY. The amount of energy production from off-site renewable energy systems that may be used to offset building energy.

BUILDING ENERGY. All energy consumed at the *building site* as measured at the site boundary. Contributions from on-site or off-site renewable energy systems shall not be considered when determining the building energy.

ENERGY UTILIZATION INTENSITY (EUI). The site energy for either the baseline building or the proposed building divided by the gross *conditioned floor area* plus any semiheated floor area of the building. For the baseline building, the EUI can be divided between regulated energy use and unregulated energy use.

OFF-SITE RENEWABLE ENERGY SYSTEM. Renewable energy system not located on the building project.

ON-SITE RENEWABLE ENERGY SYSTEM. Renewable energy systems on the building project.

RENEWABLE ENERGY SYSTEM. Photovoltaic, solar thermal, geothermal energy and wind systems used to generate energy.

SEMIHEATED SPACE. An enclosed space within a building that is heated by a heating system whose output capacity is greater than or equal to 3.4 Btu/h × ft² of floor area but is not a conditioned space.

ZERO ENERGY PERFORMANCE INDEX (ZEPI PB/EE). The ratio of the proposed building EUI without renewables to the baseline building EUI, expressed as a percentage.

SECTION CC103
MINIMUM RENEWABLE ENERGY

CC103.1 Renewable energy. On-site renewable energy systems shall be installed, or off-site renewable energy shall be procured to offset the building energy as calculated in Equation CC-1.

$$RE_{onsite} + RE_{offsite} \geq E_{building} \qquad \text{(Equation CC-1)}$$

where:

RE_{onsite} = Annual site energy production from on-site renewable energy systems (see Section CC103.2).

$RE_{offsite}$ = Adjusted annual site energy production from off-site renewable energy systems that may be credited against building energy use (see Section CC103.3).

$E_{building}$ = Building energy use without consideration of renewable energy systems.

When Section C401.2.1(1) is used for compliance with the *International Energy Conservation Code*, building energy shall be determined by multiplying the gross *conditioned floor area* plus the gross semiheated floor area of the proposed building by an EUI selected from Table CC103.1. Use a weighted average for mixed-use buildings.

When Section C401.2.1, Item 2 or Section C401.2.2 is used for compliance with the *International Energy Conservation Code*, building energy shall be determined from energy simulations.

CC103.2 Calculation of on-site renewable energy. The annual energy production from on-site renewable energy

TABLE CC103.1
ENERGY UTILIZATION INTENSITY FOR BUILDING TYPES AND CLIMATES (kBtu/ft² – yr)

BUILDING AREA TYPE	CLIMATE ZONE															
	0A/1A	0B/1B	2A	2B	3A	3B	3C	4A	4B	4C	5A	5B	5C	6A	6B	7	8
	kBtu/ft² – yr																
Healthcare/hospital (I-2)	119	120	119	113	116	109	106	116	109	106	118	110	105	126	116	131	142
Hotel/motel (R-1)	73	76	73	68	70	67	65	69	66	65	71	68	65	77	72	81	89
Multiple-family (R-2)	43	45	41	41	43	42	36	45	43	41	47	46	41	53	48	53	59
Office (B)	31	32	30	29	29	28	25	28	27	25	29	28	25	33	30	32	36
Restaurant (A-2)	389	426	411	408	444	420	395	483	437	457	531	484	484	589	538	644	750
Retail (M)	46	50	45	46	44	44	37	48	44	44	52	50	46	60	52	64	77
School (E)	42	46	42	40	40	39	36	39	40	40	39	43	37	44	40	45	54
Warehouse (S)	9	12	9	11	12	11	10	17	13	14	23	17	15	32	23	32	32
All others	55	58	54	53	53	51	48	54	52	51	57	54	50	63	57	65	73

systems shall be determined using the PVWatts software or other software approved by the code official.

CC103.3 Off-site renewable energy. Off-site energy shall comply with Sections CC103.3.1 and CC103.3.2.

CC103.3.1 Qualifying off-site procurement methods. The following are considered qualifying off-site renewable energy procurement methods:

1. Community renewables: an off-site renewable energy system for which the owner has purchased or leased renewable energy capacity along with other subscribers.

2. Renewable energy investment fund: an entity that installs renewable energy capacity on behalf of the owner.

3. Virtual power purchase agreement: a power purchase agreement for off-site renewable energy where the owner agrees to purchase renewable energy output at a fixed price schedule.

4. Direct ownership: an off-site renewable energy system owned by the building project owner.

5. Direct access to wholesale market: an agreement between the owner and a renewable energy developer to purchase renewable energy.

6. Green retail tariffs: a program by the retail electricity provider to provide 100-percent renewable energy to the owner.

7. Unbundled Renewable Energy Certificates (RECs): certificates purchased by the owner representing the environmental benefits of renewable energy generation that are sold separately from the electric power.

CC103.3.2 Requirements for all procurement methods. The following requirements shall apply to all *off-site renewable* energy procurement methods:

1. The building owner shall sign a legally binding contract to procure qualifying off-site renewable energy.

2. The procurement contract shall have duration of not less than 15 years and shall be structured to survive a partial or full transfer of ownership of the property.

3. RECs and other environmental attributes associated with the procured *off-site renewable energy* shall be assigned to the building project for the duration of the contract.

4. The renewable energy generating source shall include one or more of the following: photovoltaic systems, solar thermal power plants, geothermal power plants and wind turbines.

5. The generation source shall be located where the energy can be delivered to the building site by the same utility or distribution entity, the same independent system operator (ISO) or regional transmission organization (RTO), or within integrated ISOs (electric coordination council).

6. The *off-site renewable energy* producer shall maintain transparent accounting that clearly assigns production to the building. Records on power sent to or purchased by the building shall be retained by the building owner and made available for inspection by the code official upon request.

CC103.3.3 Adjusted off-site renewable energy. The process for calculating the adjusted *off-site renewable energy* is shown in Equation CC-2.

$$RE_{offsite} = \sum_{i=1}^{n} PF_i \times RE_i = PF_1 \times RE_1 + PF_2 \times RE_2 + \ldots + PF_n \times RE_n$$

(Equation CC-2)

where:

$RE_{offsite}$ = Adjusted off-site renewable energy.

PF_i = Procurement factor for the ith renewable energy procurement method or class taken from Table CC103.3.3.

RE_i = Annual energy production for the ith renewable energy procurement method or class.

n = The number of renewable energy procurement options or classes considered.

TABLE CC103.3.3
DEFAULT OFF-SITE RENEWABLE ENERGY PROCUREMENT METHODS, CLASSES AND COEFFICIENTS

CLASS	PROCUREMENT FACTOR (PF)	PROCUREMENT OPTIONS	ADDITIONAL REQUIREMENTS (see also Section CC103.3.2)
1	0.75	Community solar	—
		REIFs	Entity must be managed to prevent fraud or misuse of funds.
		Virtual PPA	—
		Self-owned off-site	Provisions shall prevent the generation from being sold separately from the building.
2	0.55	Green retail tariffs	The offering shall not include the purchase of unbundled RECs.
		Direct access	The offering shall not include the purchase of unbundled RECs.
3	0.20	Unbundled RECs	The vintage of the RECs shall align with building energy use.

INDEX

A

ABOVE-GRADE WALL C202
ACCESS (TO) C202
ACCESSIBLE
   Controls . . . C402.2.3, C404.6, C404.8.1, C405.2.4.1
   Defined C202
ADDITION C202
ADDITIONAL EFFICIENCY PACKAGE C406
ADDITIONS
   Defined C202
   Historic buildings C501.5
   Requirements C502
ADMINISTRATION
AIR BARRIER C402.5.1
   Access openings C402.5.6
   Assemblies C402.5.2.1.4
   Compliance options C402.5.1.2
   Construction C402.5.1.1
   Dampers C403.7.7
   Doors other than fenestration C402.5.6
   Fenestration . . . . C402.5.4, Table C402.5.4
   Materials C402.5.1.3
   Penetrations C402.5.1.1
   Recessed lighting C402.5.10
   Rooms with fuel burning appliances C402.5.5
   Testing C402.5
   Vestibules C402.5.9
AIR CONDITIONERS C403.3.2, C403.4.4
   Efficiency requirements . . . . Table C403.3.2(1),
              Table C403.3.2(3), Table C403.3.2(4),
             Table C403.3.2(8), Table C403.3.2(10),
                        Table C403.3.2(16)
AIR CURTAIN
   Defined C202
   Vestibules C402.5.9
AIR ECONOMIZERS
   Defined C202
   Requirements C403.5, C403.5.1, C403.5.2,
                      C403.5.3, C403.5.4
AIR INTAKES AND
   EXHAUSTS C402.5.7, C403.7.7
AIR SYSTEM BALANCING C408.2.2.1,
                  C408.2.2.2, C408.2.5.1
ALTERATION C104.4, C202, C401.2.1, C501.1,
             C501.2, C501.5, C503, C503.2,
               C503.3, C503.4, C503.5

ALTERNATIVE MATERIALS C102
APPROVED
   Defined C202
APPROVED AGENCY
   Defined C202
   Inspections C104.4
AREA-WEIGHTED *U*-FACTOR C402.4.3.4
AUTOMATIC
   Defined C202

B

BASEMENT WALLS (see also WALL,
   BELOW GRADE) C401.3
BELOW-GRADE WALL (see WALL,
   BELOW GRADE)
BOARD OF APPEALS C110
   Limitations on authority C110.2
   Qualifications of members C110.3
BOILER, MODULATING C202
BOILER SYSTEM C202
BOILERS
   Defined C202
   Requirements . . . . Table C403.3.2(6), C403.3.4,
               C403.4.3, C404.2, Table C404.2
   Setback controls C403.4.1.5
   Turndown controls C403.3.4
BUBBLE POINT C202
BUILDING
   Defined C202
BUILDING COMMISSIONING
   Defined C202
   Requirements C408
BUILDING ENTRANCE
   Defined C202
   Exterior lighting . . . . Table C405.5.2(2)
   Vestibules C402.5.9
BUILDING ENVELOPE
   Compliance documentation C103.2, C103.2.1
   Defined C202
   Exemptions C402.1.1, C402.1.2
   Insulation C303.1.1
   Insulation and fenestration
      criteria . . . . C402.1.3, Table C402.1.3,
                   C402.1.4, Table C402.1.4
   Performance method C407.2
   Requirements C402

BUILDING SITE C202
BUILDING THERMAL ENVELOPE
   Air leakage and barriers C402.5
   Defined C202
   Doors C402.4.5
   Low-energy buildings C402.1.1
   Performance C402.1, C402.1.3,
                             C402.1.4, C402.1.5
   Rooms with fuel-burning appliances C402.5.5
   Specific insulation C402.2

C

**CAULKING AND
  WEATHERSTRIPPING** C402.5.1.1, C402.5.6,
                          C402.5.8, C402.5.10
CAVITY INSULATION C202
C-FACTOR
   Assembly U-, C- or F-factor
     method .... Table C402.1.4, C402.1.4
   Defined C202
CHANGE OF OCCUPANCY C501.2, C505
CHILLERS .... Table C403.3.2(3), Table C407.4.1(3)
   Positive displacement chilling
     packages C403.3.2.2
   Water-cooled centrifugal chiller
     packages .... Table C403.3.2(3),
                         Table C407.4.1(5)
CIRCULATING HOT WATER SYSTEM
   Defined C202
   Requirements C404.6.1, C404.6.3
CIRCULATING PUMPS C403.4.3.3
CIRCULATING SYSTEMS C403.4.3.3, C404.6
CLIMATE TYPES
   Defined .... Table C301.3
CLIMATE ZONES .... Figure C301.1, Table C301.1,
                    C301.3, Table C301.3
   International climate zones .... Table C301.3
CODE OFFICIAL
   Approval of alternate methods C102.1
   Defined C202
   Examination of construction documents C103.3
   Inspections C105
COEFFICIENT OF PERFORMANCE (COP)
   Requirements .... Table C403.3.2(2),
                        Table C403.3.2(3)
**COEFFICENT OF PERFORMANCE
  (COP) – COOLING** C202
**COEFFICIENT OF PERFORMANCE
  (COP) – HEATING** C202

COMMERCIAL BUILDINGS
   Compliance .... Chapter 5, C101.2, C101.4.1,
                 C101.5, C401.1, C401.2
   Defined C202
   Total building performance C407
COMMISSIONING C408
COMPLIANCE AND ENFORCEMENT
   Existing buildings C501.2
   General C101.5
**COMPONENT PERFORMANCE
  APPROACH** C402.1.5
**COMPRESSOR (REFRIGERATION)
  SYSTEMS** C403.11.3.2
COMPUTER ROOM
   Air conditioning .... Table C403.3.2(10)
   Defined C202
CONDENSING UNITS
   Defined C202
   Efficiency requirements .... Table C403.3.2(1),
                       Table C403.3.2(10)
CONDITIONED FLOOR AREA
   Defined C202
   Renewable energy C406.5
CONDITIONED SPACE
   Change from nonconditioned
     or low energy C502.2
   Defined C202
   Roof solar reflectance C402.3
   Rooms containing fuel-burning
     appliances C402.5.5
CONSTRUCTION DOCUMENTS C103
   Amended C103.4
   Approval C103.3.1
   Examination C103.3
   Information required C103.2
   Phased approvals C103.3.3
   Previous approvals C103.3.2
   Retention C103.5
   Revocation C106.2
CONTINUOUS AIR BARRIER
   Defined C202
   Required C402.5.1
CONTINUOUS INSULATION (ci)
   Defined C202
   Requirements C303.2.2, C402.1.3,
           Table C402.1.3, C402.2.2, C402.2.3
CONTROLS
   Capabilities C404.6.3, C403.3.1, C403.4.1,
          C403.4.1.2, C403.4.2.1, C403.4.2.2,
        C403.4.3.3.1, C403.4.5, C403.7.1, C404.6

INDEX

Chilled water plants C403.4.5
Economizers C403.5, C403.5.1, C403.5.3.2,
    C403.5.3.3, Table C403.5.3.3
Energy recovery systems............... C403.7.4
Fan speed C403.8, C403.8.6,
    C403.8.6.1, C403.10
Freeze protection system.............. C403.13.3
Glazing................................ C402.4.3.3
Heat pump C403.4.1.1, C403.4.3.3.1
Heating and cooling ... C403.3.1, C403.4, C403.5.1
Hot water system........................ C404.6
Humidity........ C403.4.1, C403.5.1, C403.7.4.2
HVAC................... C403.4, C408.2.3.2
Hydronic systems...................... C403.4.3
Lighting......... C402.4, C402.4.1.1, C402.4.1.2,
    C402.4.2.1, C402.4.3.1, C405.2, C405.5.1
Lighting, digital........................ C406.4
Off hour............................... C403.4.2
Service water heating.................. C404.6
Shutoff dampers...................... C403.7.7
Snow melt system C403.13.2
Temperature...... C403.4.1, C403.4.2, C403.4.2.1,
    C403.4.2.2, C403.4.2.3, C403.7.7
Three-pipe system C403.4.3.1
Two-pipe changeover system C403.4.3.2
Variable air volume systems C403.5.2, C403.6
Ventilation C403.2.2
COOLING SYSTEMS
Hot gas bypass limitation C403.3.3
COOLING TOWER C403.10.3, C403.10.4
COOLING WITH OUTDOOR AIR C403.5.1
CRAWL SPACE WALLS
Defined............................... C202
Requirements....................... C303.2.1
CURTAIN WALL
Air leakage of fenestration .... Table C402.5.4
Defined................................ C202

D

DAMPERS C402.5.7, C403.7.7
DAYLIGHT RESPONSIVE CONTROL
Defined............................... C202
Required...... C402.4.1.1, C402.4.1.2, C402.4.2.1,
    C402.4.3.1, C402.4.3.2, C405.2.4.1
DAYLIGHT ZONE ... C402.4.1.1, C402.4.2, C405.2.4
Defined............................... C202
DAYLIGHT ZONE CONTROL C405.2.4
DEADBAND C403.4.1.2, C403.4.1.3,
    C403.4.3.3.1
DEGREE DAY COOLING (CDD) ...... Table C301.3

DEGREE DAY HEATING (HDD) .... Table C301.3
DEMAND CONTROL VENTILATION (DCV)
Defined............................... C202
Requirements....................... C403.7.1
DEMAND RECIRCULATION WATER SYSTEM
Defined............................... C202
Requirements..................... C404.6.1.1
DESIGN CONDITIONS C302
DIRECT EXPANSION (DX)............. C403.8.6.1
DOORS
Default U-factors .... Table C303.1.3(2)
Garage doors C303.1.3
Loading docks C402.5.8
Opaque C303.1.3
Performance requirements............. C402.5.4,
    Table C402.1.3, Table C402.1.4,
    C402.4, C402.4.5
Vestibules........................... C402.5.9
DRAIN WATER HEAT RECOVERY C404.7
DUAL DUCT VAV C403.6.3, C403.6.4
DUCT SYSTEM
Defined............................... C202
Requirements....................... C403.12.2
DUCTS
Defined............................... C202
Insulation C103.2, C403.12.1, C403.12.2,
    C403.12.2.1, C403.12.2.2, C403.12.2.3
Sealing C103.2, C403.12.1, C403.12.2.1,
    C403.12.2.2, C403.12.2.3
DWELLING UNIT
Defined............................... C202
Electrical Meter...................... C405.6
Lighting............................. C405.1
Testing.............................. C402.5.2
Vestibules........................... C402.5.9
DYNAMIC GLAZING
Defined............................... C202
Requirements...................... C402.4.3.3

E

ECONOMIZER
Air.................. Table C403.5(1), C403.5.3
Controls C403.5.1, C403.6.8, C403.6.9
Defined............................... C202
Fault detection and diagnostics (FDD).... C403.2.3
High-limit shutoff control.............. C403.5.3.3,
    Table C403.5.3.3
Requirements............. C403.5.3, C403.5.4
Water C403.5.4
ECONOMIZER, AIR C202

INDEX

ECONOMIZER, WATER C202
EFFICIENCY, ADDITIONAL C406
ELECTRICAL METERS. C405.6
ELECTRICAL MOTORS C405.8
ELECTRICAL POWER AND LIGHTING C405
ELECTRICAL TRANSFORMERS C405.7
ELEVATOR POWER C405.9.1, C405.9.2
ELEVATOR SHAFTS. C402.5.6, C402.5.7
ENCLOSED SPACE
   Defined C202
   Under skylights C402.4.2
ENERGY ANALYSIS C202, C407
ENERGY COST
   Compliance performance C401.2
   Defined C202
   Performance basis C407.2
ENERGY EFFICIENCY
   RATIO (EER) . . . . C403.3.2, Table C403.3.2(1)
ENERGY MONITORING C405.12, C406.1,
                Table C406.1(1), Table C406.1(2),
                Table C406.1(3), Table C406.1(4),
                      Table C406.1(5), C406.10
ENERGY RECOVERY VENTILATION SYSTEM
   Defined C202
   Requirements C403.1.1, C403.7.4.2
ENERGY SIMULATION TOOL
   Defined C202
   Requirements/use C101.5.1, C407,
                             C407.2, C407.5
ENTRANCE DOOR
   Air leakage . . . . Table C402.5.4
   Defined C202
   Thermal performance . . . . Table 402.4
ENVELOPE, BUILDING THERMAL
   Defined C202
ENVELOPE DESIGN METHODS C402.1.3,
                          C402.1.4, C402.1.5
EQUIPMENT BUILDINGS C402.1.2
EQUIPMENT EFFICIENCIES C103.2, C403.3.2,
                   C403.5.1, C403.8.6, C404.2
EQUIPMENT PERFORMANCE
   REQUIREMENTS C403.3.2
   Boilers . . . . Table C403.3.2(6)
   Condensing units. . . . . Table C403.3.2(10)
   Economizer exception . . . . Table C403.5(2)
   Heat rejection equipment . . . . . .Table C403.3.2(7)
   Packaged terminal air conditioners
      and heat pump. . . . . .Table C403.3.2(4)
   Unitary air conditioners and
      condensing units . . . . .Table C403.3.2(1)
   Unitary and applied
      heat pumps . . . . .Table C403.3.2(2)
   Warm air duct furnaces and
      unit heaters . . . . .Table C403.3.2(5)
   Warm air furnaces . . . . .Table C403.3.2(5)
   Warm air furnaces/air-conditioning
      units . . . . .Table C403.3.2(5)
   Water chilling packages,
      standard. . . . . .Table C403.3.2(3)
   Water heating . . . . C404.2.1, Table C404.2
EQUIPMENT ROOM . . . . . . C202, Table C405.3.2(2)
ESCALATORS. C405.9.2
EXEMPT BUILDINGS C402.1.1, C402.1.2
EXHAUSTS C402.5.7
EXISTING BUILDINGS . . . . Chapter 5
EXIT SIGNS C405.3.1
EXTERIOR LIGHTING C405.2.7, C405.5
EXTERIOR WALL C202
   Thermal performance C402, C402.2.2
EXTERNAL SHADING. . . . . .Table C407.4.1(1)

F

FAN BRAKE HORSEPOWER (BHP)
   Defined C202
FAN ENERGY INDEX (FEI) C202, C403.8.3
FAN FLOOR HORSEPOWER C403.8.1
FAN POWER LIMITATION . . . . . . Table C403.8.1(1),
                           Table C403.8.1(2)
FAN SYSTEM BHP
   Allowable. C403.8.1
   Defined C202
FAN SYSTEM DESIGN CONDITIONS
   Allowable. C403.8.1
   Defined C202
FAN SYSTEM MOTOR NAMEPLATE HP
   Defined C202
FAULT DETECTION & DIAGNOSTICS (FDD)
   Economizers C403.5.5
FEES C104
   Refunds. C104.5
   Related fees C104.4
   Schedule of permit fees. C104.2
FENESTRATION
   (see also DOORS) C303.1.3, C402.4
   Air leakage (infiltration) rate. C402.5.4,
                              Table C402.5.4
   Defined C202
   Maximum area. C402.4.1, C402.4.1.2
   Rating and labeling C303.1.3, C402.1.3
   Skylights C402.4.1.2, C402.4.2, C402.4.2.1,
                       C402.4.2.2, C402.4.3,
                       C502.3.2, C503.2.3
   Solar heat gain (SHGC) . . . Table C402.4, C402.4.3

INDEX

Vertical . . . . C402.1, Table C402.4, C402.4.1.1,
C402.4.3, C502.3.1, C503.2.2
**FENESTRATION PRODUCT,
FIELD-FABRICATED** C202
Air leakage C402.5.4
FENESTRATION PRODUCT, SITE-BUILT
Defined C202
***F*-FACTOR**
Assembly *U*-, *C*- or *F*-factor
method . . . . C402.1.4, Table C402.1.4
Defined C202
FLOOR AREA, NET
Defined C202
Fenestration increase C402.4.1.1
FLOORS
Slab on grade C402.2.4
Thermal properties C402.2.3
FREEZE PROTECTION SYSTEMS C403.13.3
FURNACE EFFICIENCY . . . . Table C403.3.2(5)

G

GENERAL LIGHTING
Additional lighting C405.3.2.2.1
Daylight controls C405.2.4
Defined C202
Interior lighting power C405.3.1
GENERAL PURPOSE ELECTRIC MOTORS
Defined C202
GLAZING AREA
Default fenestration *U*-factors . . . Table C303.1.3(1)
Dynamic C402.4.3.3
GREENHOUSE
Building envelope C402.1.1
Defined C202
GROUP R C202
GUESTROOM C403.7.6

H

HAZE FACTOR C402.4.2.2
HEAT PUMP C403.4.1.1, C403.4.3.3
HEAT RECOVERY
Drain water C404.7, C406.7.2
Economizer exemption C403.5
Kitchen exhaust C403.7.5
Service water C403.10.5
**HEAT REJECTION
EQUIPMENT** . . . . Table C403.3.2(7), C403.10
HEAT TRACE SYSTEMS C404.6.2
HEAT TRAP
Defined C202
Required C404.3, C404.4
HEATED SLAB
Defined C202
Insulation . . . . Table C402.1.3, Table C402.1.4,
C402.2.6
**HEATING AND COOLING
LOADS** C302.1, C403.1.1, C403.3.1,
C403.3.2, C403.4.1.1, C403.5
HEATING OUTSIDE A BUILDING C403.13.1
HIGH-SPEED DOOR
Air leakage . . . . Table C402.5.4
Defined C202
HISTORIC BUILDING
Compliance C501.5
Defined C202
HOT GAS BYPASS . . . . . . C403.3.3, Table C403.3.3
HOT WATER C404.2, C404.6
Efficient delivery C404.5
Piping insulation C403.12.3, C404.4
System controls C403.10.5, C404.6
HUMIDISTAT
Defined C202
Requirements C403.4.1, C403.7.4.2
HVAC EQUIPMENT
Automatic setback and shutdown C403.4.2.2
Automatic start and stop capabilities C403.4.2.3
Increased efficiency C406.2
Performance requirements C403.3.2
Supply-air temperature reset C403.6.5
System map zones . . . . Table C407.4.1(2),
C407.4.2.1
HVAC SYSTEMS C402.3, C403, C403.7.6,
C403.8, Table C407.4.1(1),
Table C407.4.1(2), C408.2.2
Manuals C408.3.2.2
Plan C408.2.1
Report C408.2.4, C408.2.5
HYDRONIC HEAT PUMP SYSTEMS C403.4.3.3

I

ICE MELT SYSTEMS C403.13.2
**IDENTIFICATION (MATERIALS,
EQUIPMENT AND SYSTEM)** C303.1
IEC DESIGN H MOTOR C202
IEC DESIGN N MOTOR C202
**INDIRECTLY CONDITIONED SPACE
(see CONDITIONED SPACE)**
INFILTRATION (see AIR BARRIER)
Defined C202
INSPECTIONS C105
Inspection agencies C105.4

INSULATED SIDING C303.1.4.1
INSULATION
   Continuous insulation C303.2.2, C402.1.3,
                          Table C402.1.3, C402.2.1
   Duct. C403.12.1
   Identification C303.1, C303.1.2
   Installation C303.1.1, C303.1.1.1,
                            C303.1.2, C303.2
   Mechanical system piping C403.12.3
   Piping C404.4
   Plenum C403.12.1
   Product rating C303.1.4
   Protection of exposed foundation C303.2.1
   Protection of piping insulation C403.12.3.1
   Radiant heating systems C402.2.6
   Requirements .... C402.1.3, Table C402.1.3,
                              C402.2, C402.2.6
INTEGRATED PART LOAD VALUE (IPLV)
   Defined C202
   Requirements .... Table C403.3.2(3),
                        Table C403.3.2(15), C403.3.2.1
INTERIOR LIGHTING POWER C405.3, C405.3.2
ISOLATION DEVICES C202

K

KITCHEN EXHAUST ...... C403.7.5, Table C403.7.5

L

LABELED
   Dampers C403.7.7
   Defined C202
   Fans C403.8.3
   Glazing, skylights C402.2.1.5, C402.4.2.2,
                                    C402.5.5
   HVAC equipment. .... Table C403.3.2(3)
   Lighting C402.5.11, C405.3.1
   Requirements .... C303.1.3, Table C403.3.2(3)
LIGHTING POWER
   Additional lighting C405.3.2.2.1, C405.5.1
   Design procedures C405.3.2, C405.3.2.1,
                                  C405.3.2.2
   Exterior connected C405.5, C405.5.1,
                            Table C405.5.2(1),
                          Table C405.5.2(2)
   Interior connected C405.3, C405.3.1,
                      C405.3.2, Table C405.5.2(1),
                      Table C405.5.2(2), C406.3
   Reduced lighting power density C406.3

LIGHTING SYSTEMS C405, C406.3
   Controls, exterior C405.2.6, C405.5.1,
                          Table C405.5.2(1),
                          Table C405.5.2(2)
   Controls, interior C405.2.2, C405.3, C405.3.1,
                       C405.3.2, Table C405.3.2(1),
                       Table C405.3.2(2), C406.4
   Daylight responsive C405.2.4, C405.2.4.1
   Dwelling and sleeping units C405.1, C405.2.2,
                       C405.2.4, C405.2.5, C405.3.1
   Existing buildings. C502.1, C502.3.6, C504.1
                             C503.1, C503.5
   Light reduction. C405.2.3.1
   Occupant sensor controls C405.2.1, C405.2.1.1,
                               C405.2.1.2
   Recessed C402.5.10
   Retail display C405.3.2.2.1
   Specific applications C405.2.5
   Time switch controls C405.2.2, C405.2.2.1
LINER SYSTEM (Ls)
   Defined C202
   Insulation .... Table C402.1.3
LISTED
   Defined C202
   Kitchen exhaust hoods C403.7.5
   Skylights C402.2.1.5
LOADING DOCK WEATHERSEALS C402.5.8
LOW-ENERGY BUILDINGS C402.1.1
LOW-SLOPED ROOF
   Defined C202
   Roof solar reflectance C402.3
LOW-VOLTAGE DRY-TYPE
DISTRIBUTION TRANSFORMER C202
LUMINAIRE
   Controls C405.2, C405.2.1, C405.2.2,
                          C405.2.4, C405.2.5
   Sealed C402.5.10
   Wattage C405.3.1, C405.5.1
LUMINAIRE-LEVEL LIGHTING CONTROLS ... C202

M

MAINTENANCE
   General C501.1.1, C501.3
   Owner responsibility C501.3
MANUAL C202
MANUALS C101.5.1
MASS
   Floor .... Table C402.1.3, Table C402.1.4
   Wall C402.2.2

INDEX

MATERIALS AND EQUIPMENT C303
MECHANICAL SYSTEMS AND EQUIPMENT C403
   Existing buildings C501.4, C502.1, C502.3.3, C503.3, C504
MECHANICAL VENTILATION C403.1, C403.2.2
METERS, ELECTRICAL C405.6
MOTOR NAMEPLATE HORSEPOWER ... C403.8.2
   Efficiency C405.8
MOVING WALKWAYS C405.9.2
MULTIFAMILY RESIDENTIAL BUILDINGS C407.4.2.3
MULTIPLE-ZONE SYSTEMS C403.6, C403.6.5

N

NAMEPLATE HORSEPOWER
   Defined C202
NEMA DESIGN A MOTOR C202
NEMA DESIGN B MOTOR C202
NEMA DESIGN C MOTOR C202
NET FLOOR AREA (see FLOOR AREA, NET)
NETWORKED GUESTROOM CONTROL SYSTEM C202
NONCONDITIONED SPACE
   Alterations C502.2
NONSTANDARD PART LOAD VALUE (NPLV) C202

O

OCCUPANCY
   Complex HVAC systems C403.6
   Compliance C101.4, C101.5
   Lighting power allowances ..... C405.2.2, C405.3.2, Table C405.3.2(1), Table C405.3.2(2), C405.3.2.1
   Mixed occupancies C101.4.1
OCCUPANT SENSOR CONTROL
   Commissioning C408.3.1.1
   Defined C202
   Lighting C406.4
   Outdoor heating C403.13.1
   Required C405.2.1, C405.2.1.1, C405.2.1.2
OFF-HOUR, CONTROLS C403.4.2, C403.7.7, C405.2.2.1
ON-SITE RENEWABLE ENERGY
   Defined C202
   Efficiency package C406.5
OPAQUE AREAS .... C402.1, Table C402.1.3, Table C402.1.4
OPAQUE DOOR C202

OPAQUE DOORS
   Defined C202
   Regulated ..... Table C402.1.3, C402.1.4, C402.4.5
OPERATING AND MAINTENANCE MANUAL C408.3.2.2
ORIENTATION
   Daylight Responsive controls C405.2.4.1
   Fenestration .... Table C402.4
   Thermostatic controls C403.4.1, C403.4.2

P

PACKAGED TERMINAL AIR CONDITIONER (PTAC)
   Requirements .... Table C403.3.2(4)
PACKAGED TERMINAL HEAT PUMP
   Requirements .... Table C403.3.2(4)
PARKING GARAGE LIGHTING CONTROL C405.2.8
PARKING GARAGE VENTILATION C403.7.2
PERFORMANCE ANALYSIS C407
PERMIT (see FEES) C104.3
PIPE INSULATION C403.12.3, C404.4
PLANS AND SPECIFICATIONS C103
PLENUMS
   Insulation and sealing C403.12.1
POOLS C404.8
   Controls C404.8.1, C404.8.2
   Covers C404.8.3
   Existing buildings C502.3.5
POWER DENSITY C406.3
POWERED ROOF/WALL VENTILATORS
   Defined C202
   Fan efficiency exemption C403.8.3
PROPOSED DESIGN
   Defined C202
   Requirements C407
PUBLIC LAVATORY C404.5.1
PUMPING SYSTEMS C403.4.3.3, C404.6, C408.2.2.2

R

RADIANT HEATING SYSTEM
   Defined C202
   Insulation C402.2.6
READILY ACCESSIBLE
   Defined C202
READY ACCESS (TO) C202
RECOOLING C403.6
REFERENCED STANDARDS ...... Chapter 6, C109
REFRIGERANT DEW POINT C202

REFRIGERATED WAREHOUSE COOLER
  Defined C202
  Requirements C403.11.1
REFRIGERATED WAREHOUSE FREEZER
  Defined C202
  Requirements C403.11.1
REFRIGERATION EQUIPMENT
  Performance . . . . C403.11, Table C403.11.1
**REFRIGERATION SYSTEM,
LOW TEMPERATURE** C202
**REFRIGERATION SYSTEM,
MEDIUM TEMPERATURE** C202
REGISTERED DESIGN PROFESSIONAL
  Commissioning C408
  Defined C202
REHEATING C403.6.5, C403.10.5
RENEWABLE ENERGY RESOURCES C202
**RENEWABLE/NONDEPLETABLE
ENERGY SOURCES** C406.5
REPAIR
  Defined C202
  Historic buildings C501.5
  Requirements C501.4, C504
REPLACEMENT MATERIALS C501.4
  Replacement fenestration C503.2.2.1
REROOFING
  Defined C202
RESET CONTROL C403.6.5
RESIDENTIAL BUILDING
  Compliance C101.2, C101.4.1, C101.5
  Defined C202
ROOF ASSEMBLY
  Air barriers C402.5.1.4
  Defined C202
  Fenestration C402.4.1, C402.4.1.2,
                              C402.4.2, C405.2.4.3
  Recover C503.1
  Reflectance and emittance options . . . Table C402.3
  Repairs C504.1
  Replacement C503.2.1
  Requirements C303.1.1.1, C402.2.1
  Solar reflectance and thermal emittance C402.3
ROOF RECOVER
  Defined C202
  Exemption C503.1
ROOF REPAIR
  Defined C202
  Exemption C504.1
ROOF REPLACEMENT
  Defined C202
  Requirements C503.2.1

ROOF VENTILATORS (see POWERED
ROOF/WALL VENTILATORS)
ROOFTOP MONITOR
  Daylight zones C405.2.4.2
  Defined C202
  Skylights required C402.4.2
**ROOMS WITH FUEL-BURNING
APPLIANCES** C402.5.6
R-VALUE (THERMAL RESISTANCE) C202

S

SATURATED CONDENSING TEMPERATURE
  Defined C202
  Refrigeration systems C403.11.3.1
SCOPE OF CODE C101.2
**SEASONAL ENERGY EFFICIENCY
RATIO (SEER)** . . . . Table C403.3.2(1),
                   Table C403.3.2(2), Table C403.3.2(8),
                               Table C403.3.2(9)
SERVICE WATER HEATING
  Defined C202
  Drain water heat recovery C404.7
  Efficiency C404.2.1, C404.5
  Existing buildings C502.3.4, C503.4, C504.1
  Reduced use C406.7
  Requirements C403.10.5, C404, C404.2,
                    C404.2.1, C404.5, C404.6, C404.7
SETBACK THERMOSTAT C403.4.2, C403.4.2.1,
                                    C403.4.2.2
SHADING C402.3, C402.4.3
SHGC (see SOLAR HEAT GAIN COEFFICIENT)
SHUTOFF DAMPERS C402.5.7, C403.7.7
**SIMULATED PERFORMANCE
ALTERNATIVE** C407
SIMULATION TOOL (see ENERGY SIMULATION
TOOL)
SINGLE ZONE C403.5
SIZING
  Equipment and system C403.3.1
SKYLIGHTS . . . . C402.1.5, C402.3, Table C402.4,
                        C402.4.3.1, C402.4.3.2
  Additions C502.3.2
  Air leakage (infiltration) . . . . Table C402.5.4
  Alterations C503.2.3
  Curb insulation C402.2.1.5
  Defined (see Fenestration) C202
  Haze factor C402.4.2.2
  Lighting controls C402.4.2.1
  Maximum area C402.4.1, C402.4.1.2
  Minimum area C402.4.2
SLAB-EDGE INSULATION C303.2.1, C402.2.4

INDEX

SLEEPING UNIT
　Defined................................C202
　Lighting.....C405.1, C405.2.2, C405.2.4, C405.2.5

SMALL ELECTRIC MOTOR
　Defined................................C202
　Minimum efficiency......................C405.8

SNOW MELT AND ICE SYSTEM CONTROLS............C403.13.2

SOLAR HEAT GAIN COEFFICIENT (SHGC)............C103.2, Table C303.1.3(3), Table C402.4, C402.4.1.1, C402.4.3, C402.4.3.1
　Defined................................C202
　Dynamic glazing........................C402.4.3.3
　Replacement products...................C503.2.2.1

SPAS...........................C404.8, C404.9

STAIRWAYS.........C402.5.7, C403.7.7, C405.2, C405.2.1.1, Table C405.3.2(2)

STANDARD REFERENCE DESIGN
　Defined................................C202
　Requirements..........C407, Table C407.4.1(1), Table C407.4.1(3)

STANDARDS, REFERENCED......Chapter 6, C109

STEEL FRAMING......Table C402.1.3, C402.1.4.2, Table C402.1.4.2

STOP WORK ORDER.....................C109
　Authority..............................C109.1
　Emergencies............................C109.3
　Failure to comply......................C109.4
　Issuance...............................C109.2

STOREFRONT
　Defined................................C202
　Glazing..........................Table C402.5.4

SUPPLY AIR TEMPERATURE RESET CONTROLS.................C403.6.5

SUSPENDED CEILINGS............C402.1.4.1.2, C402.2.1.3

SWIMMING POOLS......................C404.8

T

TEMPERATURE DEADBAND.........C403.4.3.3.1
TENANT SPACES.....................C406.1.1
THERMAL CONDUCTANCE (see *C*-FACTOR)
THERMAL MASS (see MASS)
THERMAL RESISTANCE (see *R*-VALUE)
THERMAL TRANSMITTANCE (see *U*-FACTOR)
THERMOSTAT
　Defined................................C202
　Pools and spa heaters..................C404.8.1
　Requirements.......C403.4, C403.4.1, C403.4.1.2, C403.4.1.3, C403.4.2, C403.6
　Setback capabilities...................C403.4.2

TIME SWITCH CONTROL
　Defined................................C202
　Requirements.............C405.2.2, C405.2.2.1

TOTAL BUILDING PERFORMANCE.........C407
TOWNHOUSE (see RESIDENTIAL BUILDINGS)
TRANSFORMERS, ELECTRIC............C405.7

U

***U*-FACTOR (THERMAL TRANSMITTANCE)**...C202

V

VARIABLE AIR VOLUME SYSTEMS (VAV).............C403.5.2, C403.6.1, C403.6.2, C403.6.3, C403.6.4, C403.6.6, C403.6.7, C403.6.9, Table C407.4.1(3)

VARIABLE REFRIGERANT FLOW SYSTEM
　Defined................................C202
　Increased efficiencies.................C406.2

VENTILATION........................C403.2.2
　Defined................................C202
　Demand control ventilation (DCV).......C403.7.1
　Energy recovery system........C403.7.4.2, C406.6
　Parking garages........................C403.7.2

VENTILATION AIR
　Defined................................C202
　Energy recovery.......................C403.7.4.2
　Fan controls..........................C403.8.6.1
　Kitchen exhaust........................C403.7.5

VERTICAL FENESTRATION (see FENESTRATION)
VESTIBULES..............C402.5.9, C403.4.1.4
VISIBLE TRANSMITTANCE (VT)
　Default glazed fenestration......Table C303.1.3(3)
　Defined................................C202
　Dynamic glazing.......................C402.4.3.3
　Increased fenestration................C402.4.1.1
　Skylights.................C402.4.2, C405.2.4.3

VOLTAGE DROP........................C202

W

WALK-IN COOLER
　Defined................................C202
　Requirements.....C403.11, Table C403.11.2.1(1), Table C403.11.2.1(2), Table C403.11.2.1(3)

WALK-IN FREEZER
　Defined................................C202
　Requirements..........................C403.11

WALL
- Above-grade wall, defined C202
- Below-grade wall, defined C202
- Crawl space wall, defined C202
- Exterior wall, defined C202
- Steel framed .... C402.1.4.2, Table C402.1.4.2
- Thermal resistance .... Table C402.1.3, Table C402.1.4, C402.2.2

WALL, ABOVE-GRADE C202

WALL, BELOW-GRADE C202

WALL VENTILATORS (see POWERED ROOF/WALL VENTILATORS)

WALLS (see EXTERIOR WALLS and ENVELOPE, BUILDING THERMAL)

WALLS ADJACENT TO UNCONDITIONED SPACE (see BUILDING THERMAL ENVELOPE)

WATER ECONOMIZER C403.5, C403.5.4, C403.5.4.1, C403.5.4.2

WATER HEATER
- Defined C202
- Efficiency .... Table C404.2, C404.2.1

WATER HEATING .... C404, Table C404.2

WINDOW AREA (see FENESTRATION and GLAZING AREA)

Z

ZONE (see also CLIMATE ZONES)
- Defined C202
- Requirements C403.4, C403.5, C407.4.2.1, C407.4.2.2

IECC—RESIDENTIAL PROVISIONS

TABLE OF CONTENTS

CHAPTER 1 SCOPE AND ADMINISTRATION R1-1

PART 1—SCOPE AND APPLICATION R1-1
Section
R101 Scope and General Requirements............R1-1
R102 Alternative Materials, Design and Methods of Construction and Equipment....R1-1

PART 2—ADMINISTRATION AND ENFORCEMENT R1-1
R103 Construction Documents................R1-1
R104 Fees....................R1-2
R105 Inspections R1-3
R106 Notice of Approval R1-3
R107 Validity R1-3
R108 Referenced Standards R1-3
R109 Stop Work Order....................R1-4
R110 Means of Appeals R1-4

CHAPTER 2 DEFINITIONS R2-1
Section
R201 General R2-1
R202 General Definitions..................R2-1

CHAPTER 3 GENERAL REQUIREMENTS R3-1
Section
R301 Climate Zones.....................R3-1
R302 Design Conditions....................R3-36
R303 Materials, Systems and Equipment.........R3-36

CHAPTER 4 RESIDENTIAL ENERGY EFFICIENCY.................R4-1
Section
R401 General R4-1
R402 Building Thermal Envelope R4-2
R403 Systems....................R4-9
R404 Electrical Power and Lighting Systems R4-13
R405 Total Building Performance R4-13

R406 Energy Rating Index Compliance Alternative R4-19
R407 Tropical Climate Region Compliance Path ... R4-21
R408 Additional Efficiency Package Options...... R4-21

CHAPTER 5 EXISTING BUILDINGS R5-1
Section
R501 General......................R5-1
R502 Additions R5-1
R503 Alterations R5-2
R504 Repairs.....................R5-2
R505 Change of Occupancy or Use..............R5-3

CHAPTER 6 REFERENCED STANDARDS R6-1

APPENDIX RA BOARD OF APPEALS—RESIDENTIAL ....APPENDIX RA-1
Section
RA101 General................... APPENDIX RA-1

APPENDIX RB SOLAR-READY PROVISIONS—DETACHED ONE- AND TWO-FAMILY DWELLINGS AND TOWNHOUSES....APPENDIX RB-1
Section
RB101 Scope.................... APPENDIX RB-1
RB102 General Definition........... APPENDIX RB-1
RB103 Solar-ready Zone............ APPENDIX RB-1

APPENDIX RC ZERO ENERGY RESIDENTIAL BUILDING PROVISIONS......APPENDIX RC-1
Section
RC101 Compliance .... APPENDIX RC-1
RC102 Zero Energy Residential Buildings .... APPENDIX RC-1

INDEX .... INDEX R-1

CHAPTER 1 [RE]

SCOPE AND ADMINISTRATION

User note:

About this chapter: Chapter 1 establishes the limits of applicability of this code and describes how the code is to be applied and enforced. Chapter 1 is in two parts: Part 1—Scope and Application (Sections R101–R102) and Part 2—Administration and Enforcement (Sections R103–R110). Section R101 identifies which buildings and structures come under its purview and references other I-Codes as applicable. Standards and codes are scoped to the extent referenced (see Section R108.1).

This code is intended to be adopted as a legally enforceable document, and it cannot be effective without adequate provisions for its administration and enforcement. The provisions of Chapter 1 establish the authority and duties of the code official appointed by the authority having jurisdiction and also establish the rights and privileges of the design professional, contractor and property owner.

PART 1—SCOPE AND APPLICATION

SECTION R101
SCOPE AND GENERAL REQUIREMENTS

R101.1 Title. This code shall be known as the *Energy Conservation Code* of [NAME OF JURISDICTION] and shall be cited as such. It is referred to herein as "this code."

R101.2 Scope. This code applies to *residential buildings*, *building* sites and associated systems and equipment.

R101.3 Intent. This code shall regulate the design and construction of *buildings* for the effective use and conservation of energy over the useful life of each building. This code is intended to provide flexibility to permit the use of innovative approaches and techniques to achieve this objective. This code is not intended to abridge safety, health or environmental requirements contained in other applicable codes or ordinances.

R101.4 Applicability. Where, in any specific case, different sections of this code specify different materials, methods of construction or other requirements, the most restrictive shall govern. Where there is a conflict between a general requirement and a specific requirement, the specific requirement shall govern.

R101.4.1 Mixed residential and commercial buildings. Where a *building* includes both *residential building* and *commercial building* portions, each portion shall be separately considered and meet the applicable provisions of the IECC—Commercial Provisions or IECC—Residential Provisions.

R101.5 Compliance. *Residential buildings* shall meet the provisions of IECC—Residential Provisions. *Commercial buildings* shall meet the provisions of IECC—Commercial Provisions.

R101.5.1 Compliance materials. The *code official* shall be permitted to approve specific computer software, worksheets, compliance manuals and other similar materials that meet the intent of this code.

SECTION R102
ALTERNATIVE MATERIALS, DESIGN AND METHODS OF CONSTRUCTION AND EQUIPMENT

R102.1 General. The provisions of this code are not intended to prevent the installation of any material or to prohibit any design or method of construction not specifically prescribed by this code, provided that any such alternative has been approved. The *code official* shall have the authority to approve an alternative material, design or method of construction upon the written application of the owner or the owner's authorized agent. The code official shall first find that the proposed design is satisfactory and complies with the intent of the provisions of this code, and that the material, method or work offered is, for the purpose intended, not less than the equivalent of that prescribed in this code for strength, effectiveness, fire resistance, durability, energy conservation and safety. The *code official* shall respond to the applicant, in writing, stating the reasons why the alternative was *approved* or was not *approved*.

R102.1.1 Above code programs. The *code official* or other authority having jurisdiction shall be permitted to deem a national, state or local energy-efficiency program to exceed the energy efficiency required by this code. *Buildings approved* in writing by such an energy-efficiency program shall be considered to be in compliance with this code where such buildings also meet the requirements identified in Table R405.2 and the *building thermal envelope* is greater than or equal to levels of efficiency and solar heat gain coefficients (SHGC) in Tables 402.1.1 and 402.1.3 of the 2009 *International Energy Conservation Code.*

PART 2—ADMINISTRATION AND ENFORCEMENT

SECTION R103
CONSTRUCTION DOCUMENTS

R103.1 General. Construction documents, technical reports and other supporting data shall be submitted in one or more sets, or in a digital format where allowed by the *code official*, with each application for a permit. The construction documents and technical reports shall be prepared by a registered design professional where required by the statutes of the jurisdiction in which the project is to be constructed. Where special conditions exist, the *code official* is authorized to require necessary construction documents to be prepared by a registered design professional.

> **Exception:** The *code official* is authorized to waive the requirements for construction documents or other supporting data if the *code official* determines they are not necessary to confirm compliance with this code.

R103.2 Information on construction documents. Construction documents shall be drawn to scale on suitable material. Electronic media documents are permitted to be submitted where *approved* by the *code official*. Construction documents shall be of sufficient clarity to indicate the location, nature and extent of the work proposed, and show in sufficient detail pertinent data and features of the *building*, systems and equipment as herein governed. Details shall include the following as applicable:

1. Energy compliance path.
2. Insulation materials and their *R*-values.
3. Fenestration *U*-factors and *solar heat gain coefficients* (SHGC).
4. Area-weighted *U*-factor and *solar heat gain coefficients* (SHGC) calculations.
5. Mechanical system design criteria.
6. Mechanical and service water-heating systems and equipment types, sizes and efficiencies.
7. Equipment and system controls.
8. Duct sealing, duct and pipe insulation and location.
9. Air sealing details.

R103.2.1 Building thermal envelope depiction. The *building thermal envelope* shall be represented on the construction documents.

R103.3 Examination of documents. The *code official* shall examine or cause to be examined the accompanying construction documents and shall ascertain whether the construction indicated and described is in accordance with the requirements of this code and other pertinent laws or ordinances. The *code official* is authorized to utilize a registered design professional, or other *approved* entity not affiliated with the building design or construction, in conducting the review of the plans and specifications for compliance with the code.

R103.3.1 Approval of construction documents. When the *code official* issues a permit where construction documents are required, the construction documents shall be endorsed in writing and stamped "Reviewed for Code Compliance." Such *approved* construction documents shall not be changed, modified or altered without authorization from the *code official*. Work shall be done in accordance with the *approved* construction documents.

One set of construction documents so reviewed shall be retained by the *code official*. The other set shall be returned to the applicant, kept at the site of work and shall be open to inspection by the *code official* or a duly authorized representative.

R103.3.2 Previous approvals. This code shall not require changes in the construction documents, construction or designated occupancy of a structure for which a lawful permit has been heretofore issued or otherwise lawfully authorized, and the construction of which has been pursued in good faith within 180 days after the effective date of this code and has not been abandoned.

R103.3.3 Phased approval. The *code official* shall have the authority to issue a permit for the construction of part of an energy conservation system before the construction documents for the entire system have been submitted or *approved*, provided adequate information and detailed statements have been filed complying with all pertinent requirements of this code. The holders of such permit shall proceed at their own risk without assurance that the permit for the entire energy conservation system will be granted.

R103.4 Amended construction documents. Work shall be installed in accordance with the *approved* construction documents, and any changes made during construction that are not in compliance with the *approved* construction documents shall be resubmitted for approval as an amended set of construction documents.

R103.5 Retention of construction documents. One set of *approved* construction documents shall be retained by the *code official* for a period of not less than 180 days from date of completion of the permitted work, or as required by state or local laws.

SECTION R104
FEES

R104.1 Fees. A permit shall not be issued until the fees prescribed in Section R104.2 have been paid, nor shall an amendment to a permit be released until the additional fee, if any, has been paid.

R104.2 Schedule of permit fees. Where a permit is required, a fee for each permit shall be paid as required, in accordance with the schedule as established by the applicable governing authority.

R104.3 Work commencing before permit issuance. Any person who commences any work before obtaining the necessary permits shall be subject to an additional fee established by the *code official* that shall be in addition to the required permit fees.

R104.4 Related fees. The payment of the fee for the construction, *alteration*, removal or demolition of work done in connection to or concurrently with the work or activity authorized by a permit shall not relieve the applicant or holder of the permit from the payment of other fees that are prescribed by law.

R104.5 Refunds. The *code official* is authorized to establish a refund policy.

SECTION R105
INSPECTIONS

R105.1 General. Construction or work for which a permit is required shall be subject to inspection by the *code official* or his or her designated agent, and such construction or work shall remain visible and able to be accessed for inspection purposes until *approved*. It shall be the duty of the permit applicant to cause the work to remain visible and able to be accessed for inspection purposes. Neither the *code official* nor the jurisdiction shall be liable for expense entailed in the removal or replacement of any material, product, system or building component required to allow inspection to validate compliance with this code.

R105.2 Required inspections. The *code official* or his or her designated agent, upon notification, shall make the inspections set forth in Sections R105.2.1 through R105.2.5.

> **R105.2.1 Footing and foundation inspection.** Inspections associated with footings and foundations shall verify compliance with the code as to *R*-value, location, thickness, depth of burial and protection of insulation as required by the code and *approved* plans and specifications.
>
> **R105.2.2 Framing and rough-in inspection.** Inspections at framing and rough-in shall be made before application of interior finish and shall verify compliance with the code as to: types of insulation and corresponding *R*-values and their correct location and proper installation; fenestration properties such as *U*-factor and SHGC and proper installation; air leakage controls as required by the code; and *approved* plans and specifications.
>
> **R105.2.3 Plumbing rough-in inspection.** Inspections at plumbing rough-in shall verify compliance as required by the code and *approved* plans and specifications as to types of insulation and corresponding *R*-values and protection, and required controls.
>
> **R105.2.4 Mechanical rough-in inspection.** Inspections at mechanical rough-in shall verify compliance as required by the code and *approved* plans and specifications as to installed HVAC equipment type and size, required controls, system insulation and corresponding *R*-value, system air leakage control, programmable thermostats, dampers, whole-house ventilation, and minimum fan efficiency.
>
>> **Exception:** Systems serving multiple dwelling units shall be inspected in accordance with Section C105.2.4.
>
> **R105.2.5 Final inspection.** The *building* shall have a final inspection and shall not be occupied until *approved*. The final inspection shall include verification of the installation of all required *building* systems, equipment and controls and their proper operation and the required number of high-efficacy lamps and fixtures.

R105.3 Reinspection. A *building* shall be reinspected where determined necessary by the *code official*.

R105.4 Approved inspection agencies. The *code official* is authorized to accept reports of third-party inspection agencies not affiliated with the *building* design or construction, provided that such agencies are *approved* as to qualifications and reliability relevant to the *building* components and systems that they are inspecting.

R105.5 Inspection requests. It shall be the duty of the holder of the permit or their duly authorized agent to notify the *code official* when work is ready for inspection. It shall be the duty of the permit holder to provide access to and means for inspections of such work that are required by this code.

R105.6 Reinspection and testing. Where any work or installation does not pass an initial test or inspection, the necessary corrections shall be made to achieve compliance with this code. The work or installation shall then be resubmitted to the *code official* for inspection and testing.

SECTION R106
NOTICE OF APPROVAL

R106.1 Approval. After the prescribed tests and inspections indicate that the work complies in all respects with this code, a notice of approval shall be issued by the *code official*.

R106.2 Revocation. The *code official* is authorized to, in writing, suspend or revoke a notice of approval issued under the provisions of this code wherever the certificate is issued in error, or on the basis of incorrect information supplied, or where it is determined that the *building* or structure, premise, or portion thereof is in violation of any ordinance or regulation or any of the provisions of this code.

SECTION R107
VALIDITY

R107.1 General. If a portion of this code is held to be illegal or void, such a decision shall not affect the validity of the remainder of this code.

SECTION R108
REFERENCED STANDARDS

R108.1 Referenced codes and standards. The codes and standards referenced in this code shall be those indicated in Chapter 6, and such codes and standards shall be considered as part of the requirements of this code to the prescribed extent of each such reference and as further regulated in Sections R108.1.1 and R108.1.2.

> **R108.1.1 Conflicts.** Where conflicts occur between provisions of this code and referenced codes and standards, the provisions of this code shall apply.
>
> **R108.1.2 Provisions in referenced codes and standards.** Where the extent of the reference to a referenced code or standard includes subject matter that is within the scope of this code, the provisions of this code, as applicable, shall take precedence over the provisions in the referenced code or standard.

R108.2 Application of references. References to chapter or section numbers, or to provisions not specifically identified by number, shall be construed to refer to such chapter, section or provision of this code.

R108.3 Other laws. The provisions of this code shall not be deemed to nullify any provisions of local, state or federal law.

SECTION R109
STOP WORK ORDER

R109.1 Authority. Where the *code official* finds any work regulated by this code being performed in a manner contrary to the provisions of this code or in a dangerous or unsafe manner, the *code official* is authorized to issue a stop work order.

R109.2 Issuance. The stop work order shall be in writing and shall be given to the owner of the property, the owner's authorized agent or the person performing the work. Upon issuance of a stop work order, the cited work shall immediately cease. The stop work order shall state the reason for the order and the conditions under which the cited work is authorized to resume.

R109.3 Emergencies. Where an emergency exists, the *code official* shall not be required to give a written notice prior to stopping the work.

R109.4 Failure to comply. Any person who shall continue any work after having been served with a stop work order, except such work as that person is directed to perform to remove a violation or unsafe condition, shall be subject to fines established by the authority having jurisdiction.

SECTION R110
MEANS OF APPEALS

R110.1 General. In order to hear and decide appeals of orders, decisions or determinations made by the *code official* relative to the application and interpretation of this code, there shall be and is hereby created a board of appeals. The board of appeals shall be appointed by the applicable governing authority and shall hold office at its pleasure. The board shall adopt rules of procedure for conducting its business and shall render all decisions and findings in writing to the appellant with a duplicate copy to the code official.

R110.2 Limitations on authority. An application for appeal shall be based on a claim that the true intent of this code or the rules legally adopted thereunder have been incorrectly interpreted, the provisions of this code do not fully apply or an equivalent or better form of construction is proposed. The board shall not have authority to waive requirements of this code or interpret the administration of this code.

R110.3 Qualifications. The board of appeals shall consist of members who are qualified by experience and training and are not employees of the jurisdiction.

R110.4 Administration. The code official shall take immediate action in accordance with the decision of the board.

CHAPTER 2 [RE]
DEFINITIONS

User note:

About this chapter: Codes, by their very nature, are technical documents. Every word, term and punctuation mark can add to or change the meaning of a technical requirement. It is necessary to maintain a consensus on the specific meaning of each term contained in the code. Chapter 2 performs this function by stating clearly what specific terms mean for the purpose of the code.

SECTION R201
GENERAL

R201.1 Scope. Unless stated otherwise, the following words and terms in this code shall have the meanings indicated in this chapter.

R201.2 Interchangeability. Words used in the present tense include the future; words in the masculine gender include the feminine and neuter; the singular number includes the plural and the plural includes the singular.

R201.3 Terms defined in other codes. Terms that are not defined in this code but are defined in the *International Building Code, International Fire Code, International Fuel Gas Code, International Mechanical Code, International Plumbing Code* or the *International Residential Code* shall have the meanings ascribed to them in those codes.

R201.4 Terms not defined. Terms not defined by this chapter shall have ordinarily accepted meanings such as the context implies.

SECTION R202
GENERAL DEFINITIONS

ABOVE-GRADE WALL. A wall more than 50 percent above grade and enclosing *conditioned space.* This includes between-floor spandrels, peripheral edges of floors, roof and basement knee walls, dormer walls, gable end walls, walls enclosing a mansard roof and *skylight* shafts.

ACCESS (TO). That which enables a device, appliance or equipment to be reached by *ready access* or by a means that first requires the removal or movement of a panel or similar obstruction.

ADDITION. An extension or increase in the *conditioned space* floor area, number of stories or height of a building or structure.

AIR BARRIER. One or more materials joined together in a continuous manner to restrict or prevent the passage of air through the *building thermal envelope* and its assemblies.

ALTERATION. Any construction, retrofit or renovation to an existing structure other than *repair* or *addition.* Also, a change in a building, electrical, gas, mechanical or plumbing system that involves an extension, addition or change to the arrangement, type or purpose of the original installation.

APPROVED. Acceptable to the *code official.*

APPROVED AGENCY. An established and recognized agency that is regularly engaged in conducting tests furnishing inspection services, or furnishing product certification, where such agency has been *approved* by the *code official.*

AUTOMATIC. Self-acting, operating by its own mechanism when actuated by some impersonal influence, as, for example, a change in current strength, pressure, temperature or mechanical configuration (see "*Manual*").

BASEMENT WALL. A wall 50 percent or more below grade and enclosing *conditioned space.*

BUILDING. Any structure used or intended for supporting or sheltering any use or occupancy, including any mechanical systems, service water-heating systems and electric power and lighting systems located on the building site and supporting the building.

BUILDING SITE. A contiguous area of land that is under the ownership or control of one entity.

BUILDING THERMAL ENVELOPE. The *basement walls, exterior walls,* floors, ceiling, roofs and any other *building* element assemblies that enclose *conditioned space* or provide a boundary between *conditioned space* and exempt or unconditioned space.

CAVITY INSULATION. Insulating material located between framing members.

CIRCULATING HOT WATER SYSTEM. A specifically designed water distribution system where one or more pumps are operated in the service hot water piping to circulate heated water from the water-heating equipment to fixtures and back to the water-heating equipment.

CLIMATE ZONE. A geographical region based on climatic criteria as specified in this code.

CODE OFFICIAL. The officer or other designated authority charged with the administration and enforcement of this code or a duly authorized representative.

COMMERCIAL BUILDING. For this code, all buildings that are not included in the definition of "*Residential building.*"

CONDITIONED FLOOR AREA. The horizontal projection of the floors associated with the *conditioned space.*

CONDITIONED SPACE. An area, room or space that is enclosed within the *building thermal envelope* and that is directly or indirectly heated or cooled. Spaces are indirectly heated or cooled where they communicate through openings with conditioned spaces, where they are separated from conditioned spaces by uninsulated walls, floors or ceilings, or where they contain uninsulated ducts, piping or other sources of heating or cooling.

CONTINUOUS AIR BARRIER. A combination of materials and assemblies that restrict or prevent the passage of air through the *building thermal envelope*.

CONTINUOUS INSULATION (ci). Insulating material that is continuous across all structural members without thermal bridges other than fasteners and service openings. It is installed on the interior or exterior, or is integral to any opaque surface, of the *building* envelope.

CRAWL SPACE WALL. The opaque portion of a wall that encloses a crawl space and is partially or totally below grade.

CURTAIN WALL. Fenestration products used to create an external nonload-bearing wall that is designed to separate the exterior and interior environments.

DEMAND RECIRCULATION WATER SYSTEM. A water distribution system where one or more pumps prime the service hot water piping with heated water upon demand for hot water.

DIMMER. A control device that is capable of continuously varying the light output and energy use of light sources.

DUCT. A tube or conduit utilized for conveying air. The air passages of self-contained systems are not to be construed as air ducts.

DUCT SYSTEM. A continuous passageway for the transmission of air that, in addition to ducts, includes duct fittings, dampers, plenums, fans and accessory air-handling equipment and appliances.

DWELLING UNIT. A single unit providing complete independent living facilities for one or more persons, including permanent provisions for living, sleeping, eating, cooking and sanitation.

DWELLING UNIT ENCLOSURE AREA. The sum of the area of ceiling, floors, and walls separating a *dwelling unit's conditioned space* from the exterior or from adjacent conditioned or unconditioned spaces. Wall height shall be measured from the finished floor of the *dwelling unit* to the underside of the floor above.

ENERGY ANALYSIS. A method for estimating the annual energy use of the *proposed design* and *standard reference design* based on estimates of energy use.

ENERGY COST. The total estimated annual cost for purchased energy for the building functions regulated by this code, including applicable demand charges.

ENERGY SIMULATION TOOL. An *approved* software program or calculation-based methodology that projects the annual energy use of a *building*.

ERI REFERENCE DESIGN. A version of the *rated design* that meets the minimum requirements of the 2006 *International Energy Conservation Code*.

EXTERIOR WALL. Walls including both *above-grade walls* and *basement walls*.

FENESTRATION. Products classified as either *vertical fenestration* or *skylights*.

Skylights. Glass or other transparent or translucent glazing material installed at a slope of less than 60 degrees (1.05 rad) from horizontal including unit skylights, tubular daylighting devices, and glazing materials in solariums, sunrooms, roofs and sloped walls.

Vertical fenestration. Windows that are fixed or operable, opaque doors, glazed doors, glazed block and combination opaque/glazed doors composed of glass or other transparent or translucent glazing materials and installed at a slope of not less than 60 degrees (1.05 rad) from horizontal.

FENESTRATION PRODUCT, SITE-BUILT. A fenestration designed to be made up of field-glazed or field-assembled units using specific factory cut or otherwise factory-formed framing and glazing units. Examples of site-built fenestration include storefront systems, curtain walls and atrium roof systems.

HEATED SLAB. Slab-on-grade construction in which the heating elements, hydronic tubing, or hot air distribution system is in contact with, or placed within or under, the slab.

HIGH-EFFICACY LIGHT SOURCES. Any lamp with an efficacy of not less than 65 lumens per watt, or luminaires with an efficacy of not less than 45 lumens per watt.

HISTORIC BUILDING. Any building or structure that is one or more of the following:

1. Listed, or certified as eligible for listing by the State Historic Preservation Officer or the Keeper of the National Register of Historic Places, in the National Register of Historic Places.
2. Designated as historic under an applicable state or local law.
3. Certified as a contributing resource within a National Register-listed, state-designated or locally designated historic district.

INFILTRATION. The uncontrolled inward air leakage into a *building* caused by the pressure effects of wind or the effect of differences in the indoor and outdoor air density or both.

INSULATED SIDING. A type of continuous insulation with manufacturer-installed insulating material as an integral part of the cladding product having an *R*-value of not less than R-2.

LABELED. Equipment, materials or products to which have been affixed a label, seal, symbol or other identifying mark of a nationally recognized testing laboratory, *approved* agency or other organization concerned with product evaluation that maintains periodic inspection of the production of such labeled items and whose labeling indicates either that the equipment, material or product meets identified standards or has been tested and found suitable for a specified purpose.

LISTED. Equipment, materials, products or services included in a list published by an organization acceptable to the *code official* and concerned with evaluation of products or services that maintains periodic inspection of production of *listed* equipment or materials or periodic evaluation of services and whose listing states either that the equipment, material, product or service meets identified standards or has been tested and found suitable for a specified purpose.

MANUAL. Capable of being operated by personal intervention (see "*Automatic*").

OCCUPANT SENSOR CONTROL. An automatic control device that detects the presence or absence of people within an area and causes lighting, equipment or appliances to be regulated accordingly.

ON-SITE RENEWABLE ENERGY. Energy from renewable energy resources harvested at the building site.

OPAQUE DOOR. A door that is not less than 50-percent opaque in surface area.

PROPOSED DESIGN. A description of the proposed *building* used to estimate annual energy use for determining compliance based on total building performance.

RATED DESIGN. A description of the proposed *building* used to determine the energy rating index.

READY ACCESS (TO). That which enables a device, appliance or equipment to be directly reached without requiring the removal or movement of any panel or similar obstruction.

RENEWABLE ENERGY CERTIFICATE (REC). An instrument that represents the environmental attributes of one megawatt hour of renewable energy; also known as an energy attribute certificate (EAC).

RENEWABLE ENERGY RESOURCES. Energy derived from solar radiation, wind, waves, tides, landfill gas, biogas, biomass or extracted from hot fluid or steam heated within the earth.

REPAIR. The reconstruction or renewal of any part of an existing *building* for the purpose of its maintenance or to correct damage.

REROOFING. The process of recovering or replacing an existing roof covering. See "*Roof recover*" and "*Roof replacement*."

RESIDENTIAL BUILDING. For this code, includes detached one- and two-family dwellings and townhouses as well as *Group R*-2, *R*-3 and *R*-4 buildings three stories or less in height above grade plane.

ROOF ASSEMBLY. A system designed to provide weather protection and resistance to design loads. The system consists of a roof covering and roof deck or a single component serving as both the roof covering and the roof deck. A roof assembly includes the roof covering, underlayment and roof deck and can also include a thermal barrier, an ignition barrier, insulation or a vapor retarder.

ROOF RECOVER. The process of installing an additional roof covering over an existing roof covering without removing the existing roof covering.

ROOF REPAIR. Reconstruction or renewal of any part of an existing roof for the purposes of its maintenance.

ROOF REPLACEMENT. The process of removing the existing roof covering, repairing any damaged substrate and installing a new roof covering.

***R*-VALUE (THERMAL RESISTANCE).** The inverse of the time rate of heat flow through a body from one of its bounding surfaces to the other surface for a unit temperature difference between the two surfaces, under steady state conditions, per unit area (h × ft^2 × °F/Btu) [(m^2 × K)/W].

SERVICE WATER HEATING. Supply of hot water for purposes other than comfort heating.

SOLAR HEAT GAIN COEFFICIENT (SHGC). The ratio of the solar heat gain entering the space through the fenestration assembly to the incident solar radiation. Solar heat gain includes directly transmitted solar heat and absorbed solar radiation that is then reradiated, conducted or convected into the space.

STANDARD REFERENCE DESIGN. A version of the *proposed design* that meets the minimum requirements of this code and is used to determine the maximum annual energy use requirement for compliance based on total building performance.

SUNROOM. A one-story structure attached to a dwelling with a glazing area in excess of 40 percent of the gross area of the structure's *exterior walls* and roof.

THERMAL DISTRIBUTION EFFICIENCY (TDE). The resistance to changes in air heat as air is conveyed through a distance of air duct. TDE is a heat loss calculation evaluating the difference in the heat of the air between the air duct inlet and outlet caused by differences in temperatures between the air in the duct and the duct material. TDE is expressed as a percent difference between the inlet and outlet heat in the duct.

THERMAL ISOLATION. Physical and space conditioning separation from *conditioned spaces*. The *conditioned spaces* shall be controlled as separate zones for heating and cooling or conditioned by separate equipment.

THERMOSTAT. An automatic control device used to maintain temperature at a fixed or adjustable setpoint.

***U*-FACTOR (THERMAL TRANSMITTANCE).** The coefficient of heat transmission (air to air) through a building component or assembly, equal to the time rate of heat flow per unit area and unit temperature difference between the warm side and cold side air films (Btu/h × ft^2 × °F) [W/(m^2 × K)].

VENTILATION. The natural or mechanical process of supplying conditioned or unconditioned air to, or removing such air from, any space.

VENTILATION AIR. That portion of supply air that comes from outside (outdoors) plus any recirculated air that has been treated to maintain the desired quality of air within a designated space.

VISIBLE TRANSMITTANCE (VT). The ratio of visible light entering the space through the fenestration product assembly to the incident visible light. Visible Transmittance includes the effects of glazing material and frame and is expressed as a number between 0 and 1.

DEFINITIONS

VENTILATION AIR. That portion of supply air that comes from outside (outdoors) plus any recirculated air that has been treated to maintain the desired quality of air within a designated space.

VISIBLE TRANSMITTANCE (VT). The ratio of visible light entering the space through the fenestration product assembly to the incident visible light. Visible Transmittance includes the effects of glazing material and frame and is expressed as a number between 0 and 1.

WHOLE HOUSE MECHANICAL VENTILATION SYSTEM. An exhaust system, supply system, or combination thereof that is designed to mechanically exchange indoor air with outdoor air when operating continuously or through a programmed intermittent schedule to satisfy the whole house ventilation rates.

ZONE. A space or group of spaces within a *building* with heating or cooling requirements that are sufficiently similar so that desired conditions can be maintained throughout using a single controlling device.

CHAPTER 3 [RE]
GENERAL REQUIREMENTS

User note:

About this chapter: *Chapter 3 addresses broadly applicable requirements that would not be at home in other chapters having more specific coverage of subject matter. This chapter establishes climate zone by US counties and territories and includes methodology for determining climate zones elsewhere. It also contains product rating, marking and installation requirements for materials such as insulation, windows, doors and siding.*

SECTION R301
CLIMATE ZONES

R301.1 General. *Climate zones* from Figure R301.1 or Table R301.1 shall be used for determining the applicable requirements from Chapter 4. Locations not indicated in Table R301.1 shall be assigned a *climate zone* in accordance with Section R301.3.

R301.2 Warm Humid counties. In Table R301.1, Warm Humid counties are identified by an asterisk.

GENERAL REQUIREMENTS

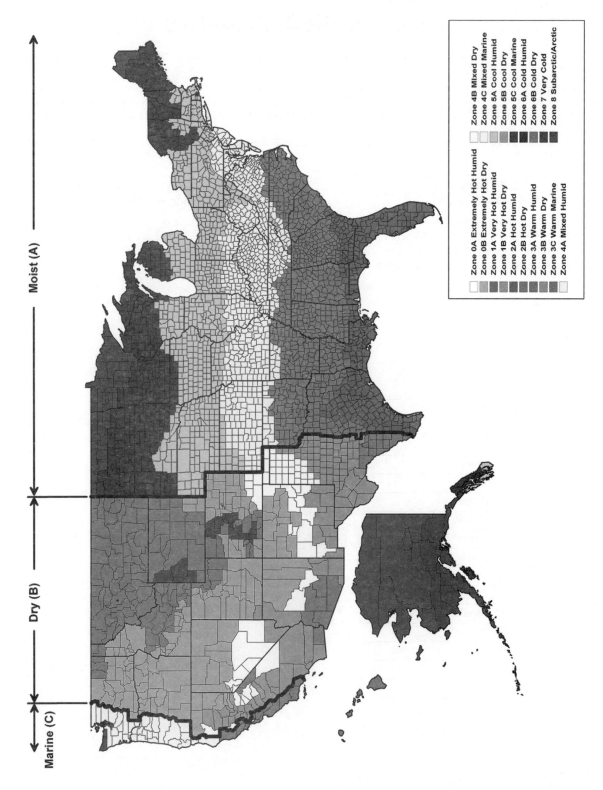

FIGURE R301.1
CLIMATE ZONES

TABLE R301.1
CLIMATE ZONES, MOISTURE REGIMES, AND WARM HUMID DESIGNATIONS BY STATE, COUNTY AND TERRITORY[a]

US STATES	
ALABAMA	3A Lowndes*
3A Autauga*	3A Macon*
2A Baldwin*	3A Madison
3A Barbour*	3A Marengo*
3A Bibb	3A Marion
3A Blount	3A Marshall
3A Bullock*	2A Mobile*
3A Butler*	3A Monroe*
3A Calhoun	3A Montgomery*
3A Chambers	3A Morgan
3A Cherokee	3A Perry*
3A Chilton	3A Pickens
3A Choctaw*	3A Pike*
3A Clarke*	3A Randolph
3A Clay	3A Russell*
3A Cleburne	3A Shelby
2A Coffee*	3A St. Clair
3A Colbert	3A Sumter
3A Conecuh*	3A Talladega
3A Coosa	3A Tallapoosa
2A Covington*	3A Tuscaloosa
3A Crenshaw*	3A Walker
3A Cullman	3A Washington*
2A Dale*	3A Wilcox*
3A Dallas*	3A Winston
3A DeKalb	**ALASKA**
3A Elmore*	7 Aleutians East
2A Escambia*	7 Aleutians West
3A Etowah	7 Anchorage
3A Fayette	7 Bethel
3A Franklin	7 Bristol Bay
2A Geneva*	8 Denali
3A Greene	7 Dillingham
3A Hale	8 Fairbanks North Star
2A Henry*	6A Haines
2A Houston*	6A Juneau
3A Jackson	7 Kenai Peninsula
3A Jefferson	5C Ketchikan Gateway
3A Lamar	6A Kodiak Island
3A Lauderdale	7 Lake and Peninsula
3A Lawrence	7 Matanuska-Susitna
3A Lee	8 Nome
3A Limestone	8 North Slope
	8 Northwest Arctic

(continued)

GENERAL REQUIREMENTS

TABLE R301.1—continued
CLIMATE ZONES, MOISTURE REGIMES, AND WARM HUMID DESIGNATIONS BY STATE, COUNTY AND TERRITORY[a]

US STATES—continued	
ALASKA *(continued)*	3A Crittenden
5C Prince of Wales Outer Ketchikan	3A Cross
5C Sitka	3A Dallas
6A Skagway-Hoonah-Angoon	3A Desha
8 Southeast Fairbanks	3A Drew
7 Valdez-Cordova	3A Faulkner
8 Wade Hampton	3A Franklin
6A Wrangell-Petersburg	4A Fulton
7 Yakutat	3A Garland
8 Yukon-Koyukuk	3A Grant
ARIZONA	3A Greene
5B Apache	3A Hempstead*
3B Cochise	3A Hot Spring
5B Coconino	3A Howard
4B Gila	3A Independence
3B Graham	4A Izard
3B Greenlee	3A Jackson
2B La Paz	3A Jefferson
2B Maricopa	3A Johnson
3B Mohave	3A Lafayette*
5B Navajo	3A Lawrence
2B Pima	3A Lee
2B Pinal	3A Lincoln
3B Santa Cruz	3A Little River*
4B Yavapai	3A Logan
2B Yuma	3A Lonoke
ARKANSAS	4A Madison
3A Arkansas	4A Marion
3A Ashley	3A Miller*
4A Baxter	3A Mississippi
4A Benton	3A Monroe
4A Boone	3A Montgomery
3A Bradley	3A Nevada
3A Calhoun	4A Newton
4A Carroll	3A Ouachita
3A Chicot	3A Perry
3A Clark	3A Phillips
3A Clay	3A Pike
3A Cleburne	3A Poinsett
3A Cleveland	3A Polk
3A Columbia*	3A Pope
3A Conway	3A Prairie
3A Craighead	3A Pulaski
3A Crawford	3A Randolph

(continued)

TABLE R301.1—continued
CLIMATE ZONES, MOISTURE REGIMES, AND WARM HUMID DESIGNATIONS BY STATE, COUNTY AND TERRITORY[a]

US STATES—continued
ARKANSAS (continued)
3A Saline
3A Scott
4A Searcy
3A Sebastian
3A Sevier*
3A Sharp
3A St. Francis
4A Stone
3A Union*
3A Van Buren
4A Washington
3A White
3A Woodruff
3A Yell
CALIFORNIA
3C Alameda
6B Alpine
4B Amador
3B Butte
4B Calaveras
3B Colusa
3B Contra Costa
4C Del Norte
4B El Dorado
3B Fresno
3B Glenn
4C Humboldt
2B Imperial
4B Inyo
3B Kern
3B Kings
4B Lake
5B Lassen
3B Los Angeles
3B Madera
3C Marin
4B Mariposa
3C Mendocino
3B Merced
5B Modoc
6B Mono
3C Monterey
3C Napa
5B Nevada
3B Orange
3B Placer
5B Plumas
3B Riverside
3B Sacramento
3C San Benito
3B San Bernardino
3B San Diego
3C San Francisco
3B San Joaquin
3C San Luis Obispo
3C San Mateo
3C Santa Barbara
3C Santa Clara
3C Santa Cruz
3B Shasta
5B Sierra
5B Siskiyou
3B Solano
3C Sonoma
3B Stanislaus
3B Sutter
3B Tehama
4B Trinity
3B Tulare
4B Tuolumne
3C Ventura
3B Yolo
3B Yuba
COLORADO
5B Adams
6B Alamosa
5B Arapahoe
6B Archuleta
4B Baca
4B Bent
5B Boulder
5B Broomfield
6B Chaffee
5B Cheyenne
7 Clear Creek
6B Conejos
6B Costilla
5B Crowley

(continued)

GENERAL REQUIREMENTS

TABLE R301.1—continued
CLIMATE ZONES, MOISTURE REGIMES, AND WARM HUMID DESIGNATIONS BY STATE, COUNTY AND TERRITORY[a]

US STATES—continued
COLORADO *(continued)*
5B Custer
5B Delta
5B Denver
6B Dolores
5B Douglas
6B Eagle
5B Elbert
5B El Paso
5B Fremont
5B Garfield
5B Gilpin
7 Grand
7 Gunnison
7 Hinsdale
5B Huerfano
7 Jackson
5B Jefferson
5B Kiowa
5B Kit Carson
7 Lake
5B La Plata
5B Larimer
4B Las Animas
5B Lincoln
5B Logan
5B Mesa
7 Mineral
6B Moffat
5B Montezuma
5B Montrose
5B Morgan
4B Otero
6B Ouray
7 Park
5B Phillips
7 Pitkin
4B Prowers
5B Pueblo
6B Rio Blanco
7 Rio Grande
7 Routt
6B Saguache
7 San Juan

6B San Miguel
5B Sedgwick
7 Summit
5B Teller
5B Washington
5B Weld
5B Yuma
CONNECTICUT
5A (all)
DELAWARE
4A (all)
DISTRICT OF COLUMBIA
4A (all)
FLORIDA
2A Alachua*
2A Baker*
2A Bay*
2A Bradford*
2A Brevard*
1A Broward*
2A Calhoun*
2A Charlotte*
2A Citrus*
2A Clay*
2A Collier*
2A Columbia*
2A DeSoto*
2A Dixie*
2A Duval*
2A Escambia*
2A Flagler*
2A Franklin*
2A Gadsden*
2A Gilchrist*
2A Glades*
2A Gulf*
2A Hamilton*
2A Hardee*
2A Hendry*
2A Hernando*
2A Highlands*
2A Hillsborough*
2A Holmes*
2A Indian River*
2A Jackson*

(continued)

TABLE R301.1—continued
CLIMATE ZONES, MOISTURE REGIMES, AND WARM HUMID DESIGNATIONS BY STATE, COUNTY AND TERRITORY[a]

US STATES—continued	
FLORIDA *(continued)*	3A Barrow
2A Jefferson*	3A Bartow
2A Lafayette*	3A Ben Hill*
2A Lake*	2A Berrien*
2A Lee*	3A Bibb
2A Leon*	3A Bleckley*
2A Levy*	2A Brantley*
2A Liberty*	2A Brooks*
2A Madison*	2A Bryan*
2A Manatee*	3A Bulloch*
2A Marion*	3A Burke
2A Martin*	3A Butts
1A Miami-Dade*	2A Calhoun*
1A Monroe*	2A Camden*
2A Nassau*	3A Candler*
2A Okaloosa*	3A Carroll
2A Okeechobee*	3A Catoosa
2A Orange*	2A Charlton*
2A Osceola*	2A Chatham*
1A Palm Beach*	3A Chattahoochee*
2A Pasco*	3A Chattooga
2A Pinellas*	3A Cherokee
2A Polk*	3A Clarke
2A Putnam*	3A Clay*
2A Santa Rosa*	3A Clayton
2A Sarasota*	2A Clinch*
2A Seminole*	3A Cobb
2A St. Johns*	2A Coffee*
2A St. Lucie*	2A Colquitt*
2A Sumter*	3A Columbia
2A Suwannee*	2A Cook*
2A Taylor*	3A Coweta
2A Union*	3A Crawford
2A Volusia*	3A Crisp*
2A Wakulla*	3A Dade
2A Walton*	3A Dawson
2A Washington*	2A Decatur*
GEORGIA	3A DeKalb
2A Appling*	3A Dodge*
2A Atkinson*	3A Dooly*
2A Bacon*	2A Dougherty*
2A Baker*	3A Douglas
3A Baldwin	2A Early*
3A Banks	2A Echols*
	2A Effingham*

(continued)

GENERAL REQUIREMENTS

TABLE R301.1—continued
CLIMATE ZONES, MOISTURE REGIMES, AND WARM HUMID DESIGNATIONS BY STATE, COUNTY AND TERRITORY[a]

US STATES—continued	
GEORGIA *(continued)*	3A Madison
3A Elbert	3A Marion*
3A Emanuel*	3A McDuffie
2A Evans*	2A McIntosh*
3A Fannin	3A Meriwether
3A Fayette	2A Miller*
3A Floyd	2A Mitchell*
3A Forsyth	3A Monroe
3A Franklin	3A Montgomery*
3A Fulton	3A Morgan
3A Gilmer	3A Murray
3A Glascock	3A Muscogee
2A Glynn*	3A Newton
3A Gordon	3A Oconee
2A Grady*	3A Oglethorpe
3A Greene	3A Paulding
3A Gwinnett	3A Peach*
3A Habersham	3A Pickens
3A Hall	2A Pierce*
3A Hancock	3A Pike
3A Haralson	3A Polk
3A Harris	3A Pulaski*
3A Hart	3A Putnam
3A Heard	3A Quitman*
3A Henry	3A Rabun
3A Houston*	3A Randolph*
3A Irwin*	3A Richmond
3A Jackson	3A Rockdale
3A Jasper	3A Schley*
2A Jeff Davis*	3A Screven*
3A Jefferson	2A Seminole*
3A Jenkins*	3A Spalding
3A Johnson*	3A Stephens
3A Jones	3A Stewart*
3A Lamar	3A Sumter*
2A Lanier*	3A Talbot
3A Laurens*	3A Taliaferro
3A Lee*	2A Tattnall*
2A Liberty*	3A Taylor*
3A Lincoln	3A Telfair*
2A Long*	3A Terrell*
2A Lowndes*	2A Thomas*
3A Lumpkin	2A Tift*
3A Macon*	2A Toombs*
	3A Towns

(continued)

GENERAL REQUIREMENTS

TABLE R301.1—continued
CLIMATE ZONES, MOISTURE REGIMES, AND WARM HUMID DESIGNATIONS BY STATE, COUNTY AND TERRITORY[a]

US STATES—continued	
GEORGIA *(continued)*	6B Franklin
3A Treutlen*	6B Fremont
3A Troup	5B Gem
3A Turner*	5B Gooding
3A Twiggs*	5B Idaho
3A Union	6B Jefferson
3A Upson	5B Jerome
3A Walker	5B Kootenai
3A Walton	5B Latah
2A Ware*	6B Lemhi
3A Warren	5B Lewis
3A Washington	5B Lincoln
2A Wayne*	6B Madison
3A Webster*	5B Minidoka
3A Wheeler*	5B Nez Perce
3A White	6B Oneida
3A Whitfield	5B Owyhee
3A Wilcox*	5B Payette
3A Wilkes	5B Power
3A Wilkinson	5B Shoshone
2A Worth*	6B Teton
HAWAII	5B Twin Falls
1A (all)*	6B Valley
IDAHO	5B Washington
5B Ada	**ILLINOIS**
6B Adams	5A Adams
6B Bannock	4A Alexander
6B Bear Lake	4A Bond
5B Benewah	5A Boone
6B Bingham	5A Brown
6B Blaine	5A Bureau
6B Boise	4A Calhoun
6B Bonner	5A Carroll
6B Bonneville	5A Cass
6B Boundary	5A Champaign
6B Butte	4A Christian
6B Camas	4A Clark
5B Canyon	4A Clay
6B Caribou	4A Clinton
5B Cassia	4A Coles
6B Clark	5A Cook
5B Clearwater	4A Crawford
6B Custer	4A Cumberland
5B Elmore	5A DeKalb
	5A De Witt

(continued)

TABLE R301.1—continued
CLIMATE ZONES, MOISTURE REGIMES, AND WARM HUMID DESIGNATIONS BY STATE, COUNTY AND TERRITORY[a]

US STATES—continued	
ILLINOIS (continued)	5A Menard
5A Douglas	5A Mercer
5A DuPage	4A Monroe
5A Edgar	4A Montgomery
4A Edwards	5A Morgan
4A Effingham	5A Moultrie
4A Fayette	5A Ogle
5A Ford	5A Peoria
4A Franklin	4A Perry
5A Fulton	5A Piatt
4A Gallatin	5A Pike
4A Greene	4A Pope
5A Grundy	4A Pulaski
4A Hamilton	5A Putnam
5A Hancock	4A Randolph
4A Hardin	4A Richland
5A Henderson	5A Rock Island
5A Henry	4A Saline
5A Iroquois	5A Sangamon
4A Jackson	5A Schuyler
4A Jasper	5A Scott
4A Jefferson	4A Shelby
4A Jersey	5A Stark
5A Jo Daviess	4A St. Clair
4A Johnson	5A Stephenson
5A Kane	5A Tazewell
5A Kankakee	4A Union
5A Kendall	5A Vermilion
5A Knox	4A Wabash
5A Lake	5A Warren
5A La Salle	4A Washington
4A Lawrence	4A Wayne
5A Lee	4A White
5A Livingston	5A Whiteside
5A Logan	5A Will
5A Macon	4A Williamson
4A Macoupin	5A Winnebago
4A Madison	5A Woodford
4A Marion	**INDIANA**
5A Marshall	5A Adams
5A Mason	5A Allen
4A Massac	4A Bartholomew
5A McDonough	5A Benton
5A McHenry	5A Blackford
5A McLean	5A Boone
	4A Brown

(continued)

TABLE R301.1—continued
CLIMATE ZONES, MOISTURE REGIMES, AND WARM HUMID DESIGNATIONS BY STATE, COUNTY AND TERRITORY[a]

US STATES—continued	
INDIANA *(continued)*	4A Martin
5A Carroll	5A Miami
5A Cass	4A Monroe
4A Clark	5A Montgomery
4A Clay	4A Morgan
5A Clinton	5A Newton
4A Crawford	5A Noble
4A Daviess	4A Ohio
4A Dearborn	4A Orange
4A Decatur	4A Owen
5A De Kalb	5A Parke
5A Delaware	4A Perry
4A Dubois	4A Pike
5A Elkhart	5A Porter
4A Fayette	4A Posey
4A Floyd	5A Pulaski
5A Fountain	4A Putnam
4A Franklin	5A Randolph
5A Fulton	4A Ripley
4A Gibson	4A Rush
5A Grant	4A Scott
4A Greene	4A Shelby
5A Hamilton	4A Spencer
5A Hancock	5A Starke
4A Harrison	5A Steuben
4A Hendricks	5A St. Joseph
5A Henry	4A Sullivan
5A Howard	4A Switzerland
5A Huntington	5A Tippecanoe
4A Jackson	5A Tipton
5A Jasper	4A Union
5A Jay	4A Vanderburgh
4A Jefferson	5A Vermillion
4A Jennings	4A Vigo
4A Johnson	5A Wabash
4A Knox	5A Warren
5A Kosciusko	4A Warrick
5A LaGrange	4A Washington
5A Lake	5A Wayne
5A LaPorte	5A Wells
4A Lawrence	5A White
5A Madison	5A Whitley
4A Marion	**IOWA**
5A Marshall	5A Adair
	5A Adams

(continued)

GENERAL REQUIREMENTS

TABLE R301.1—continued
CLIMATE ZONES, MOISTURE REGIMES, AND WARM HUMID DESIGNATIONS BY STATE, COUNTY AND TERRITORY[a]

US STATES—continued	
IOWA *(continued)*	5A Humboldt
5A Allamakee	5A Ida
5A Appanoose	5A Iowa
5A Audubon	5A Jackson
5A Benton	5A Jasper
6A Black Hawk	5A Jefferson
5A Boone	5A Johnson
5A Bremer	5A Jones
5A Buchanan	5A Keokuk
5A Buena Vista	6A Kossuth
5A Butler	5A Lee
5A Calhoun	5A Linn
5A Carroll	5A Louisa
5A Cass	5A Lucas
5A Cedar	6A Lyon
6A Cerro Gordo	5A Madison
5A Cherokee	5A Mahaska
5A Chickasaw	5A Marion
5A Clarke	5A Marshall
6A Clay	5A Mills
5A Clayton	6A Mitchell
5A Clinton	5A Monona
5A Crawford	5A Monroe
5A Dallas	5A Montgomery
5A Davis	5A Muscatine
5A Decatur	6A O'Brien
5A Delaware	6A Osceola
5A Des Moines	5A Page
6A Dickinson	6A Palo Alto
5A Dubuque	5A Plymouth
6A Emmet	5A Pocahontas
5A Fayette	5A Polk
5A Floyd	5A Pottawattamie
5A Franklin	5A Poweshiek
5A Fremont	5A Ringgold
5A Greene	5A Sac
5A Grundy	5A Scott
5A Guthrie	5A Shelby
5A Hamilton	6A Sioux
6A Hancock	5A Story
5A Hardin	5A Tama
5A Harrison	5A Taylor
5A Henry	5A Union
5A Howard	5A Van Buren
	5A Wapello

(continued)

TABLE R301.1—continued
CLIMATE ZONES, MOISTURE REGIMES, AND WARM HUMID DESIGNATIONS BY STATE, COUNTY AND TERRITORY[a]

US STATES—continued	
IOWA *(continued)*	4A Grant
5A Warren	4A Gray
5A Washington	5A Greeley
5A Wayne	4A Greenwood
5A Webster	4A Hamilton
6A Winnebago	4A Harper
5A Winneshiek	4A Harvey
5A Woodbury	4A Haskell
6A Worth	4A Hodgeman
5A Wright	4A Jackson
KANSAS	4A Jefferson
4A Allen	5A Jewell
4A Anderson	4A Johnson
4A Atchison	4A Kearny
4A Barber	4A Kingman
4A Barton	4A Kiowa
4A Bourbon	4A Labette
4A Brown	4A Lane
4A Butler	4A Leavenworth
4A Chase	4A Lincoln
4A Chautauqua	4A Linn
4A Cherokee	5A Logan
5A Cheyenne	4A Lyon
4A Clark	4A Marion
4A Clay	4A Marshall
4A Cloud	4A McPherson
4A Coffey	4A Meade
4A Comanche	4A Miami
4A Cowley	4A Mitchell
4A Crawford	4A Montgomery
5A Decatur	4A Morris
4A Dickinson	4A Morton
4A Doniphan	4A Nemaha
4A Douglas	4A Neosho
4A Edwards	4A Ness
4A Elk	5A Norton
4A Ellis	4A Osage
4A Ellsworth	4A Osborne
4A Finney	4A Ottawa
4A Ford	4A Pawnee
4A Franklin	5A Phillips
4A Geary	4A Pottawatomie
5A Gove	4A Pratt
4A Graham	5A Rawlins
	4A Reno

(continued)

GENERAL REQUIREMENTS

TABLE R301.1—continued
CLIMATE ZONES, MOISTURE REGIMES, AND WARM HUMID DESIGNATIONS BY STATE, COUNTY AND TERRITORY[a]

US STATES—continued	
KANSAS (continued)	3A Claiborne*
5A Republic	3A Concordia*
4A Rice	3A De Soto*
4A Riley	2A East Baton Rouge*
4A Rooks	3A East Carroll
4A Rush	2A East Feliciana*
4A Russell	2A Evangeline*
4A Saline	3A Franklin*
5A Scott	3A Grant*
4A Sedgwick	2A Iberia*
4A Seward	2A Iberville*
4A Shawnee	3A Jackson*
5A Sheridan	2A Jefferson*
5A Sherman	2A Jefferson Davis*
5A Smith	2A Lafayette*
4A Stafford	2A Lafourche*
4A Stanton	3A La Salle*
4A Stevens	3A Lincoln*
4A Sumner	2A Livingston*
5A Thomas	3A Madison*
4A Trego	3A Morehouse
4A Wabaunsee	3A Natchitoches*
5A Wallace	2A Orleans*
4A Washington	3A Ouachita*
5A Wichita	2A Plaquemines*
4A Wilson	2A Pointe Coupee*
4A Woodson	2A Rapides*
4A Wyandotte	3A Red River*
KENTUCKY	3A Richland*
4A (all)	3A Sabine*
LOUISIANA	2A St. Bernard*
2A Acadia*	2A St. Charles*
2A Allen*	2A St. Helena*
2A Ascension*	2A St. James*
2A Assumption*	2A St. John the Baptist*
2A Avoyelles*	2A St. Landry*
2A Beauregard*	2A St. Martin*
3A Bienville*	2A St. Mary*
3A Bossier*	2A St. Tammany*
3A Caddo*	2A Tangipahoa*
2A Calcasieu*	3A Tensas*
3A Caldwell*	2A Terrebonne*
2A Cameron*	3A Union*
3A Catahoula*	2A Vermilion*
	3A Vernon*

(continued)

GENERAL REQUIREMENTS

TABLE R301.1—continued
CLIMATE ZONES, MOISTURE REGIMES, AND WARM HUMID DESIGNATIONS BY STATE, COUNTY AND TERRITORY[a]

US STATES—continued	
LOUISIANA *(continued)*	4A St. Mary's
2A Washington*	4A Talbot
3A Webster*	4A Washington
2A West Baton Rouge*	4A Wicomico
3A West Carroll	4A Worcester
2A West Feliciana*	**MASSACHUSETTS**
3A Winn*	5A (all)
MAINE	**MICHIGAN**
6A Androscoggin	6A Alcona
7 Aroostook	6A Alger
6A Cumberland	5A Allegan
6A Franklin	6A Alpena
6A Hancock	6A Antrim
6A Kennebec	6A Arenac
6A Knox	6A Baraga
6A Lincoln	5A Barry
6A Oxford	5A Bay
6A Penobscot	6A Benzie
6A Piscataquis	5A Berrien
6A Sagadahoc	5A Branch
6A Somerset	5A Calhoun
6A Waldo	5A Cass
6A Washington	6A Charlevoix
6A York	6A Cheboygan
MARYLAND	6A Chippewa
5A Allegany	6A Clare
4A Anne Arundel	5A Clinton
4A Baltimore	6A Crawford
4A Baltimore (city)	6A Delta
4A Calvert	6A Dickinson
4A Caroline	5A Eaton
4A Carroll	6A Emmet
4A Cecil	5A Genesee
4A Charles	6A Gladwin
4A Dorchester	6A Gogebic
4A Frederick	6A Grand Traverse
5A Garrett	5A Gratiot
4A Harford	5A Hillsdale
4A Howard	6A Houghton
4A Kent	5A Huron
4A Montgomery	5A Ingham
4A Prince George's	5A Ionia
4A Queen Anne's	6A Iosco
4A Somerset	6A Iron
	6A Isabella

(continued)

2021 INTERNATIONAL ENERGY CONSERVATION CODE®   R3-15

GENERAL REQUIREMENTS

TABLE R301.1—continued
CLIMATE ZONES, MOISTURE REGIMES, AND WARM HUMID DESIGNATIONS BY STATE, COUNTY AND TERRITORY[a]

US STATES—continued
MICHIGAN *(continued)*
5A Jackson
5A Kalamazoo
6A Kalkaska
5A Kent
7 Keweenaw
6A Lake
5A Lapeer
6A Leelanau
5A Lenawee
5A Livingston
6A Luce
6A Mackinac
5A Macomb
6A Manistee
7 Marquette
6A Mason
6A Mecosta
6A Menominee
5A Midland
6A Missaukee
5A Monroe
5A Montcalm
6A Montmorency
5A Muskegon
6A Newaygo
5A Oakland
6A Oceana
6A Ogemaw
6A Ontonagon
6A Osceola
6A Oscoda
6A Otsego
5A Ottawa
6A Presque Isle
6A Roscommon
5A Saginaw
5A Sanilac
6A Schoolcraft
5A Shiawassee
5A St. Clair
5A St. Joseph
5A Tuscola
5A Van Buren

5A Washtenaw
5A Wayne
6A Wexford
MINNESOTA
7 Aitkin
6A Anoka
6A Becker
7 Beltrami
6A Benton
6A Big Stone
6A Blue Earth
6A Brown
7 Carlton
6A Carver
7 Cass
6A Chippewa
6A Chisago
6A Clay
7 Clearwater
7 Cook
6A Cottonwood
7 Crow Wing
6A Dakota
6A Dodge
6A Douglas
6A Faribault
5A Fillmore
6A Freeborn
6A Goodhue
6A Grant
6A Hennepin
5A Houston
7 Hubbard
6A Isanti
7 Itasca
6A Jackson
6A Kanabec
6A Kandiyohi
7 Kittson
7 Koochiching
6A Lac qui Parle
7 Lake
7 Lake of the Woods
6A Le Sueur
6A Lincoln

(continued)

GENERAL REQUIREMENTS

TABLE R301.1—continued
CLIMATE ZONES, MOISTURE REGIMES, AND WARM HUMID DESIGNATIONS BY STATE, COUNTY AND TERRITORY[a]

US STATES—continued	
MINNESOTA *(continued)*	5A Winona
6A Lyon	6A Wright
7 Mahnomen	6A Yellow Medicine
7 Marshall	**MISSISSIPPI**
6A Martin	3A Adams*
6A McLeod	3A Alcorn
6A Meeker	3A Amite*
6A Mille Lacs	3A Attala
6A Morrison	3A Benton
6A Mower	3A Bolivar
6A Murray	3A Calhoun
6A Nicollet	3A Carroll
6A Nobles	3A Chickasaw
7 Norman	3A Choctaw
6A Olmsted	3A Claiborne*
6A Otter Tail	3A Clarke
7 Pennington	3A Clay
7 Pine	3A Coahoma
6A Pipestone	3A Copiah*
7 Polk	3A Covington*
6A Pope	3A DeSoto
6A Ramsey	3A Forrest*
7 Red Lake	3A Franklin*
6A Redwood	2A George*
6A Renville	3A Greene*
6A Rice	3A Grenada
6A Rock	2A Hancock*
7 Roseau	2A Harrison*
6A Scott	3A Hinds*
6A Sherburne	3A Holmes
6A Sibley	3A Humphreys
6A Stearns	3A Issaquena
6A Steele	3A Itawamba
6A Stevens	2A Jackson*
7 St. Louis	3A Jasper
6A Swift	3A Jefferson*
6A Todd	3A Jefferson Davis*
6A Traverse	3A Jones*
6A Wabasha	3A Kemper
7 Wadena	3A Lafayette
6A Waseca	3A Lamar*
6A Washington	3A Lauderdale
6A Watonwan	3A Lawrence*
6A Wilkin	3A Leake
	3A Lee

(continued)

TABLE R301.1—continued
CLIMATE ZONES, MOISTURE REGIMES, AND WARM HUMID DESIGNATIONS BY STATE, COUNTY AND TERRITORY[a]

US STATES—continued	
MISSISSIPPI *(continued)*	5A Andrew
3A Leflore	5A Atchison
3A Lincoln*	4A Audrain
3A Lowndes	4A Barry
3A Madison	4A Barton
3A Marion*	4A Bates
3A Marshall	4A Benton
3A Monroe	4A Bollinger
3A Montgomery	4A Boone
3A Neshoba	4A Buchanan
3A Newton	4A Butler
3A Noxubee	4A Caldwell
3A Oktibbeha	4A Callaway
3A Panola	4A Camden
2A Pearl River*	4A Cape Girardeau
3A Perry*	4A Carroll
3A Pike*	4A Carter
3A Pontotoc	4A Cass
3A Prentiss	4A Cedar
3A Quitman	4A Chariton
3A Rankin*	4A Christian
3A Scott	5A Clark
3A Sharkey	4A Clay
3A Simpson*	4A Clinton
3A Smith*	4A Cole
2A Stone*	4A Cooper
3A Sunflower	4A Crawford
3A Tallahatchie	4A Dade
3A Tate	4A Dallas
3A Tippah	5A Daviess
3A Tishomingo	5A DeKalb
3A Tunica	4A Dent
3A Union	4A Douglas
3A Walthall*	3A Dunklin
3A Warren*	4A Franklin
3A Washington	4A Gasconade
3A Wayne*	5A Gentry
3A Webster	4A Greene
3A Wilkinson*	5A Grundy
3A Winston	5A Harrison
3A Yalobusha	4A Henry
3A Yazoo	4A Hickory
MISSOURI	5A Holt
5A Adair	4A Howard
	4A Howell

(continued)

GENERAL REQUIREMENTS

TABLE R301.1—continued
CLIMATE ZONES, MOISTURE REGIMES, AND WARM HUMID DESIGNATIONS BY STATE, COUNTY AND TERRITORY[a]

US STATES—continued	
MISSOURI (continued)	4A Reynolds
4A Iron	4A Ripley
4A Jackson	4A Saline
4A Jasper	5A Schuyler
4A Jefferson	5A Scotland
4A Johnson	4A Scott
5A Knox	4A Shannon
4A Laclede	5A Shelby
4A Lafayette	4A St. Charles
4A Lawrence	4A St. Clair
5A Lewis	4A St. Francois
4A Lincoln	4A St. Louis
5A Linn	4A St. Louis (city)
5A Livingston	4A Ste. Genevieve
5A Macon	4A Stoddard
4A Madison	4A Stone
4A Maries	5A Sullivan
5A Marion	4A Taney
4A McDonald	4A Texas
5A Mercer	4A Vernon
4A Miller	4A Warren
4A Mississippi	4A Washington
4A Moniteau	4A Wayne
4A Monroe	4A Webster
4A Montgomery	5A Worth
4A Morgan	4A Wright
4A New Madrid	**MONTANA**
4A Newton	6B (all)
5A Nodaway	**NEBRASKA**
4A Oregon	5A (all)
4A Osage	**NEVADA**
4A Ozark	4B Carson City (city)
3A Pemiscot	5B Churchill
4A Perry	3B Clark
4A Pettis	4B Douglas
4A Phelps	5B Elko
5A Pike	4B Esmeralda
4A Platte	5B Eureka
4A Polk	5B Humboldt
4A Pulaski	5B Lander
5A Putnam	4B Lincoln
5A Ralls	4B Lyon
4A Randolph	4B Mineral
4A Ray	4B Nye
	5B Pershing

(continued)

TABLE R301.1—continued
CLIMATE ZONES, MOISTURE REGIMES, AND WARM HUMID DESIGNATIONS BY STATE, COUNTY AND TERRITORY[a]

US STATES—continued
NEVADA *(continued)*
5B Storey
5B Washoe
5B White Pine
NEW HAMPSHIRE
6A Belknap
6A Carroll
5A Cheshire
6A Coos
6A Grafton
5A Hillsborough
5A Merrimack
5A Rockingham
5A Strafford
6A Sullivan
NEW JERSEY
4A Atlantic
5A Bergen
4A Burlington
4A Camden
4A Cape May
4A Cumberland
4A Essex
4A Gloucester
4A Hudson
5A Hunterdon
4A Mercer
4A Middlesex
4A Monmouth
5A Morris
4A Ocean
5A Passaic
4A Salem
5A Somerset
5A Sussex
4A Union
5A Warren
NEW MEXICO
4B Bernalillo
4B Catron
3B Chaves
4B Cibola
5B Colfax
4B Curry

4B DeBaca
3B Doña Ana
3B Eddy
4B Grant
4B Guadalupe
5B Harding
3B Hidalgo
3B Lea
4B Lincoln
5B Los Alamos
3B Luna
5B McKinley
5B Mora
3B Otero
4B Quay
5B Rio Arriba
4B Roosevelt
5B Sandoval
5B San Juan
5B San Miguel
5B Santa Fe
3B Sierra
4B Socorro
5B Taos
5B Torrance
4B Union
4B Valencia
NEW YORK
5A Albany
5A Allegany
4A Bronx
5A Broome
5A Cattaraugus
5A Cayuga
5A Chautauqua
5A Chemung
6A Chenango
6A Clinton
5A Columbia
5A Cortland
6A Delaware
5A Dutchess
5A Erie
6A Essex
6A Franklin

(continued)

TABLE R301.1—continued
CLIMATE ZONES, MOISTURE REGIMES, AND WARM HUMID DESIGNATIONS BY STATE, COUNTY AND TERRITORY[a]

US STATES—continued	
NEW YORK *(continued)*	
6A Fulton	5A Wyoming
5A Genesee	5A Yates
5A Greene	**NORTH CAROLINA**
6A Hamilton	3A Alamance
6A Herkimer	3A Alexander
6A Jefferson	5A Alleghany
4A Kings	3A Anson
6A Lewis	5A Ashe
5A Livingston	5A Avery
6A Madison	3A Beaufort
5A Monroe	3A Bertie
6A Montgomery	3A Bladen
4A Nassau	3A Brunswick*
4A New York	4A Buncombe
5A Niagara	4A Burke
6A Oneida	3A Cabarrus
5A Onondaga	4A Caldwell
5A Ontario	3A Camden
5A Orange	3A Carteret*
5A Orleans	3A Caswell
5A Oswego	3A Catawba
6A Otsego	3A Chatham
5A Putnam	3A Cherokee
4A Queens	3A Chowan
5A Rensselaer	3A Clay
4A Richmond	3A Cleveland
5A Rockland	3A Columbus*
5A Saratoga	3A Craven
5A Schenectady	3A Cumberland
5A Schoharie	3A Currituck
5A Schuyler	3A Dare
5A Seneca	3A Davidson
5A Steuben	3A Davie
6A St. Lawrence	3A Duplin
4A Suffolk	3A Durham
6A Sullivan	3A Edgecombe
5A Tioga	3A Forsyth
5A Tompkins	3A Franklin
6A Ulster	3A Gaston
6A Warren	3A Gates
5A Washington	4A Graham
5A Wayne	3A Granville
4A Westchester	3A Greene
	3A Guilford
	3A Halifax

(continued)

GENERAL REQUIREMENTS

TABLE R301.1—continued
CLIMATE ZONES, MOISTURE REGIMES, AND WARM HUMID DESIGNATIONS BY STATE, COUNTY AND TERRITORY[a]

US STATES—continued	
NORTH CAROLINA *(continued)*	4A Surry
3A Harnett	4A Swain
4A Haywood	4A Transylvania
4A Henderson	3A Tyrrell
3A Hertford	3A Union
3A Hoke	3A Vance
3A Hyde	3A Wake
3A Iredell	3A Warren
4A Jackson	3A Washington
3A Johnston	5A Watauga
3A Jones	3A Wayne
3A Lee	3A Wilkes
3A Lenoir	3A Wilson
3A Lincoln	4A Yadkin
4A Macon	5A Yancey
4A Madison	**NORTH DAKOTA**
3A Martin	6A Adams
4A McDowell	6A Barnes
3A Mecklenburg	7 Benson
4A Mitchell	6A Billings
3A Montgomery	7 Bottineau
3A Moore	6A Bowman
3A Nash	7 Burke
3A New Hanover*	6A Burleigh
3A Northampton	6A Cass
3A Onslow*	7 Cavalier
3A Orange	6A Dickey
3A Pamlico	7 Divide
3A Pasquotank	6A Dunn
3A Pender*	6A Eddy
3A Perquimans	6A Emmons
3A Person	6A Foster
3A Pitt	6A Golden Valley
3A Polk	7 Grand Forks
3A Randolph	6A Grant
3A Richmond	6A Griggs
3A Robeson	6A Hettinger
3A Rockingham	6A Kidder
3A Rowan	6A LaMoure
3A Rutherford	6A Logan
3A Sampson	7 McHenry
3A Scotland	6A McIntosh
3A Stanly	6A McKenzie
4A Stokes	6A McLean
	6A Mercer

(continued)

TABLE R301.1—continued
CLIMATE ZONES, MOISTURE REGIMES, AND WARM HUMID DESIGNATIONS BY STATE, COUNTY AND TERRITORY[a]

US STATES—continued	
NORTH DAKOTA *(continued)*	5A Darke
6A Morton	5A Defiance
6A Mountrail	5A Delaware
7 Nelson	5A Erie
6A Oliver	5A Fairfield
7 Pembina	4A Fayette
7 Pierce	4A Franklin
7 Ramsey	5A Fulton
6A Ransom	4A Gallia
7 Renville	5A Geauga
6A Richland	4A Greene
7 Rolette	5A Guernsey
6A Sargent	4A Hamilton
6A Sheridan	5A Hancock
6A Sioux	5A Hardin
6A Slope	5A Harrison
6A Stark	5A Henry
6A Steele	4A Highland
6A Stutsman	4A Hocking
7 Towner	5A Holmes
6A Traill	5A Huron
7 Walsh	4A Jackson
7 Ward	5A Jefferson
6A Wells	5A Knox
6A Williams	5A Lake
OHIO	4A Lawrence
4A Adams	5A Licking
5A Allen	5A Logan
5A Ashland	5A Lorain
5A Ashtabula	5A Lucas
4A Athens	4A Madison
5A Auglaize	5A Mahoning
5A Belmont	5A Marion
4A Brown	5A Medina
4A Butler	4A Meigs
5A Carroll	5A Mercer
5A Champaign	5A Miami
5A Clark	5A Monroe
4A Clermont	5A Montgomery
4A Clinton	5A Morgan
5A Columbiana	5A Morrow
5A Coshocton	5A Muskingum
5A Crawford	5A Noble
5A Cuyahoga	5A Ottawa
	5A Paulding

(continued)

GENERAL REQUIREMENTS

TABLE R301.1—continued
CLIMATE ZONES, MOISTURE REGIMES, AND WARM HUMID DESIGNATIONS BY STATE, COUNTY AND TERRITORY[a]

US STATES—continued	
OHIO (continued)	4A Craig
5A Perry	3A Creek
4A Pickaway	3A Custer
4A Pike	4A Delaware
5A Portage	3A Dewey
5A Preble	4A Ellis
5A Putnam	4A Garfield
5A Richland	3A Garvin
4A Ross	3A Grady
5A Sandusky	4A Grant
4A Scioto	3A Greer
5A Seneca	3A Harmon
5A Shelby	4A Harper
5A Stark	3A Haskell
5A Summit	3A Hughes
5A Trumbull	3A Jackson
5A Tuscarawas	3A Jefferson
5A Union	3A Johnston
5A Van Wert	4A Kay
4A Vinton	3A Kingfisher
4A Warren	3A Kiowa
4A Washington	3A Latimer
5A Wayne	3A Le Flore
5A Williams	3A Lincoln
5A Wood	3A Logan
5A Wyandot	3A Love
OKLAHOMA	4A Major
3A Adair	3A Marshall
4A Alfalfa	3A Mayes
3A Atoka	3A McClain
4B Beaver	3A McCurtain
3A Beckham	3A McIntosh
3A Blaine	3A Murray
3A Bryan	3A Muskogee
3A Caddo	3A Noble
3A Canadian	4A Nowata
3A Carter	3A Okfuskee
3A Cherokee	3A Oklahoma
3A Choctaw	3A Okmulgee
4B Cimarron	4A Osage
3A Cleveland	4A Ottawa
3A Coal	3A Pawnee
3A Comanche	3A Payne
3A Cotton	3A Pittsburg
	3A Pontotoc

(continued)

TABLE R301.1—continued
CLIMATE ZONES, MOISTURE REGIMES, AND WARM HUMID DESIGNATIONS BY STATE, COUNTY AND TERRITORY[a]

US STATES—continued	
OKLAHOMA *(continued)*	5B Sherman
3A Pottawatomie	4C Tillamook
3A Pushmataha	5B Umatilla
3A Roger Mills	5B Union
3A Rogers	5B Wallowa
3A Seminole	5B Wasco
3A Sequoyah	4C Washington
3A Stephens	5B Wheeler
4B Texas	4C Yamhill
3A Tillman	**PENNSYLVANIA**
3A Tulsa	4A Adams
3A Wagoner	5A Allegheny
4A Washington	5A Armstrong
3A Washita	5A Beaver
4A Woods	5A Bedford
4A Woodward	4A Berks
OREGON	5A Blair
5B Baker	5A Bradford
4C Benton	4A Bucks
4C Clackamas	5A Butler
4C Clatsop	5A Cambria
4C Columbia	5A Cameron
4C Coos	5A Carbon
5B Crook	5A Centre
4C Curry	4A Chester
5B Deschutes	5A Clarion
4C Douglas	5A Clearfield
5B Gilliam	5A Clinton
5B Grant	5A Columbia
5B Harney	5A Crawford
5B Hood River	4A Cumberland
4C Jackson	4A Dauphin
5B Jefferson	4A Delaware
4C Josephine	5A Elk
5B Klamath	5A Erie
5B Lake	5A Fayette
4C Lane	5A Forest
4C Lincoln	4A Franklin
4C Linn	5A Fulton
5B Malheur	5A Greene
4C Marion	5A Huntingdon
5B Morrow	5A Indiana
4C Multnomah	5A Jefferson
4C Polk	5A Juniata
	5A Lackawanna

(continued)

GENERAL REQUIREMENTS

TABLE R301.1—continued
CLIMATE ZONES, MOISTURE REGIMES, AND WARM HUMID DESIGNATIONS BY STATE, COUNTY AND TERRITORY[a]

US STATES—continued	
PENNSYLVANIA *(continued)*	3A Calhoun
4A Lancaster	3A Charleston*
5A Lawrence	3A Cherokee
4A Lebanon	3A Chester
5A Lehigh	3A Chesterfield
5A Luzerne	3A Clarendon
5A Lycoming	3A Colleton*
5A McKean	3A Darlington
5A Mercer	3A Dillon
5A Mifflin	3A Dorchester*
5A Monroe	3A Edgefield
4A Montgomery	3A Fairfield
5A Montour	3A Florence
5A Northampton	3A Georgetown*
5A Northumberland	3A Greenville
4A Perry	3A Greenwood
4A Philadelphia	3A Hampton*
5A Pike	3A Horry*
5A Potter	2A Jasper*
5A Schuylkill	3A Kershaw
5A Snyder	3A Lancaster
5A Somerset	3A Laurens
5A Sullivan	3A Lee
5A Susquehanna	3A Lexington
5A Tioga	3A Marion
5A Union	3A Marlboro
5A Venango	3A McCormick
5A Warren	3A Newberry
5A Washington	3A Oconee
5A Wayne	3A Orangeburg
5A Westmoreland	3A Pickens
5A Wyoming	3A Richland
4A York	3A Saluda
RHODE ISLAND	3A Spartanburg
5A (all)	3A Sumter
SOUTH CAROLINA	3A Union
3A Abbeville	3A Williamsburg
3A Aiken	3A York
3A Allendale*	**SOUTH DAKOTA**
3A Anderson	6A Aurora
3A Bamberg*	6A Beadle
3A Barnwell*	5A Bennett
2A Beaufort*	5A Bon Homme
3A Berkeley*	6A Brookings
	6A Brown

(continued)

TABLE R301.1—continued
CLIMATE ZONES, MOISTURE REGIMES, AND WARM HUMID DESIGNATIONS BY STATE, COUNTY AND TERRITORY[a]

US STATES—continued	
SOUTH DAKOTA *(continued)*	6A Moody
5A Brule	6A Pennington
6A Buffalo	6A Perkins
6A Butte	6A Potter
6A Campbell	6A Roberts
5A Charles Mix	6A Sanborn
6A Clark	6A Shannon
5A Clay	6A Spink
6A Codington	5A Stanley
6A Corson	6A Sully
6A Custer	5A Todd
6A Davison	5A Tripp
6A Day	6A Turner
6A Deuel	5A Union
6A Dewey	6A Walworth
5A Douglas	5A Yankton
6A Edmunds	6A Ziebach
6A Fall River	**TENNESSEE**
6A Faulk	4A Anderson
6A Grant	3A Bedford
5A Gregory	4A Benton
5A Haakon	4A Bledsoe
6A Hamlin	4A Blount
6A Hand	4A Bradley
6A Hanson	4A Campbell
6A Harding	4A Cannon
6A Hughes	4A Carroll
5A Hutchinson	4A Carter
6A Hyde	4A Cheatham
5A Jackson	3A Chester
6A Jerauld	4A Claiborne
5A Jones	4A Clay
6A Kingsbury	4A Cocke
6A Lake	3A Coffee
6A Lawrence	3A Crockett
6A Lincoln	4A Cumberland
5A Lyman	3A Davidson
6A Marshall	3A Decatur
6A McCook	4A DeKalb
6A McPherson	4A Dickson
6A Meade	3A Dyer
5A Mellette	3A Fayette
6A Miner	4A Fentress
6A Minnehaha	3A Franklin
	3A Gibson

(continued)

GENERAL REQUIREMENTS

TABLE R301.1—continued
CLIMATE ZONES, MOISTURE REGIMES, AND WARM HUMID DESIGNATIONS BY STATE, COUNTY AND TERRITORY[a]

US STATES—continued	
TENNESSEE *(continued)*	4A Putnam
3A Giles	4A Rhea
4A Grainger	4A Roane
4A Greene	4A Robertson
3A Grundy	3A Rutherford
4A Hamblen	4A Scott
3A Hamilton	4A Sequatchie
4A Hancock	4A Sevier
3A Hardeman	3A Shelby
3A Hardin	4A Smith
4A Hawkins	4A Stewart
3A Haywood	4A Sullivan
3A Henderson	4A Sumner
4A Henry	3A Tipton
3A Hickman	4A Trousdale
4A Houston	4A Unicoi
4A Humphreys	4A Union
4A Jackson	4A Van Buren
4A Jefferson	4A Warren
4A Johnson	4A Washington
4A Knox	3A Wayne
4A Lake	4A Weakley
3A Lauderdale	4A White
3A Lawrence	3A Williamson
3A Lewis	4A Wilson
3A Lincoln	**TEXAS**
4A Loudon	2A Anderson*
4A Macon	3B Andrews
3A Madison	2A Angelina*
3A Marion	2A Aransas*
3A Marshall	3A Archer
3A Maury	4B Armstrong
4A McMinn	2A Atascosa*
3A McNairy	2A Austin*
4A Meigs	4B Bailey
4A Monroe	2B Bandera
4A Montgomery	2A Bastrop*
3A Moore	3B Baylor
4A Morgan	2A Bee*
4A Obion	2A Bell*
4A Overton	2A Bexar*
3A Perry	3A Blanco*
4A Pickett	3B Borden
4A Polk	2A Bosque*
	3A Bowie*

(continued)

TABLE R301.1—continued
CLIMATE ZONES, MOISTURE REGIMES, AND WARM HUMID DESIGNATIONS BY STATE, COUNTY AND TERRITORY[a]

US STATES—continued
TEXAS *(continued)*
2A Brazoria*
2A Brazos*
3B Brewster
4B Briscoe
2A Brooks*
3A Brown*
2A Burleson*
3A Burnet*
2A Caldwell*
2A Calhoun*
3B Callahan
1A Cameron*
3A Camp*
4B Carson
3A Cass*
4B Castro
2A Chambers*
2A Cherokee*
3B Childress
3A Clay
4B Cochran
3B Coke
3B Coleman
3A Collin*
3B Collingsworth
2A Colorado*
2A Comal*
3A Comanche*
3B Concho
3A Cooke
2A Coryell*
3B Cottle
3B Crane
3B Crockett
3B Crosby
3B Culberson
4B Dallam
2A Dallas*
3B Dawson
4B Deaf Smith
3A Delta
3A Denton*
2A DeWitt*
3B Dickens
2B Dimmit
4B Donley
2A Duval*
3A Eastland
3B Ector
2B Edwards
2A Ellis*
3B El Paso
3A Erath*
2A Falls*
3A Fannin
2A Fayette*
3B Fisher
4B Floyd
3B Foard
2A Fort Bend*
3A Franklin*
2A Freestone*
2B Frio
3B Gaines
2A Galveston*
3B Garza
3A Gillespie*
3B Glasscock
2A Goliad*
2A Gonzales*
4B Gray
3A Grayson
3A Gregg*
2A Grimes*
2A Guadalupe*
4B Hale
3B Hall
3A Hamilton*
4B Hansford
3B Hardeman
2A Hardin*
2A Harris*
3A Harrison*
4B Hartley
3B Haskell
2A Hays*
3B Hemphill
3A Henderson*

(continued)

GENERAL REQUIREMENTS

TABLE R301.1—continued
CLIMATE ZONES, MOISTURE REGIMES, AND WARM HUMID DESIGNATIONS BY STATE, COUNTY AND TERRITORY[a]

US STATES—continued	
TEXAS *(continued)*	3B Loving
1A Hidalgo*	3B Lubbock
2A Hill*	3B Lynn
4B Hockley	2A Madison*
3A Hood*	3A Marion*
3A Hopkins*	3B Martin
2A Houston*	3B Mason
3B Howard	2A Matagorda*
3B Hudspeth	2B Maverick
3A Hunt*	3B McCulloch
4B Hutchinson	2A McLennan*
3B Irion	2A McMullen*
3A Jack	2B Medina
2A Jackson*	3B Menard
2A Jasper*	3B Midland
3B Jeff Davis	2A Milam*
2A Jefferson*	3A Mills*
2A Jim Hogg*	3B Mitchell
2A Jim Wells*	3A Montague
2A Johnson*	2A Montgomery*
3B Jones	4B Moore
2A Karnes*	3A Morris*
3A Kaufman*	3B Motley
3A Kendall*	3A Nacogdoches*
2A Kenedy*	2A Navarro*
3B Kent	2A Newton*
3B Kerr	3B Nolan
3B Kimble	2A Nueces*
3B King	4B Ochiltree
2B Kinney	4B Oldham
2A Kleberg*	2A Orange*
3B Knox	3A Palo Pinto*
3A Lamar*	3A Panola*
4B Lamb	3A Parker*
3A Lampasas*	4B Parmer
2B La Salle	3B Pecos
2A Lavaca*	2A Polk*
2A Lee*	4B Potter
2A Leon*	3B Presidio
2A Liberty*	3A Rains*
2A Limestone*	4B Randall
4B Lipscomb	3B Reagan
2A Live Oak*	2B Real
3A Llano*	3A Red River*
	3B Reeves

(continued)

GENERAL REQUIREMENTS

TABLE R301.1—continued
CLIMATE ZONES, MOISTURE REGIMES, AND WARM HUMID DESIGNATIONS BY STATE, COUNTY AND TERRITORY[a]

US STATES—continued	
TEXAS *(continued)*	2A Washington*
2A Refugio*	2B Webb
4B Roberts	2A Wharton*
2A Robertson*	3B Wheeler
3A Rockwall*	3A Wichita
3B Runnels	3B Wilbarger
3A Rusk*	1A Willacy*
3A Sabine*	2A Williamson*
3A San Augustine*	2A Wilson*
2A San Jacinto*	3B Winkler
2A San Patricio*	3A Wise
3A San Saba*	3A Wood*
3B Schleicher	4B Yoakum
3B Scurry	3A Young
3B Shackelford	2B Zapata
3A Shelby*	2B Zavala
4B Sherman	**UTAH**
3A Smith*	5B Beaver
3A Somervell*	5B Box Elder
2A Starr*	5B Cache
3A Stephens	5B Carbon
3B Sterling	6B Daggett
3B Stonewall	5B Davis
3B Sutton	6B Duchesne
4B Swisher	5B Emery
2A Tarrant*	5B Garfield
3B Taylor	5B Grand
3B Terrell	5B Iron
3B Terry	5B Juab
3B Throckmorton	5B Kane
3A Titus*	5B Millard
3B Tom Green	6B Morgan
2A Travis*	5B Piute
2A Trinity*	6B Rich
2A Tyler*	5B Salt Lake
3A Upshur*	5B San Juan
3B Upton	5B Sanpete
2B Uvalde	5B Sevier
2B Val Verde	6B Summit
3A Van Zandt*	5B Tooele
2A Victoria*	6B Uintah
2A Walker*	5B Utah
2A Waller*	6B Wasatch
3B Ward	3B Washington
	5B Wayne

(continued)

TABLE R301.1—continued
CLIMATE ZONES, MOISTURE REGIMES, AND WARM HUMID DESIGNATIONS BY STATE, COUNTY AND TERRITORY[a]

US STATES—continued	
UTAH *(continued)*	4C Grays Harbor
5B Weber	5C Island
VERMONT	4C Jefferson
6A (all)	4C King
VIRGINIA	5C Kitsap
4A (all except as follows:)	5B Kittitas
5A Alleghany	5B Klickitat
5A Bath	4C Lewis
3A Brunswick	5B Lincoln
3A Chesapeake	4C Mason
5A Clifton Forge	5B Okanogan
5A Covington	4C Pacific
3A Emporia	6B Pend Oreille
3A Franklin	4C Pierce
3A Greensville	5C San Juan
3A Halifax	4C Skagit
3A Hampton	5B Skamania
5A Highland	4C Snohomish
3A Isle of Wight	5B Spokane
3A Mecklenburg	6B Stevens
3A Newport News	4C Thurston
3A Norfolk	4C Wahkiakum
3A Pittsylvania	5B Walla Walla
3A Portsmouth	4C Whatcom
3A South Boston	5B Whitman
3A Southampton	5B Yakima
3A Suffolk	**WEST VIRGINIA**
3A Surry	5A Barbour
3A Sussex	4A Berkeley
3A Virginia Beach	4A Boone
WASHINGTON	4A Braxton
5B Adams	5A Brooke
5B Asotin	4A Cabell
5B Benton	4A Calhoun
5B Chelan	4A Clay
5C Clallam	4A Doddridge
4C Clark	4A Fayette
5B Columbia	4A Gilmer
4C Cowlitz	5A Grant
5B Douglas	4A Greenbrier
6B Ferry	5A Hampshire
5B Franklin	5A Hancock
5B Garfield	5A Hardy
5B Grant	5A Harrison
	4A Jackson

(continued)

GENERAL REQUIREMENTS

TABLE R301.1—continued
CLIMATE ZONES, MOISTURE REGIMES, AND WARM HUMID DESIGNATIONS BY STATE, COUNTY AND TERRITORY[a]

US STATES—continued
WEST VIRGINIA (continued)
4A Jefferson
4A Kanawha
4A Lewis
4A Lincoln
4A Logan
5A Marion
5A Marshall
4A Mason
4A McDowell
4A Mercer
5A Mineral
4A Mingo
5A Monongalia
4A Monroe
4A Morgan
4A Nicholas
5A Ohio
5A Pendleton
4A Pleasants
5A Pocahontas
5A Preston
4A Putnam
4A Raleigh
5A Randolph
4A Ritchie
4A Roane
4A Summers
5A Taylor
5A Tucker
4A Tyler
4A Upshur
4A Wayne
4A Webster
5A Wetzel
4A Wirt
4A Wood
4A Wyoming
WISCONSIN
5A Adams
6A Ashland
6A Barron
6A Bayfield
6A Brown
6A Buffalo
6A Burnett
5A Calumet
6A Chippewa
6A Clark
5A Columbia
5A Crawford
5A Dane
5A Dodge
6A Door
6A Douglas
6A Dunn
6A Eau Claire
6A Florence
5A Fond du Lac
6A Forest
5A Grant
5A Green
5A Green Lake
5A Iowa
6A Iron
6A Jackson
5A Jefferson
5A Juneau
5A Kenosha
6A Kewaunee
5A La Crosse
5A Lafayette
6A Langlade
6A Lincoln
6A Manitowoc
6A Marathon
6A Marinette
6A Marquette
6A Menominee
5A Milwaukee
5A Monroe
6A Oconto
6A Oneida
5A Outagamie
5A Ozaukee
6A Pepin
6A Pierce
6A Polk
6A Portage

(continued)

GENERAL REQUIREMENTS

TABLE R301.1—continued
CLIMATE ZONES, MOISTURE REGIMES, AND WARM HUMID DESIGNATIONS BY STATE, COUNTY AND TERRITORY[a]

US STATES—continued	
WISCONSIN *(continued)*	6B Fremont
6A Price	5B Goshen
5A Racine	6B Hot Springs
5A Richland	6B Johnson
5A Rock	5B Laramie
6A Rusk	7 Lincoln
5A Sauk	6B Natrona
6A Sawyer	6B Niobrara
6A Shawano	6B Park
6A Sheboygan	5B Platte
6A St. Croix	6B Sheridan
6A Taylor	7 Sublette
6A Trempealeau	6B Sweetwater
5A Vernon	7 Teton
6A Vilas	6B Uinta
5A Walworth	6B Washakie
6A Washburn	6B Weston
5A Washington	**US TERRITORIES**
5A Waukesha	**AMERICAN SAMOA**
6A Waupaca	1A (all)*
5A Waushara	**GUAM**
5A Winnebago	1A (all)*
6A Wood	**NORTHERN MARIANA ISLANDS**
WYOMING	1A (all)*
6B Albany	**PUERTO RICO**
6B Big Horn	1A (all except as follows:)*
6B Campbell	2B Barraquitas
6B Carbon	2B Cayey
6B Converse	**VIRGIN ISLANDS**
6B Crook	1A (all)*

a. Key: A – Moist, B – Dry, C – Marine. Absence of moisture designation indicates moisture regime is irrelevant. Asterisk (*) indicates a Warm Humid location.

R301.3 Climate zone definitions. To determine the climate zones for locations not listed in this code, use the following information to determine climate zone numbers and letters in accordance with Items 1 through 5.

1. Determine the thermal climate zone, 0 through 8, from Table R301.3 using the heating (HDD) and cooling degree-days (CDD) for the location.
2. Determine the moisture zone (Marine, Dry or Humid) in accordance with Items 2.1 through 2.3.
    2.1. If monthly average temperature and precipitation data are available, use the Marine, Dry and Humid definitions to determine the moisture zone (C, B or A).
    2.2. If annual average temperature information (including degree-days) and annual precipitation (i.e., annual mean) are available, use Items 2.2.1 through 2.2.3 to determine the moisture zone. If the moisture zone is not Marine, then use the Dry definition to determine whether Dry or Humid.
        2.2.1. If thermal climate zone is 3 and CDD50°F $\leq$ 4,500 (CDD10°C $\leq$ 2500), climate zone is Marine (3C).
        2.2.2. If thermal climate zone is 4 and CDD50°F $\leq$ 2,700 (CDD10°C $\leq$ 1500), climate zone is Marine (4C).
        2.2.3. If thermal climate zone is 5 and CDD50°F $\leq$ 1,800 (CDD10°C $\leq$ 1000), climate zone is Marine (5C).
    2.3. If only degree-day information is available, use Items 2.3.1 through 2.3.3 to determine the moisture zone. If the moisture zone is not Marine, then it is not possible to assign Humid or Dry moisture zone for this location.
        2.3.1. If thermal climate zone is 3 and CDD50°F $\leq$ 4,500 (CDD10°C $\leq$ 2500), climate zone is Marine (3C).
        2.3.2. If thermal climate zone is 4 and CDD50°F $\leq$ 2,700 (CDD10°C $\leq$ 1500), climate zone is Marine (4C).
        2.3.3. If thermal climate zone is 5 and CDD50°F $\leq$ 1,800 (CDD10°C $\leq$ 1000), climate zone is Marine (5C).
3. Marine (C) Zone definition: Locations meeting all the criteria in Items 3.1 through 3.4.
    3.1. Mean temperature of coldest month between 27°F (-3°C) and 65°F (18°C).
    3.2. Warmest month mean < 72°F (22°C).
    3.3. Not fewer than four months with mean temperatures over 50°F (10°C).
    3.4. Dry season in summer. The month with the heaviest precipitation in the cold season has at least three times as much precipitation as the month with the least precipitation in the rest of the year. The cold season is October through March in the Northern Hemisphere and April through September in the Southern Hemisphere.
4. Dry (B) definition: Locations meeting the criteria in Items 4.1 through 4.4.
    4.1. Not Marine (C).
    4.2. If 70 percent or more of the precipitation, P, occurs during the high sun period, defined as April through September in the Northern Hemisphere and October through March in the Southern Hemisphere, then the dry/humid threshold is in accordance with Equation 3-1.

    $P < 0.44 \times (T - 7)$
    $[P < 20.0 \times (T + 14)$ in SI units]
    (Equation 3-1)
    where:
    P = Annual precipitation, inches (mm).
    T = Annual mean temperature, °F (°C).

    4.3. If between 30 and 70 percent of the precipitation, P, occurs during the high sun period, defined as April through September in the Northern Hemisphere and October through March in the Southern Hemisphere, then the dry/humid threshold is in accordance with Equation 3-2.

    $P < 0.44 \times (T - 19.5)$
    $[P < 20.0 \times (T + 7)$ in SI units]
    (Equation 3-2)
    where:
    P = Annual precipitation, inches (mm).
    T = Annual mean temperature, °F (°C).

    4.4. If 30 percent or less of the precipitation, P, occurs during the high sun period, defined as April through September in the Northern Hemisphere and October through March in the Southern Hemisphere, then the dry/humid threshold is in accordance with Equation 3-3.

    $P < 0.44 \times (T - 32)$
    $[P < 20.0 \times T$ in SI units]
    (Equation 3-3)
    where:
    P = Annual precipitation, inches (mm).
    T = Annual mean temperature, °F (°C).

5. Humid (A) definition: Locations that are not Marine (C) or Dry (B).

GENERAL REQUIREMENTS

TABLE R301.3
THERMAL CLIMATE ZONE DEFINITIONS

ZONE NUMBER	THERMAL CRITERIA	
	IP Units	SI Units
0	10,800 < CDD50°F	6000 < CDD10°C
1	9,000 < CDD50°F < 10,800	5000 < CDD10°C < 6000
2	6,300 < CDD50°F ≤ 9,000	3500 < CDD10°C ≤ 5000
3	CDD50°F ≤ 6,300 AND HDD65°F ≤ 3,600	CDD10°C < 3500 AND HDD18°C ≤ 2000
4	CDD50°F ≤ 6,300 AND 3,600 < HDD65°F ≤ 5,400	CDD10°C < 3500 AND 2000 < HDD18°C ≤ 3000
5	CDD50°F < 6,300 AND 5,400 < HDD65°F ≤ 7,200	CDD10°C < 3500 AND 3000 < HDD18°C ≤ 4000
6	7,200 < HDD65°F ≤ 9,000	4000 < HDD18°C ≤ 5000
7	9,000 < HDD65°F ≤ 12,600	5000 < HDD18°C ≤ 7000
8	12,600 < HDD65°F	7000 < HDD18°C

For SI: °C = [(°F) - 32]/1.8.

R301.4 Tropical climate region. The tropical region shall be defined as:

1. Hawaii, Puerto Rico, Guam, American Samoa, U.S. Virgin Islands, Commonwealth of Northern Mariana Islands; and
2. Islands in the area between the Tropic of Cancer and the Tropic of Capricorn.

SECTION R302
DESIGN CONDITIONS

R302.1 Interior design conditions. The interior design temperatures used for heating and cooling load calculations shall be a maximum of 72°F (22°C) for heating and minimum of 75°F (24°C) for cooling.

SECTION R303
MATERIALS, SYSTEMS AND EQUIPMENT

R303.1 Identification. Materials, systems and equipment shall be identified in a manner that will allow a determination of compliance with the applicable provisions of this code.

R303.1.1 Building thermal envelope insulation. An R-value identification mark shall be applied by the manufacturer to each piece of *building thermal envelope* insulation that is 12 inches (305 mm) or greater in width. Alternatively, the insulation installers shall provide a certification that indicates the type, manufacturer and R-value of insulation installed in each element of the *building thermal envelope*. For blown-in or sprayed fiberglass and cellulose insulation, the initial installed thickness, settled thickness, settled R-value, installed density, coverage area and number of bags installed shall be indicated on the certification. For sprayed polyurethane foam (SPF) insulation, the installed thickness of the areas covered and the R-value of the installed thickness shall be indicated on the certification. For insulated siding, the R-value shall be on a label on the product's package and shall be indicated on the certification. The insulation installer shall sign, date and post the certification in a conspicuous location on the job site.

> **Exception:** For roof insulation installed above the deck, the R-value shall be labeled as required by the material standards specified in Table 1508.2 of the *International Building Code* or Table R906.2 of the *International Residential Code*, as applicable.

R303.1.1.1 Blown-in or sprayed roof and ceiling insulation. The thickness of blown-in or sprayed fiberglass and cellulose roof and ceiling insulation shall be written in inches (mm) on markers that are installed at not less than one for every 300 square feet (28 m^2) throughout the attic space. The markers shall be affixed to the trusses or joists and marked with the minimum initial installed thickness with numbers not less than 1 inch (25 mm) in height. Each marker shall face the attic access opening. The thickness and installed R-value of sprayed polyurethane foam insulation shall be indicated on the certification provided by the insulation installer.

R303.1.2 Insulation mark installation. Insulating materials shall be installed such that the manufacturer's R-value mark is readily observable at inspection. For insulation materials that are installed without an observable manufacturer's R-value mark, such as blown or draped products, an insulation certificate complying with Section R303.1.1 shall be left immediately after installation by the installer, in a conspicuous location within the building, to certify the installed R-value of the insulation material.

R303.1.3 Fenestration product rating. U-factors of fenestration products such as windows, doors and *skylights* shall be determined in accordance with NFRC 100.

> **Exception:** Where required, garage door U-factors shall be determined in accordance with either NFRC 100 or ANSI/DASMA 105.

U-factors shall be determined by an accredited, independent laboratory, and labeled and certified by the manufacturer.

Products lacking such a labeled U-factor shall be assigned a default U-factor from Table R303.1.3(1) or Table R303.1.3(2). The *solar heat gain coefficient* (SHGC) and *visible transmittance* (VT) of glazed fenestration products such as windows, glazed doors and *skylights* shall be determined in accordance with NFRC 200 by an accredited, independent laboratory, and labeled and certified by the manufacturer. Products lacking such a labeled SHGC or VT shall be assigned a default SHGC or VT from Table R303.1.3(3).

TABLE R303.1.3(1)
DEFAULT GLAZED WINDOW, GLASS DOOR AND SKYLIGHT *U*-FACTORS

FRAME TYPE	WINDOW AND GLASS DOOR		SKYLIGHT	
	Single pane	Double pane	Single	Double
Metal	1.20	0.80	2.00	1.30
Metal with Thermal Break	1.10	0.65	1.90	1.10
Nonmetal or Metal Clad	0.95	0.55	1.75	1.05
Glazed Block	0.60			

TABLE R303.1.3(2)
DEFAULT OPAQUE DOOR *U*-FACTORS

DOOR TYPE	OPAQUE *U*-FACTOR
Uninsulated Metal	1.20
Insulated Metal	0.60
Wood	0.50
Insulated, nonmetal edge, not exceeding 45% glazing, any glazing double pane	0.35

TABLE R303.1.3(3)
DEFAULT GLAZED FENESTRATION SHGC AND VT

	SINGLE GLAZED		DOUBLE GLAZED		GLAZED BLOCK
	Clear	Tinted	Clear	Tinted	
SHGC	0.8	0.7	0.7	0.6	0.6
VT	0.6	0.3	0.6	0.3	0.6

R303.1.4 Insulation product rating. The thermal resistance, *R*-value, of insulation shall be determined in accordance with Part 460 of US-FTC CFR Title 16 in units of h × ft^2 × °F/Btu at a mean temperature of 75°F (24°C).

R303.1.4.1 Insulated siding. The thermal resistance, *R*-value, of insulated siding shall be determined in accordance with ASTM C1363. Installation for testing shall be in accordance with the manufacturer's instructions.

R303.1.5 Air-impermeable insulation. Insulation having an air permeability not greater than 0.004 cubic feet per minute per square foot [0.002 L/(s × m^2)] under pressure differential of 0.3 inch water gauge (75 Pa) when tested in accordance with ASTM E2178 shall be determined air-impermeable insulation.

R303.2 Installation. Materials, systems and equipment shall be installed in accordance with the manufacturer's instructions and the *International Building Code* or the *International Residential Code*, as applicable.

R303.2.1 Protection of exposed foundation insulation. Insulation applied to the exterior of *basement walls*, crawl space walls and the perimeter of slab-on-grade floors shall have a rigid, opaque and weather-resistant protective covering to prevent the degradation of the insulation's thermal performance. The protective covering shall cover the exposed exterior insulation and extend not less than 6 inches (153 mm) below grade.

R303.3 Maintenance information. Maintenance instructions shall be furnished for equipment and systems that require preventive maintenance. Required regular maintenance actions shall be clearly stated and incorporated on a readily visible label. The label shall include the title or publication number for the operation and maintenance manual for that particular model and type of product.

CHAPTER 4 [RE]
RESIDENTIAL ENERGY EFFICIENCY

User note:

About this chapter: Chapter 4 presents the paths and options for compliance with the energy efficiency provisions. Chapter 4 contains energy efficiency provisions for the building envelope, mechanical and water heating systems, lighting and additional efficiency requirements. A performance alternative, energy rating alternative, and tropical regional alternative are also provided to allow for energy code compliance other than by the prescriptive method.

SECTION R401
GENERAL

R401.1 Scope. This chapter applies to residential buildings.

R401.2 Application. Residential buildings shall comply with Section R401.2.5 and either Sections R401.2.1, R401.2.2, R401.2.3 or R401.2.4.

Exception: Additions, *alterations*, repairs and changes of occupancy to existing buildings complying with Chapter 5.

R401.2.1 Prescriptive Compliance Option. The Prescriptive Compliance Option requires compliance with Sections R401 through R404.

R401.2.2 Total Building Performance Option. The Total Building Performance Option requires compliance with Section R405.

R401.2.3 Energy Rating Index Option. The Energy Rating Index (ERI) Option requires compliance with Section R406.

R401.2.4 Tropical Climate Region Option. The Tropical Climate Region Option requires compliance with Section R407.

R401.2.5 Additional energy efficiency. This section establishes additional requirements applicable to all compliance approaches to achieve additional energy efficiency.

1. For buildings complying with Section R401.2.1, one of the additional efficiency package options shall be installed according to Section R408.2.
2. For buildings complying with Section R401.2.2, the building shall meet one of the following:
    2.1. One of the additional efficiency package options in Section R408.2 shall be installed without including such measures in the proposed design under Section R405; or
    2.2. The proposed design of the building under Section R405.2 shall have an annual energy cost that is less than or equal to 95 percent of the annual energy cost of the standard reference design.
3. For buildings complying with the Energy Rating Index alternative Section R401.2.3, the Energy Rating Index value shall be at least 5 percent less than the Energy Rating Index target specified in Table R406.5.

The option selected for compliance shall be identified in the certificate required by Section R401.3.

R401.3 Certificate. A permanent certificate shall be completed by the builder or other *approved* party and posted on a wall in the space where the furnace is located, a utility room or an *approved* location inside the *building*. Where located on an electrical panel, the certificate shall not cover or obstruct the visibility of the circuit directory *label*, service disconnect *label* or other required labels. The certificate shall indicate the following:

1. The predominant *R*-values of insulation installed in or on ceilings, roofs, walls, foundation components such as slabs, *basement walls*, *crawl space walls* and floors and ducts outside *conditioned spaces*.
2. *U*-factors of fenestration and the *solar heat gain coefficient* (SHGC) of fenestration. Where there is more than one value for any component of the building envelope, the certificate shall indicate both the value covering the largest area and the area weighted average value if available.
3. The results from any required duct system and building envelope air leakage testing performed on the building.
4. The types, sizes and efficiencies of heating, cooling and service water-heating equipment. Where a gas-fired unvented room heater, electric furnace or baseboard electric heater is installed in the residence, the certificate shall indicate "gas-fired unvented room heater," "electric furnace" or "baseboard electric heater," as appropriate. An efficiency shall not be indicated for gas-fired unvented room heaters, electric furnaces and electric baseboard heaters.
5. Where on-site *photovoltaic panel* systems have been installed, the array capacity, inverter efficiency, panel tilt and orientation shall be noted on the certificate.
6. For buildings where an Energy Rating Index score is determined in accordance with Section R406, the Energy Rating Index score, both with and without

any on-site generation, shall be listed on the certificate.

7. The code edition under which the structure was permitted, and the compliance path used.

SECTION R402
BUILDING THERMAL ENVELOPE

R402.1 General. The *building thermal envelope* shall comply with the requirements of Sections R402.1.1 through R402.1.5.

Exceptions:

1. The following low-energy *buildings*, or portions thereof, separated from the remainder of the building by *building thermal envelope* assemblies complying with this section shall be exempt from the *building thermal envelope* provisions of Section R402.

   1.1. Those with a peak design rate of energy usage less than 3.4 Btu/h × ft^2 (10.7 W/m^2) or 1.0 watt/ft^2 of floor area for space-conditioning purposes.

   1.2. Those that do not contain *conditioned space*.

2. Log homes designed in accordance with ICC 400.

R402.1.1 Vapor retarder. Wall assemblies in the *building thermal envelope* shall comply with the vapor retarder requirements of Section R702.7 of the *International Residential Code* or Section 1404.3 of the *International Building Code*, as applicable.

R402.1.2 Insulation and fenestration criteria. The *building thermal envelope* shall meet the requirements of Table R402.1.2, based on the *climate zone* specified in Chapter 3. Assemblies shall have a *U*-factor equal to or less than that specified in Table R402.1.2. Fenestration shall have a *U*-factor and glazed fenestration SHGC equal to or less than that specified in Table R402.1.2.

R402.1.3 *R*-value alternative. Assemblies with *R*-value of insulation materials equal to or greater than that specified in Table R402.1.3 shall be an alternative to the *U*-factor in Table R402.1.2

R402.1.4 *R*-value computation. Cavity insulation alone shall be used to determine compliance with the cavity insulation *R*-value requirements in Table R402.1.3. Where cavity insulation is installed in multiple layers, the *R*-values of the cavity insulation layers shall be summed to determine compliance with the cavity insulation *R*-value requirements. The manufacturer's settled *R*-value shall be used for blown-in insulation. Continuous insulation (ci) alone shall be used to determine compliance with the continuous insulation *R*-value requirements in Table R402.1.3. Where continuous insulation is installed in multiple layers, the *R*-values of the continuous insulation layers shall be summed to determine compliance with the continuous insulation *R*-value requirements. Cavity insulation *R*-values shall not be used to determine compliance with the continuous insulation *R*-value requirements in Table R402.1.3. Computed *R*-values shall not include an

TABLE R402.1.2
MAXIMUM ASSEMBLY *U*-FACTORS[a] AND FENESTRATION REQUIREMENTS

CLIMATE ZONE	FENESTRATION *U*-FACTOR[f]	SKYLIGHT *U*-FACTOR	GLAZED FENESTRATION SHGC[d, e]	CEILING *U*-FACTOR	WOOD FRAME WALL *U*-FACTOR	MASS WALL *U*-FACTOR[b]	FLOOR *U*-FACTOR	BASEMENT WALL *U*-FACTOR	CRAWL SPACE WALL *U*-FACTOR
0	0.50	0.75	0.25	0.035	0.084	0.197	0.064	0.360	0.477
1	0.50	0.75	0.25	0.035	0.084	0.197	0.064	0.360	0.477
2	0.40	0.65	0.25	0.026	0.084	0.165	0.064	0.360	0.477
3	0.30	0.55	0.25	0.026	0.060	0.098	0.047	0.091[c]	0.136
4 except Marine	0.30	0.55	0.40	0.024	0.045	0.098	0.047	0.059	0.065
5 and Marine 4	0.30	0.55	0.40	0.024	0.045	0.082	0.033	0.050	0.055
6	0.30	0.55	NR	0.024	0.045	0.060	0.033	0.050	0.055
7 and 8	0.30	0.55	NR	0.024	0.045	0.057	0.028	0.050	0.055

For SI: 1 foot = 304.8 mm.

a. Nonfenestration *U*-factors shall be obtained from measurement, calculation or an approved source.

b. Mass walls shall be in accordance with Section R402.2.5. Where more than half the insulation is on the interior, the mass wall *U*-factors shall not exceed 0.17 in Climate Zones 0 and 1, 0.14 in Climate Zone 2, 0.12 in Climate Zone 3, 0.087 in Climate Zone 4 except Marine, 0.065 in Climate Zone 5 and Marine 4, and 0.057 in Climate Zones 6 through 8.

c. In Warm Humid locations as defined by Figure R301.1 and Table R301.1, the basement wall *U*-factor shall not exceed 0.360.

d. The SHGC column applies to all glazed fenestration.
   Exception: In Climate Zones 0 through 3, skylights shall be permitted to be excluded from glazed fenestration SHGC requirements provided that the SHGC for such skylights does not exceed 0.30.

e. There are no SHGC requirements in the Marine Zone.

f. A maximum *U*-factor of 0.32 shall apply in Marine Climate Zone 4 and Climate Zones 5 through 8 to vertical fenestration products installed in buildings located either:

   1. Above 4,000 feet in elevation above sea level, or

   2. In windborne debris regions where protection of openings is required by Section R301.2.1.2 of the *International Residential Code*.

TABLE R402.1.3
INSULATION MINIMUM *R*-VALUES AND FENESTRATION REQUIREMENTS BY COMPONENT[a]

CLIMATE ZONE	FENESTRATION U-FACTOR[b, i]	SKYLIGHT[b] U-FACTOR	GLAZED FENESTRATION SHGC[b, e]	CEILING R-VALUE	WOOD FRAME WALL R-VALUE[g]	MASS WALL R-VALUE[h]	FLOOR R-VALUE	BASEMENT[c, g] WALL R-VALUE	SLAB[d] R-VALUE & DEPTH	CRAWL SPACE[c, g] WALL R-VALUE
0	NR	0.75	0.25	30	13 or 0&10ci	3/4	13	0	0	0
1	NR	0.75	0.25	30	13 or 0&10ci	3/4	13	0	0	0
2	0.40	0.65	0.25	49	13 or 0&10ci	4/6	13	0	0	0
3	.30	0.55	0.25	49	20 or 13&5ci[h] or 0&15ci[h]	8/13	19	5ci or 13[f]	10ci, 2 ft	5ci or 13[f]
4 except Marine	.30	0.55	0.40	60	30 or 20&5ci[h] or 13&10ci[h] or 0&20ci[h]	8/13	19	10ci or 13	10ci, 4 ft	10ci or 13
5 and Marine 4	0.30[i]	0.55	0.40	60	30 or 20&5ci[h] or 13&10ci[h] or 0&20ci[h]	13/17	30	15ci or 19 or 13&5ci	10ci, 4 ft	15ci or 19 or 13&5ci
6	0.30[i]	0.55	NR	60	30 or 20&5ci[h] or 13&10ci[h] or 0&20ci[h]	15/20	30	15ci or 19 or 13&5ci	10ci, 4 ft	15ci or 19 or 13&5ci
7 and 8	0.30[i]	0.55	NR	60	30 or 20&5ci[h] or 13&10ci[h] or 0&20ci[h]	19/21	38	15ci or 19 or 13&5ci	10ci, 4 ft	15ci or 19 or 13&5ci

For SI: 1 foot = 304.8 mm.
NR = Not Required.
ci = continuous insulation.

a. *R*-values are minimums. *U*-factors and SHGC are maximums. Where insulation is installed in a cavity that is less than the label or design thickness of the insulation, the installed *R*-value of the insulation shall be not less than the *R*-value specified in the table.
b. The fenestration *U*-factor column excludes skylights. The SHGC column applies to all glazed fenestration.
   Exception: In Climate Zones 0 through 3, skylights shall be permitted to be excluded from glazed fenestration SHGC requirements provided that the SHGC for such skylights does not exceed 0.30.
c. "5ci or 13" means R-5 continuous insulation (ci) on the interior or exterior surface of the wall or R-13 cavity insulation on the interior side of the wall. "10ci or 13" means R-10 continuous insulation (ci) on the interior or exterior surface of the wall or R-13 cavity insulation on the interior side of the wall. "15ci or 19 or 13&5ci" means R-15 continuous insulation (ci) on the interior or exterior surface of the wall; or R-19 cavity insulation on the interior side of the wall; or R-13 cavity insulation on the interior of the wall in addition to R-5 continuous insulation on the interior or exterior surface of the wall.
d. R-5 insulation shall be provided under the full slab area of a heated slab in addition to the required slab edge insulation *R*-value for slabs, as indicated in the table. The slab-edge insulation for heated slabs shall not be required to extend below the slab.
e. There are no SHGC requirements in the Marine Zone.
f. Basement wall insulation is not required in Warm Humid locations as defined by Figure R301.1 and Table R301.1.
g. The first value is cavity insulation; the second value is continuous insulation. Therefore, as an example, "13&5" means R-13 cavity insulation plus R-5 continuous insulation.
h. Mass walls shall be in accordance with Section R402.2.5. The second *R*-value applies where more than half of the insulation is on the interior of the mass wall.
i. A maximum *U*-factor of 0.32 shall apply in Climate Zones 3 through 8 to vertical fenestration products installed in buildings located either:
   1. Above 4,000 feet in elevation, or
   2. In windborne debris regions where protection of openings is required by Section R301.2.1.2 of the *International Residential Code*.

R-value for other building materials or air films. Where insulated siding is used for the purpose of complying with the continuous insulation requirements of Table R402.1.3, the manufacturer's labeled *R*-value for the insulated siding shall be reduced by R-0.6.

R402.1.5 Total UA alternative. Where the total *building thermal envelope* UA, the sum of *U*-factor times assembly area, is less than or equal to the total UA resulting from multiplying the *U*-factors in Table R402.1.2 by the same assembly area as in the proposed *building*, the *building* shall be considered to be in compliance with Table R402.1.2. The UA calculation shall be performed using a method consistent with the ASHRAE *Handbook of Fundamentals* and shall include the thermal bridging effects of framing materials. In addition to UA compliance, the

SHGC requirements of Table R402.1.2 and the maximum fenestration *U*-factors of Section R402.5 shall be met.

R402.2 Specific insulation requirements. In addition to the requirements of Section R402.1, insulation shall meet the specific requirements of Sections R402.2.1 through R402.2.12.

R402.2.1 Ceilings with attics. Where Section R402.1.3 requires R-49 insulation in the ceiling or attic, installing R-38 over 100 percent of the ceiling or attic area requiring insulation shall satisfy the requirement for R-49 insulation wherever the full height of uncompressed R-38 insulation extends over the wall top plate at the eaves. Where Section R402.1.3 requires R-60 insulation in the ceiling or attic, installing R-49 over 100 percent of the ceiling or attic area requiring insulation shall satisfy the requirement for R-60 insulation wherever the full height of uncompressed R-49 insulation extends over the wall top plate at the eaves. This reduction shall not apply to the insulation and fenestration criteria in Section R402.1.2 and the Total UA alternative in Section R402.1.5.

R402.2.2 Ceilings without attics. Where Section R402.1.3 requires insulation *R*-values greater than R-30 in the interstitial space above a ceiling and below the structural roof deck, and the design of the roof/ceiling assembly does not allow sufficient space for the required insulation, the minimum required insulation *R*-value for such roof/ceiling assemblies shall be R-30. Insulation shall extend over the top of the wall plate to the outer edge of such plate and shall not be compressed. This reduction of insulation from the requirements of Section R402.1.3 shall be limited to 500 square feet (46 m^2) or 20 percent of the total insulated ceiling area, whichever is less. This reduction shall not apply to the Total UA alternative in Section R402.1.5.

R402.2.3 Eave baffle. For air-permeable insulation in vented attics, a baffle shall be installed adjacent to soffit and eave vents. Baffles shall maintain a net free area opening equal to or greater than the size of the vent. The baffle shall extend over the top of the attic insulation. The baffle shall be permitted to be any solid material. The baffle shall be installed to the outer edge of the *exterior wall* top plate so as to provide maximum space for attic insulation coverage over the top plate. Where soffit venting is not continuous, baffles shall be installed continuously to prevent ventilation air in the eave soffit from bypassing the baffle.

R402.2.4 Access hatches and doors. Access hatches and doors from conditioned to unconditioned spaces such as attics and crawl spaces shall be insulated to the same *R*-value required by Table R402.1.3 for the wall or ceiling in which they are installed.

Exceptions:

1. Vertical doors providing access from conditioned spaces to unconditioned spaces that comply with the fenestration requirements of Table R402.1.3 based on the applicable climate zone specified in Chapter 3.

2. Horizontal pull-down, stair-type access hatches in ceiling assemblies that provide access from conditioned to unconditioned spaces in Climate Zones 0 through 4 shall not be required to comply with the insulation level of the surrounding surfaces provided the hatch meets all of the following:

    2.1. The average *U*-factor of the hatch shall be less than or equal to U-0.10 or have an average insulation *R*-value of R-10 or greater.

    2.2. Not less than 75 percent of the panel area shall have an insulation *R*-value of R-13 or greater.

    2.3. The net area of the framed opening shall be less than or equal to 13.5 square feet (1.25 m^2).

    2.4. The perimeter of the hatch edge shall be weatherstripped.

The reduction shall not apply to the total UA alternative in Section R402.1.5.

R402.2.4.1 Access hatches and door insulation installation and retention. Vertical or horizontal access hatches and doors from *conditioned spaces* to *unconditioned spaces* such as attics and crawl spaces shall be weatherstripped. Access that prevents damaging or compressing the insulation shall be provided to all equipment. Where loose-fill insulation is installed, a wood-framed or equivalent baffle, retainer, or dam shall be installed to prevent loose-fill insulation from spilling into living space from higher to lower sections of the attic and from attics covering conditioned spaces to unconditioned spaces. The baffle or retainer shall provide a permanent means of maintaining the installed *R*-value of the loose-fill insulation.

R402.2.5 Mass walls. Mass walls where used as a component of the *building thermal envelope* shall be one of the following:

1. Above-ground walls of concrete block, concrete, insulated concrete form, masonry cavity, brick but not brick veneer, adobe, compressed earth block, rammed earth, solid timber, mass timber or solid logs.

2. Any wall having a heat capacity greater than or equal to 6 Btu/ft^2 × °F (123 kJ/m^2 × K).

R402.2.6 Steel-frame ceilings, walls and floors. Steel-frame ceilings, walls, and floors shall comply with the insulation requirements of Table R402.2.6 or the *U*-factor requirements of Table R402.1.2. The calculation of the *U*-factor for a steel-frame envelope assembly shall use a series-parallel path calculation method.

TABLE R402.2.6
STEEL-FRAME CEILING, WALL AND FLOOR INSULATION R-VALUES

WOOD FRAME R-VALUE REQUIREMENT	COLD-FORMED STEEL-FRAME EQUIVALENT R-VALUE[a]
Steel Truss Ceilings[b]	
R-30	R-38 or R-30 + 3 or R-26 + 5
R-38	R-49 or R-38 + 3
R-49	R-38 + 5
Steel Joist Ceilings[b]	
R-30	R-38 in 2 × 4 or 2 × 6 or 2 × 8 R-49 in any framing
R-38	R-49 in 2 × 4 or 2 × 6 or 2 × 8 or 2 × 10
Steel-frame Wall, 16 inches on center	
R-13	R-13 + 4.2 or R-21 + 2.8 or R-0 + 9.3 or R-15 + 3.8 or R-21 + 3.1
R-13 + 5	R-0 + 15 or R-13 + 9 or R-15 + 8.5 or R-19 + 8 or R-21 + 7
R-13 + 10	R-0 + 20 or R-13 + 15 or R-15 + 14 or R-19 + 13 or R-21 + 13
R-20	R-0 + 14.0 or R-13 + 8.9 or R-15 + 8.5 or R-19 + 7.8 or R-21 + 7.5
R-20 + 5	R-13 + 12.7 or R-15 + 12.3 or R-19 + 11.6 or R-21 + 11.3 or R-25 + 10.9
R-21	R-0 + 14.6 or R-13 + 9.5 or R-15 + 9.1 or R-19 + 8.4 or R-21 + 8.1 or R-25 + 7.7
Steel-frame Wall, 24 inches on center	
R-13	R-0 + 9.3 or R-13 + 3.0 or R-15 + 2.4
R-13 + 5	R-0 + 15 or R-13 + 7.5 or R-15 + 7 or R-19 + 6 or R-21 + 6
R-13 + 10	R-0 + 20 or R-13 + 13 or R-15 + 12 or R-19 + 11 or R-21 + 11
R-20	R-0 + 14.0 or R-13 + 7.7 or R-15 + 7.1 or R-19 + 6.3 or R-21 + 5.9
R-20 + 5	R-13 + 11.5 or R-15 + 10.9 or R-19 + 10.1 or R-21 + 9.7 or R-25 + 9.1
R-21	R-0 + 14.6 or R-13 + 8.3 or R-15 + 7.7 or R-19 + 6.9 or R-21 + 6.5 or R-25 + 5.9
Steel Joist Floor	
R-13	R-19 in 2 × 6, or R-19 + 6 in 2 × 8 or 2 × 10
R-19	R-19 + 6 in 2 × 6, or R-19 + 12 in 2 × 8 or 2 × 10

a. The first value is cavity insulation R-value; the second value is continuous insulation R-value. Therefore, for example, "R-30 + 3" means R-30 cavity insulation plus R-3 continuous insulation.
b. Insulation exceeding the height of the framing shall cover the framing.

R402.2.7 Floors. Floor *cavity insulation* shall comply with one of the following:

1. Installation shall be installed to maintain permanent contact with the underside of the subfloor decking in accordance with manufacturer instructions to maintain required R-value or readily fill the available cavity space.

2. Floor framing cavity insulation shall be permitted to be in contact with the top side of sheathing separating the cavity and the unconditioned space below. Insulation shall extend from the bottom to the top of all perimeter floor framing members and the framing members shall be air sealed.

3. A combination of cavity and continuous insulation shall be installed so that the cavity insulation is in contact with the top side of the continuous insulation that is installed on the underside of the floor framing separating the cavity and the unconditioned space below. The combined R-value of the cavity and continuous insulation shall equal the required R-value for floors. Insulation shall extend from the bottom to the top of all perimeter floor framing members and the framing members shall be air sealed.

R402.2.8 Basement walls. Basement walls shall be insulated in accordance with Table R402.1.3.

Exception: Basement walls associated with unconditioned basements where all of the following requirements are met:

1. The floor overhead, including the underside stairway stringer leading to the basement, is insulated in accordance with Section R402.1.3 and applicable provisions of Sections R402.2 and R402.2.7.

2. There are no uninsulated duct, domestic hot water, or hydronic heating surfaces exposed to the basement.

3. There are no HVAC supply or return diffusers serving the basement.

4. The walls surrounding the stairway and adjacent to conditioned space are insulated in accordance with Section R402.1.3 and applicable provisions of Section R402.2.

5. The door(s) leading to the basement from conditioned spaces are insulated in accordance with Section R402.1.3 and applicable provisions of Section R402.2, and weatherstripped in accordance with Section R402.4.

6. The building thermal envelope separating the basement from adjacent conditioned spaces complies with Section R402.4.

R402.2.8.1 Basement wall insulation installation. Where *basement walls* are insulated, the insulation shall be installed from the top of the *basement wall* down to 10 feet (3048 mm) below grade or to the basement floor, whichever is less.

R402.2.9 Slab-on-grade floors. Slab-on-grade floors with a floor surface less than 12 inches (305 mm) below grade shall be insulated in accordance with Table R402.1.3.

Exception: Slab-edge insulation is not required in jurisdictions designated by the code official as having a very heavy termite infestation.

R402.2.9.1 Slab-on-grade floor insulation installation. Where installed, the insulation shall extend downward from the top of the slab on the outside or inside of the foundation wall. Insulation located below grade shall be extended the distance provided in Table R402.1.3 or the distance of the proposed design, as applicable, by any combination of vertical insulation, insulation extending under the slab or insulation extending out from the building. Insulation extending away from the building shall be protected by pavement or by not less than 10 inches (254 mm) of soil. The top edge of the insulation installed between the *exterior wall* and the edge of the interior slab shall be permitted to be cut at a 45-degree (0.79 rad) angle away from the *exterior wall*.

R402.2.10 Crawl space walls. Crawl space walls shall be insulated in accordance with Table R402.1.3.

> **Exception:** Crawl space walls associated with a crawl space that is vented to the outdoors and the floor overhead is insulated in accordance with Table R402.1.3 and Section R402.2.7.

R402.2.10.1 Crawl space wall insulation installations. Where crawl space wall insulation is installed, it shall be permanently fastened to the wall and shall extend downward from the floor to the finished grade elevation and then vertically or horizontally for not less than an additional 24 inches (610 mm). Exposed earth in unvented crawl space foundations shall be covered with a continuous Class I vapor retarder in accordance with the *International Building Code* or *International Residential Code*, as applicable. Joints of the vapor retarder shall overlap by 6 inches (153 mm) and be sealed or taped. The edges of the vapor retarder shall extend not less than 6 inches (153 mm) up stem walls and shall be attached to the stem walls.

R402.2.11 Masonry veneer. Insulation shall not be required on the horizontal portion of a foundation that supports a masonry veneer.

R402.2.12 Sunroom and heated garage insulation. *Sunrooms* enclosing *conditioned space* and heated garages shall meet the insulation requirements of this code.

> **Exception:** For *sunrooms* and heated garages provided *thermal isolation*, and enclosed *conditioned space*, the following exceptions to the insulation requirements of this code shall apply:
> 1. The minimum ceiling insulation *R*-values shall be R-19 in *Climate Zones* 0 through 4 and R-24 in *Climate Zones* 5 through 8.
> 2. The minimum wall insulation *R*-value shall be R-13 in all *climate zones*. Walls separating a *sunroom* or heated garage with *thermal isolation* from *conditioned space* shall comply with the *building thermal envelope* requirements of this code.

R402.3 Fenestration. In addition to the requirements of Section R402, fenestration shall comply with Sections R402.3.1 through R402.3.5.

R402.3.1 *U*-factor. An area-weighted average of fenestration products shall be permitted to satisfy the *U*-factor requirements.

R402.3.2 Glazed fenestration SHGC. An area-weighted average of fenestration products more than 50-percent glazed shall be permitted to satisfy the SHGC requirements.

Dynamic glazing shall be permitted to satisfy the SHGC requirements of Table R402.1.2 provided that the ratio of the higher to lower labeled SHGC is greater than or equal to 2.4, and the *dynamic glazing* is automatically controlled to modulate the amount of solar gain into the space in multiple steps. *Dynamic glazing* shall be considered separately from other fenestration, and area-weighted averaging with other fenestration that is not dynamic glazing shall be prohibited.

> **Exception:** Dynamic glazing shall not be required to comply with this section where both the lower and higher labeled SHGC comply with the requirements of Table R402.1.2.

R402.3.3 Glazed fenestration exemption. Not greater than 15 square feet (1.4 m²) of glazed fenestration per *dwelling unit* shall be exempt from the *U*-factor and SHGC requirements in Section R402.1.2. This exemption shall not apply to the Total UA alternative in Section R402.1.5.

R402.3.4 Opaque door exemption. One side-hinged opaque door assembly not greater than 24 square feet (2.22 m²) in area shall be exempt from the *U*-factor requirement in Section R402.1.2. This exemption shall not apply to the Total UA alternative in Section R402.1.5.

R402.3.5 Sunroom and heated garage fenestration. *Sunrooms* and heated garages enclosing *conditioned space* shall comply with the fenestration requirements of this code.

> **Exception:** In Climate Zones 2 through 8, for *sunrooms* and heated garages with *thermal isolation* and enclosing *conditioned space*, the fenestration *U*-factor shall not exceed 0.45 and the skylight *U*-factor shall not exceed 0.70.

New fenestration separating a *sunroom* or heated garage with *thermal isolation* from *conditioned space* shall comply with the *building thermal envelope* requirements of this code.

R402.4 Air leakage. The *building thermal envelope* shall be constructed to limit air leakage in accordance with the requirements of Sections R402.4.1 through R402.4.5.

R402.4.1 Building thermal envelope. The *building thermal envelope* shall comply with Sections R402.4.1.1 through R402.4.1.3. The sealing methods between dissimilar materials shall allow for differential expansion and contraction.

R402.4.1.1 Installation. The components of the *building thermal envelope* as indicated in Table R402.4.1.1 shall be installed in accordance with the manufacturer's instructions and the criteria indicated in Table R402.4.1.1, as applicable to the method of construction. Where required by the *code official*, an *approved* third party shall inspect all components and verify compliance.

TABLE R402.4.1.1
AIR BARRIER, AIR SEALING AND INSULATION INSTALLATION[a]

COMPONENT	AIR BARRIER CRITERIA	INSULATION INSTALLATION CRITERIA
General requirements	A continuous air barrier shall be installed in the building envelope. Breaks or joints in the air barrier shall be sealed.	Air-permeable insulation shall not be used as a sealing material.
Ceiling/attic	The air barrier in any dropped ceiling or soffit shall be aligned with the insulation and any gaps in the air barrier shall be sealed. Access openings, drop down stairs or knee wall doors to unconditioned attic spaces shall be sealed.	The insulation in any dropped ceiling/soffit shall be aligned with the air barrier.
Walls	The junction of the foundation and sill plate shall be sealed. The junction of the top plate and the top of exterior walls shall be sealed. Knee walls shall be sealed.	Cavities within corners and headers of frame walls shall be insulated by completely filling the cavity with a material having a thermal resistance, *R*-value, of not less than R-3 per inch. Exterior thermal envelope insulation for framed walls shall be installed in substantial contact and continuous alignment with the air barrier.
Windows, skylights and doors	The space between framing and skylights, and the jambs of windows and doors, shall be sealed.	—
Rim joists	Rim joists shall include an exterior air barrier.[b] The junctions of the rim board to the sill plate and the rim board and the subfloor shall be air sealed.	Rim joists shall be insulated so that the insulation maintains permanent contact with the exterior rim board.[b]
Floors, including cantilevered floors and floors above garages	The air barrier shall be installed at any exposed edge of insulation.	Floor framing cavity insulation shall be installed to maintain permanent contact with the underside of subfloor decking. Alternatively, floor framing cavity insulation shall be in contact with the top side of sheathing, or continuous insulation installed on the underside of floor framing and extending from the bottom to the top of all perimeter floor framing members.
Basement crawl space and slab foundations	Exposed earth in unvented crawl spaces shall be covered with a Class I vapor retarder/air barrier in accordance with Section R402.2.10. Penetrations through concrete foundation walls and slabs shall be air sealed. Class 1 vapor retarders shall not be used as an air barrier on below-grade walls and shall be installed in accordance with Section R702.7 of the *International Residential Code*.	Crawl space insulation, where provided instead of floor insulation, shall be installed in accordance with Section R402.2.10. Conditioned basement foundation wall insulation shall be installed in accordance with Section R402.2.8.1. Slab-on-grade floor insulation shall be installed in accordance with Section R402.2.10.
Shafts, penetrations	Duct and flue shafts to exterior or unconditioned space shall be sealed. Utility penetrations of the air barrier shall be caulked, gasketed or otherwise sealed and shall allow for expansion, contraction of materials and mechanical vibration.	Insulation shall be fitted tightly around utilities passing through shafts and penetrations in the building thermal envelope to maintain required *R*-value.
Narrow cavities	Narrow cavities of 1 inch or less that are not able to be insulated shall be air sealed.	Batts to be installed in narrow cavities shall be cut to fit or narrow cavities shall be filled with insulation that on installation readily conforms to the available cavity space.
Garage separation	Air sealing shall be provided between the garage and conditioned spaces.	Insulated portions of the garage separation assembly shall be installed in accordance with Sections R303 and R402.2.7.

(continued)

TABLE R402.4.1.1—continued
AIR BARRIER, AIR SEALING AND INSULATION INSTALLATION[a]

COMPONENT	AIR BARRIER CRITERIA	INSULATION INSTALLATION CRITERIA
Recessed lighting	Recessed light fixtures installed in the building thermal envelope shall be air sealed in accordance with Section R402.4.5.	Recessed light fixtures installed in the building thermal envelope shall be airtight and IC rated, and shall be buried or surrounded with insulation.
Plumbing, wiring or other obstructions	All holes created by wiring, plumbing or other obstructions in the air barrier assembly shall be air sealed.	Insulation shall be installed to fill the available space and surround wiring, plumbing, or other obstructions, unless the required R-value can be met by installing insulation and air barrier systems completely to the exterior side of the obstructions.
Shower/tub on exterior wall	The air barrier installed at exterior walls adjacent to showers and tubs shall separate the wall from the shower or tub.	Exterior walls adjacent to showers and tubs shall be insulated.
Electrical/phone box on exterior walls	The air barrier shall be installed behind electrical and communication boxes. Alternatively, air-sealed boxes shall be installed.	—
HVAC register boots	HVAC supply and return register boots that penetrate building thermal envelope shall be sealed to the subfloor, wall covering or ceiling penetrated by the boot.	—
Concealed sprinklers	Where required to be sealed, concealed fire sprinklers shall only be sealed in a manner that is recommended by the manufacturer. Caulking or other adhesive sealants shall not be used to fill voids between fire sprinkler cover plates and walls or ceilings.	—

a. Inspection of log walls shall be in accordance with the provisions of ICC 400.
b. Air barrier and insulation full enclosure is not required in unconditioned/ventilated attic spaces and at rim joists.

R402.4.1.2 Testing. The *building* or *dwelling unit* shall be tested for air leakage. The maximum air leakage rate for any *building* or *dwelling unit* under any compliance path shall not exceed 5.0 air changes per hour or 0.28 cubic feet per minute (CFM) per square foot [0.0079 m^3/(s × m^2)] of dwelling unit enclosure area. Testing shall be conducted in accordance with ANSI/RESNET/ICC 380, ASTM E779 or ASTM E1827 and reported at a pressure of 0.2 inch w.g. (50 Pascals). Where required by the *code official*, testing shall be conducted by an *approved* third party. A written report of the results of the test shall be signed by the party conducting the test and provided to the *code official*. Testing shall be performed at any time after creation of all penetrations of the *building thermal envelope* have been sealed.

> **Exception:** For heated, attached private garages and heated, detached private garages accessory to one- and two-family dwellings and townhouses not more than three stories above *grade plane* in height, building envelope tightness and insulation installation shall be considered acceptable where the items in Table R402.4.1.1, applicable to the method of construction, are field verified. Where required by the code official, an *approved* third party independent from the installer shall inspect both air barrier and insulation installation criteria. Heated, attached private garage space and heated, detached private garage space shall be thermally isolated from all other habitable, *conditioned spaces* in accordance with Sections R402.2.12 and R402.3.5, as applicable.

During testing:

1. Exterior windows and doors, fireplace and stove doors shall be closed, but not sealed, beyond the intended weatherstripping or other infiltration control measures.
2. Dampers including exhaust, intake, makeup air, backdraft and flue dampers shall be closed, but not sealed beyond intended infiltration control measures.
3. Interior doors, where installed at the time of the test, shall be open.
4. Exterior or interior terminations for continuous ventilation systems shall be sealed.
5. Heating and cooling systems, where installed at the time of the test, shall be turned off.
6. Supply and return registers, where installed at the time of the test, shall be fully open.

> **Exception:** When testing individual *dwelling units*, an air leakage rate not exceeding 0.30 cubic feet per minute per square foot [0.008 m^3/(s × m^2)] of the dwelling unit enclosure area, tested in accordance with ANSI/RESNET/ICC 380, ASTM E779 or ASTM E1827 and reported at a pressure of

0.2 inch w.g. (50 Pa), shall be permitted in all climate zones for:

1. Attached single and multiple-family building *dwelling units*.
2. Buildings or *dwelling units* that are 1,500 square feet (139.4 m^2) or smaller.

Mechanical ventilation shall be provided in accordance with Section M1505 of the *International Residential Code* or Section 403.3.2 of the *International Mechanical Code*, as applicable, or with other *approved* means of ventilation.

R402.4.1.3 Leakage rate. When complying with Section R401.2.1, the building or dwelling unit shall have an air leakage rate not exceeding 5.0 air changes per hour in Climate Zones 0, 1 and 2, and 3.0 air changes per hour in Climate Zones 3 through 8, when tested in accordance with Section R402.4.1.2.

R402.4.2 Fireplaces. New wood-burning fireplaces shall have tight-fitting flue dampers or doors, and outdoor combustion air. Where using tight-fitting doors on factory-built fireplaces *listed* and *labeled* in accordance with UL 127, the doors shall be tested and *listed* for the fireplace.

R402.4.3 Fenestration air leakage. Windows, *skylights* and sliding glass doors shall have an air infiltration rate of not greater than 0.3 cfm per square foot (1.5 L/s/m^2), and for swinging doors, not greater than 0.5 cfm per square foot (2.6 L/s/m^2), when tested in accordance with NFRC 400 or AAMA/WDMA/CSA 101/I.S.2/A440 by an accredited, independent laboratory and *listed* and *labeled* by the manufacturer.

Exception: Site-built windows, *skylights* and doors.

R402.4.4 Rooms containing fuel-burning appliances. In Climate Zones 3 through 8, where open combustion air ducts provide combustion air to open combustion fuel-burning appliances, the appliances and combustion air opening shall be located outside the *building thermal envelope* or enclosed in a room that is isolated from inside the thermal envelope. Such rooms shall be sealed and insulated in accordance with the envelope requirements of Table R402.1.3, where the walls, floors and ceilings shall meet a minimum of the *basement wall R*-value requirement. The door into the room shall be fully gasketed and any water lines and ducts in the room insulated in accordance with Section R403. The combustion air duct shall be insulated where it passes through *conditioned space* to an *R*-value of not less than R-8.

Exceptions:

1. Direct vent appliances with both intake and exhaust pipes installed continuous to the outside.
2. Fireplaces and stoves complying with Section R402.4.2 and Section R1006 of the *International Residential Code*.

R402.4.5 Recessed lighting. Recessed luminaires installed in the *building thermal envelope* shall be sealed to limit air leakage between conditioned and *unconditioned spaces*. Recessed luminaires shall be IC-rated and *labeled* as having an air leakage rate of not greater than 2.0 cfm (0.944 L/s) when tested in accordance with ASTM E283 at a pressure differential of 1.57 psf (75 Pa). Recessed luminaires shall be sealed with a gasket or caulked between the housing and the interior wall or ceiling covering.

R402.4.6 Electrical and communication outlet boxes (air-sealed boxes). Electrical and communication outlet boxes installed in the building thermal envelope shall be sealed to limit air leakage between conditioned and unconditioned spaces. Electrical and communication outlet boxes shall be tested in accordance with NEMA OS 4, *Requirements for Air-Sealed Boxes for Electrical and Communication Applications*, and shall have an air leakage rate of not greater than 2.0 cubic feet per minute (0.944 L/s) at a pressure differential of 1.57 psf (75 Pa). Electrical and communication outlet boxes shall be marked "NEMA OS 4" or "OS 4" in accordance with NEMA OS 4. Electrical and communication outlet boxes shall be installed per the manufacturer's instructions and with any supplied components required to achieve compliance with NEMA OS 4.

R402.5 Maximum fenestration U-factor and SHGC. The area-weighted average maximum fenestration *U*-factor permitted using tradeoffs from Section R402.1.5 or R405 shall be 0.48 in Climate Zones 4 and 5 and 0.40 in Climate Zones 6 through 8 for vertical fenestration, and 0.75 in Climate Zones 4 through 8 for skylights. The area-weighted average maximum fenestration SHGC permitted using tradeoffs from Section R405 in *Climate Zones* 0 through 3 shall be 0.40.

Exception: The maximum *U*-factor and solar heat gain coefficient (SHGC) for fenestration shall not be required in storm shelters complying with ICC 500.

SECTION R403
SYSTEMS

R403.1 Controls. Not less than one thermostat shall be provided for each separate heating and cooling system.

R403.1.1 Programmable thermostat. The thermostat controlling the primary heating or cooling system of the *dwelling unit* shall be capable of controlling the heating and cooling system on a daily schedule to maintain different temperature set points at different times of day and different days of the week. This thermostat shall include the capability to set back or temporarily operate the system to maintain *zone* temperatures of not less than 55°F (13°C) to not greater than 85°F (29°C). The thermostat shall be programmed initially by the manufacturer with a heating temperature setpoint of not greater than 70°F (21°C) and a cooling temperature setpoint of not less than 78°F (26°C).

R403.1.2 Heat pump supplementary heat. Heat pumps having supplementary electric-resistance heat shall have controls that, except during defrost, prevent supplemental heat operation when the heat pump compressor can meet the heating load.

R403.2 Hot water boiler temperature reset. The manufacturer shall equip each gas, oil and electric boiler (other than a boiler equipped with a tankless domestic water heating coil) with automatic means of adjusting the water temperature supplied by the boiler to ensure incremental change of the inferred heat load will cause an incremental change in the temperature of the water supplied by the boiler. This can be accomplished with outdoor reset, indoor reset or water temperature sensing.

R403.3 Ducts. Ducts and air handlers shall be installed in accordance with Sections R403.3.1 through R403.3.7.

R403.3.1 Ducts located outside conditioned space. Supply and return ducts located outside *conditioned space* shall be insulated to an *R*-value of not less than R-8 for ducts 3 inches (76 mm) in diameter and larger and not less than R-6 for ducts smaller than 3 inches (76 mm) in diameter. Ducts buried beneath a building shall be insulated as required per this section or have an equivalent *thermal distribution efficiency*. Underground ducts utilizing the *thermal distribution efficiency* method shall be listed and *labeled* to indicate the *R*-value equivalency.

R403.3.2 Ducts located in conditioned space. For ductwork to be considered inside a *conditioned space*, it shall comply with one of the following:

1. The duct system shall be located completely within the *continuous air barrier* and within the building thermal envelope.
2. Ductwork in ventilated attic spaces shall be buried within ceiling insulation in accordance with Section R403.3.3 and all of the following conditions shall exist:
   2.1. The air handler is located completely within the *continuous air barrier* and within the *building thermal envelope*.
   2.2. The duct leakage, as measured either by a rough-in test of the ducts or a post-construction total system leakage test to outside the *building thermal envelope* in accordance with Section R403.3.6, is less than or equal to 1.5 cubic feet per minute (42.5 L/min) per 100 square feet (9.29 m^2) of *conditioned floor area* served by the duct system.
   2.3. The ceiling insulation *R*-value installed against and above the insulated duct is greater than or equal to the proposed ceiling insulation *R*-value, less the *R*-value of the insulation on the duct.
3. Ductwork in floor cavities located over unconditioned space shall comply with all of the following:
   3.1. A *continuous air barrier* installed between unconditioned space and the duct.
   3.2. Insulation installed in accordance with Section R402.2.7.
   3.3. A minimum R-19 insulation installed in the cavity width separating the duct from unconditioned space.
4. Ductwork located within *exterior walls* of the *building thermal envelope* shall comply with the following:
   4.1. A *continuous air barrier* installed between unconditioned space and the duct.
   4.2. Minimum R-10 insulation installed in the cavity width separating the duct from the outside sheathing.
   4.3. The remainder of the cavity insulation shall be fully insulated to the drywall side.

R403.3.3 Ducts buried within ceiling insulation. Where supply and return air ducts are partially or completely buried in ceiling insulation, such ducts shall comply with all of the following:

1. The supply and return ducts shall have an insulation *R*-value not less than R-8.
2. At all points along each duct, the sum of the ceiling insulation *R*-value against and above the top of the duct, and against and below the bottom of the duct, shall be not less than R-19, excluding the *R*-value of the duct insulation.
3. In Climate Zones 0A, 1A, 2A and 3A, the supply ducts shall be completely buried within ceiling insulation, insulated to an *R*-value of not less than R-13 and in compliance with the vapor retarder requirements of Section 604.11 of the *International Mechanical Code* or Section M1601.4.6 of the *International Residential Code*, as applicable.

   Exception: Sections of the supply duct that are less than 3 feet (914 mm) from the supply outlet shall not be required to comply with these requirements.

R403.3.3.1 Effective *R*-value of deeply buried ducts. Where using the Total Building Performance Compliance Option in accordance with Section R401.2.2, sections of ducts that are installed in accordance with Section R403.3.3, located directly on or within 5.5 inches (140 mm) of the ceiling, surrounded with blown-in attic insulation having an *R*-value of R-30 or greater and located such that the top of the duct is not less than 3.5 inches (89 mm) below the top of the insulation, shall be considered as having an effective duct insulation *R*-value of R-25.

R403.3.4 Sealing. Ducts, air handlers and filter boxes shall be sealed. Joints and seams shall comply with either the *International Mechanical Code* or *International Residential Code*, as applicable.

R403.3.4.1 Sealed air handler. Air handlers shall have a manufacturer's designation for an air leakage of not greater than 2 percent of the design airflow rate when tested in accordance with ASHRAE 193.

R403.3.5 Duct testing. Ducts shall be pressure tested in accordance with ANSI/RESNET/ICC 380 or ASTM E1554 to determine air leakage by one of the following methods:

1. Rough-in test: Total leakage shall be measured with a pressure differential of 0.1 inch w.g. (25 Pa) across the system, including the manufacturer's air handler enclosure if installed at the time of the test. Registers shall be taped or otherwise sealed during the test.

2. Postconstruction test: Total leakage shall be measured with a pressure differential of 0.1 inch w.g. (25 Pa) across the entire system, including the manufacturer's air handler enclosure. Registers shall be taped or otherwise sealed during the test.

A written report of the results of the test shall be signed by the party conducting the test and provided to the *code official*.

Exception: A duct air-leakage test shall not be required for ducts serving ventilation systems that are not integrated with ducts serving heating or cooling systems.

R403.3.6 Duct leakage. The total leakage of the ducts, where measured in accordance with Section R403.3.5, shall be as follows:

1. Rough-in test: The total leakage shall be less than or equal to 4.0 cubic feet per minute (113.3 L/min) per 100 square feet (9.29 m^2) of *conditioned floor area* where the air handler is installed at the time of the test. Where the air handler is not installed at the time of the test, the total leakage shall be less than or equal to 3.0 cubic feet per minute (85 L/min) per 100 square feet (9.29 m^2) of *conditioned floor area*.

2. Postconstruction test: Total leakage shall be less than or equal to 4.0 cubic feet per minute (113.3 L/min) per 100 square feet (9.29 m^2) of *conditioned floor area*.

3. Test for ducts within thermal envelope: Where all ducts and air handlers are located entirely within the *building thermal envelope*, total leakage shall be less than or equal to 8.0 cubic feet per minute (226.6 L/min) per 100 square feet (9.29 m^2) of *conditioned floor area*.

R403.3.7 Building cavities. *Building* framing cavities shall not be used as ducts or plenums.

R403.4 Mechanical system piping insulation. Mechanical system piping capable of carrying fluids greater than 105°F (41°C) or less than 55°F (13°C) shall be insulated to an *R*-value of not less than R-3.

R403.4.1 Protection of piping insulation. Piping insulation exposed to weather shall be protected from damage, including that caused by sunlight, moisture, equipment maintenance and wind. The protection shall provide shielding from solar radiation that can cause degradation of the material. Adhesive tape shall be prohibited.

R403.5 Service hot water systems. Energy conservation measures for service hot water systems shall be in accordance with Sections R403.5.1 through R403.5.3.

R403.5.1 Heated water circulation and temperature maintenance systems. Heated water circulation systems shall be in accordance with Section R403.5.1.1. Heat trace temperature maintenance systems shall be in accordance with Section R403.5.1.2. Automatic controls, temperature sensors and pumps shall be in a location with access. Manual controls shall be in a location with *ready access*.

R403.5.1.1 Circulation systems. Heated water circulation systems shall be provided with a circulation pump. The system return pipe shall be a dedicated return pipe or a cold water supply pipe. Gravity and thermosyphon circulation systems shall be prohibited. Controls for circulating hot water system pumps shall automatically turn off the pump when the water in the circulation loop is at the desired temperature and when there is no demand for hot water. The controls shall limit the temperature of the water entering the cold water piping to not greater than 104°F (40°C).

R403.5.1.1.1 Demand recirculation water systems. Where installed, *demand recirculation water systems* shall have controls that start the pump upon receiving a signal from the action of a user of a fixture or appliance, sensing the presence of a user of a fixture or sensing the flow of hot or tempered water to a fixture fitting or appliance.

R403.5.1.2 Heat trace systems. Electric heat trace systems shall comply with IEEE 515.1 or UL 515. Controls for such systems shall automatically adjust the energy input to the heat tracing to maintain the desired water temperature in the piping in accordance with the times when heated water is used in the occupancy.

R403.5.2 Hot water pipe insulation. Insulation for service hot water piping with a thermal resistance, *R*-value, of not less than R-3 shall be applied to the following:

1. Piping $^3/_4$ inch (19.1 mm) and larger in nominal diameter located inside the *conditioned space*.
2. Piping serving more than one dwelling unit.
3. Piping located outside the *conditioned space*.
4. Piping from the water heater to a distribution manifold.
5. Piping located under a floor slab.
6. Buried piping.
7. Supply and return piping in circulation and recirculation systems other than cold water pipe return demand recirculation systems.

R403.5.3 Drain water heat recovery units. Where installed, drain water heat recovery units shall comply with CSA B55.2. Drain water heat recovery units shall be tested in accordance with CSA B55.1. Potable water-side pressure loss of drain water heat recovery units shall be less than 3 psi (20.7 kPa) for individual units connected to one or two showers. Potable water-side pressure loss of drain water heat recovery units shall be less than 2 psi (13.8 kPa) for individual units connected to three or more showers.

R403.6 Mechanical ventilation. The *buildings* complying with Section R402.4.1 shall be provided with ventilation that complies with the requirements of Section M1505 of the *International Residential Code* or *International Mechanical Code*, as applicable, or with other *approved* means of *ventilation*. Outdoor air intakes and exhausts shall have automatic or gravity dampers that close when the *ventilation* system is not operating.

R403.6.1 Heat or energy recovery ventilation. *Dwelling units* shall be provided with a heat recovery or energy recovery ventilation system in Climate Zones 7 and 8. The system shall be balanced with a minimum sensible heat recovery efficiency of 65 percent at 32°F (0°C) at a flow greater than or equal to the design airflow.

R403.6.2 Whole-dwelling mechanical ventilation system fan efficacy. Fans used to provide whole-dwelling mechanical ventilation shall meet the efficacy requirements of Table R403.6.2 at one or more rating points. Fans shall be tested in accordance with HVI 916 and listed. The airflow shall be reported in the product listing or on the label. Fan efficacy shall be reported in the product listing or shall be derived from the input power and airflow values reported in the product listing or on the label. Fan efficacy for fully ducted HRV, ERC, balanced, and in-line fans shall be determined at a static pressure of not less than 0.2 inch w.c. (49.85 Pa). Fan efficacy for ducted range hoods, bathroom and utility room fans shall be determined at a static pressure of not less than 0.1 inch w.c. (24.91 Pa).

TABLE R403.6.2
WHOLE-DWELLING MECHANICAL VENTILATION SYSTEM FAN EFFICACY[a]

FAN LOCATION	AIRFLOW RATE MINIMUM (CFM)	MINIMUM EFFICACY (CFM/WATT)
HRV, ERV	Any	1.2 cfm/watt
In-line supply or exhaust fan	Any	3.8 cfm/watt
Other exhaust fan	< 90	2.8 cfm/watt
Other exhaust fan	≥ 90	3.5 cfm/watt
Air-handler that is integrated to tested and *listed* HVAC equipment	Any	1.2 cfm/watt

For SI: 1 cubic foot per minute = 28.3 L/min.
a. Design outdoor airflow rate/watts of fan used.

R403.6.3 Testing. Mechanical ventilation systems shall be tested and verified to provide the minimum ventilation flow rates required by Section R403.6. Testing shall be performed according to the ventilation *equipment* manufacturer's instructions, or by using a flow hood or box, flow grid, or other airflow measuring device at the mechanical ventilation fan's inlet terminals or grilles, outlet terminals or grilles, or in the connected ventilation ducts. Where required by the code official, testing shall be conducted by an *approved* third party. A written report of the results of the test shall be signed by the party conducting the test and provided to the code official.

Exception: Kitchen range hoods that are ducted to the outside with 6-inch (152 mm) or larger duct and not more than one 90-degree (1.57 rad) elbow or equivalent in the duct run.

R403.7 Equipment sizing and efficiency rating. Heating and cooling *equipment* shall be sized in accordance with ACCA Manual S based on *building* loads calculated in accordance with ACCA Manual J or other *approved* heating and cooling calculation methodologies. New or replacement heating and cooling *equipment* shall have an efficiency rating equal to or greater than the minimum required by federal law for the geographic location where the *equipment* is installed.

R403.8 Systems serving multiple dwelling units. Systems serving multiple *dwelling units* shall comply with Sections C403 and C404 of the *International Energy Conservation Code*—Commercial Provisions instead of Section R403.

R403.9 Snow melt and ice system controls. Snow- and ice-melting systems, supplied through energy service to the building, shall include automatic controls capable of shutting off the system when the pavement temperature is greater than 50°F (10°C) and precipitation is not falling, and an automatic or manual control that will allow shutoff when the outdoor temperature is greater than 40°F (4.8°C).

R403.10 Energy consumption of pools and spas. The energy consumption of pools and permanent spas shall be controlled by the requirements in Sections R403.10.1 through R403.10.3.

R403.10.1 Heaters. The electric power to heaters shall be controlled by an on-off switch that is an integral part of the heater mounted on the exterior of the heater in a location with *ready access*, or external to and within 3 feet (914 mm) of the heater. Operation of such switch shall not change the setting of the heater thermostat. Such switches shall be in addition to a circuit breaker for the power to the heater. Gas-fired heaters shall not be equipped with continuously burning ignition pilots.

R403.10.2 Time switches. Time switches or other control methods that can automatically turn heaters and pump motors off and on according to a preset schedule shall be installed for heaters and pump motors. Heaters and pump motors that have built-in time switches shall be in compliance with this section.

Exceptions:

1. Where public health standards require 24-hour pump operation.
2. Pumps that operate solar- and waste-heat-recovery pool heating systems.

R403.10.3 Covers. Outdoor heated pools and outdoor permanent spas shall be provided with a vapor-retardant cover or other *approved* vapor-retardant means.

Exception: Where more than 75 percent of the energy for heating, computed over an operation season of not fewer than 3 calendar months, is from a heat pump or an on-site renewable energy system, covers or other vapor-retardant means shall not be required.

R403.11 Portable spas. The energy consumption of electric-powered portable spas shall be controlled by the requirements of APSP 14.

R403.12 Residential pools and permanent residential spas. Where installed, the energy consumption of residential swimming pools and permanent residential spas shall be controlled in accordance with the requirements of APSP 15.

SECTION R404
ELECTRICAL POWER AND LIGHTING SYSTEMS

R404.1 Lighting equipment. All permanently installed lighting fixtures, excluding kitchen appliance lighting fixtures, shall contain only high-efficacy lighting sources.

R404.1.1 Exterior lighting. Connected exterior lighting for residential buildings shall comply with Section C405.5.

Exceptions:
1. Detached one- and two- family dwellings.
2. Townhouses.
3. Solar-powered lamps not connected to any electrical service.
4. Luminaires controlled by a motion sensor.
5. Lamps and luminaires that comply with Section R404.1.

R404.1.2 Fuel gas lighting equipment. Fuel gas lighting systems shall not have continuously burning pilot lights.

R404.2 Interior lighting controls. Permanently installed lighting fixtures shall be controlled with either a dimmer, an occupant sensor control or other control that is installed or built into the fixture.

Exception: Lighting controls shall not be required for the following:
1. Bathrooms.
2. Hallways.
3. Exterior lighting fixtures.
4. Lighting designed for safety or security.

R404.3 Exterior lighting controls. Where the total permanently installed exterior lighting power is greater than 30 watts, the permanently installed exterior lighting shall comply with the following:

1. Lighting shall be controlled by a manual on and off switch which permits automatic shut-off actions.

    Exception: Lighting serving multiple *dwelling units*.

2. Lighting shall be automatically shut off when daylight is present and satisfies the lighting needs.
3. Controls that override automatic shut-off actions shall not be allowed unless the override automatically returns automatic control to its normal operation within 24 hours.

SECTION R405
TOTAL BUILDING PERFORMANCE

R405.1 Scope. This section establishes criteria for compliance using total building performance analysis. Such analysis shall include heating, cooling, mechanical ventilation and service water-heating energy only.

R405.2 Performance-based compliance. Compliance based on total building performance requires that a *proposed design* meets all of the following:

1. The requirements of the sections indicated within Table R405.2.
2. The building thermal envelope shall be greater than or equal to levels of efficiency and solar heat gain coefficients in Table R402.1.1 or R402.1.3 of the 2009 *International Energy Conservation Code*.
3. An annual energy cost that is less than or equal to the annual energy cost of the *standard reference design*. Energy prices shall be taken from a source *approved* by the *code official*, such as the Department of Energy, Energy Information Administration's State Energy Data System Prices and Expenditures reports. Code officials shall be permitted to require time-of-use pricing in energy cost calculations.

Exception: The energy use based on source energy expressed in Btu or Btu per square foot of conditioned floor area shall be permitted to be substituted for the energy cost. The source energy multiplier for electricity shall be 3.16. The source energy multiplier for fuels other than electricity shall be 1.1.

R405.3 Documentation. Documentation of the software used for the performance design and the parameters for the *building* shall be in accordance with Sections R405.3.1 through R405.3.2.2.

R405.3.1 Compliance software tools. Documentation verifying that the methods and accuracy of the compliance software tools conform to the provisions of this section shall be provided to the *code official*.

R405.3.2 Compliance report. Compliance software tools shall generate a report that documents that the *proposed design* complies with Section R405.3. A compliance report on the *proposed design* shall be submitted with the application for the building permit. Upon completion of the building, a confirmed compliance report based on the confirmed condition of the building shall be submitted to the *code official* before a certificate of occupancy is issued.

Compliance reports shall include information in accordance with Sections R405.3.2.1 and R405.3.2.2.

**TABLE R405.2
REQUIREMENTS FOR TOTAL BUILDING PERFORMANCE**

SECTION[a]	TITLE
General	
R401.2.5	Additional energy efficiency
R401.3	Certificate
Building Thermal Envelope	
R402.1.1	Vapor retarder
R402.2.3	Eave baffle
R402.2.4.1	Access hatches and doors
R402.2.10.1	Crawl space wall insulation installations
R402.4.1.1	Installation
R402.4.1.2	Testing
R402.5	Maximum fenestration U-factor and SHGC
Mechanical	
R403.1	Controls
R403.3, including R403.3.1, except Sections R403.3.2, R403.3.3 and R403.3.6	Ducts
R403.4	Mechanical system piping insulation
R403.5.1	Heated water circulation and temperature maintenance systems
R403.5.3	Drain water heat recovery units
R403.6	Mechanical ventilation
R403.7	Equipment sizing and efficiency rating
R403.8	Systems serving multiple dwelling units
R403.9	Snow melt and ice systems
R403.10	Energy consumption of pools and spas
R403.11	Portable spas
R403.12	Residential pools and permanent residential spas
Electrical Power and Lighting Systems	
R404.1	Lighting equipment
R404.2	Interior lighting controls

a. Reference to a code section includes all the relative subsections except as indicated in the table.

R405.3.2.1 Compliance report for permit application. A compliance report submitted with the application for building permit shall include the following:

1. Building street address, or other *building site* identification.
2. The name of the individual performing the analysis and generating the compliance report.
3. The name and version of the compliance software tool.
4. Documentation of all inputs entered into the software used to produce the results for the reference design and/or the rated home.
5. A certificate indicating that the proposed design complies with Section R405.3. The certificate shall document the building components' energy specifications that are included in the calculation including: component-level insulation R-values or U-factors; duct system and building envelope air leakage testing assumptions; and the type and rated efficiencies of proposed heating, cooling, mechanical ventilation and service water-heating equipment to be installed. If on-site renewable energy systems will be installed, the certificate shall report the type and production size of the proposed system.
6. Where a site-specific report is not generated, the proposed design shall be based on the worst-case orientation and configuration of the rated home.

R405.3.2.2 Compliance report for certificate of occupancy. A compliance report submitted for obtaining the certificate of occupancy shall include the following:

1. Building street address, or other building site identification.
2. Declaration of the total building performance path on the title page of the energy report and the title page of the building plans.
3. A statement, bearing the name of the individual performing the analysis and generating the report, indicating that the as-built building complies with Section R405.3.
4. The name and version of the compliance software tool.
5. A site-specific energy analysis report that is in compliance with Section R405.3.
6. A final confirmed certificate indicating compliance based on inspection, and a statement indicating that the confirmed rated design of the built home complies with Section R405.3. The certificate shall report the energy features that were confirmed to be in the home, including component-level insulation R-values or U-factors; results from any required duct system and building envelope air leakage testing; and the type and rated efficiencies of the heating, cooling, mechanical ventilation and service water-heating equipment installed.
7. When on-site renewable energy systems have been installed, the certificate shall report the type and production size of the installed system.

R405.4 Calculation procedure. Calculations of the performance design shall be in accordance with Sections R405.4.1 and R405.4.2.

R405.4.1 General. Except as specified by this section, the *standard reference design* and *proposed design* shall be configured and analyzed using identical methods and techniques.

R405.4.2 Residence specifications. The *standard reference design* and *proposed design* shall be configured and analyzed as specified by Table R405.4.2(1). Table R405.4.2(1) shall include, by reference, all notes contained in Table R402.1.3.

TABLE R405.4.2(1)
SPECIFICATIONS FOR THE STANDARD REFERENCE AND PROPOSED DESIGNS

BUILDING COMPONENT	STANDARD REFERENCE DESIGN	PROPOSED DESIGN
Above-grade walls	Type: mass where the proposed wall is a mass wall; otherwise wood frame.	As proposed
	Gross area: same as proposed.	As proposed
	U-factor: as specified in Table R402.1.2.	As proposed
	Solar absorptance = 0.75.	As proposed
	Emittance = 0.90.	As proposed
Basement and crawl space walls	Type: same as proposed.	As proposed
	Gross area: same as proposed.	As proposed
	U-factor: as specified in Table R402.1.2, with the insulation layer on the interior side of the walls.	As proposed
Above-grade floors	Type: wood frame.	As proposed
	Gross area: same as proposed.	As proposed
	U-factor: as specified in Table R402.1.2.	As proposed
Ceilings	Type: wood frame.	As proposed
	Gross area: same as proposed.	As proposed
	U-factor: as specified in Table R402.1.2.	As proposed
Roofs	Type: composition shingle on wood sheathing.	As proposed
	Gross area: same as proposed.	As proposed
	Solar absorptance = 0.75.	As proposed
	Emittance = 0.90.	As proposed
Attics	Type: vented with an aperture of 1 ft^2 per 300 ft^2 of ceiling area.	As proposed
Foundations	Type: same as proposed.	As proposed
	Foundation wall area above and below grade and soil characteristics: same as proposed.	As proposed
Opaque doors	Area: 40 ft^2.	As proposed
	Orientation: North.	As proposed
	U-factor: same as fenestration as specified in Table R402.1.2.	As proposed
Vertical fenestration other than opaque doors	Total area^h = (a) The proposed glazing area, where the proposed glazing area is less than 15 percent of the conditioned floor area. (b) 15 percent of the conditioned floor area, where the proposed glazing area is 15 percent or more of the conditioned floor area.	As proposed
	Orientation: equally distributed to four cardinal compass orientations (N, E, S & W).	As proposed
	U-factor: as specified in Table R402.1.2.	As proposed
	SHGC: as specified in Table R402.1.2 except for climate zones without an SHGC requirement, the SHGC shall be equal to 0.40.	As proposed
	Interior shade fraction: 0.92 − (0.21 × SHGC for the standard reference design).	Interior shade fraction: 0.92 − (0.21 × SHGC as proposed)
	External shading: none	As proposed

(continued)

TABLE R405.4.2(1)—continued
SPECIFICATIONS FOR THE STANDARD REFERENCE AND PROPOSED DESIGNS

BUILDING COMPONENT	STANDARD REFERENCE DESIGN	PROPOSED DESIGN
Skylights	None	As proposed
Thermally isolated sunrooms	None	As proposed
Air exchange rate	The air leakage rate at a pressure of 0.2 inch w.g. (50 Pa) shall be Climate Zones 0 through 2: 5.0 air changes per hour. Climate Zones 3 through 8: 3.0 air changes per hour.	The measured air exchange rate.[a]
	The mechanical ventilation rate shall be in addition to the air leakage rate and shall be the same as in the proposed design, but not greater than $0.01 \times CFA + 7.5 \times (N_{br} + 1)$ where: CFA = conditioned floor area, ft². N_{br} = number of bedrooms. The mechanical ventilation system type shall be the same as in the proposed design. Energy recovery shall not be assumed for mechanical ventilation.	The mechanical ventilation rate[b] shall be in addition to the air leakage rate and shall be as proposed.
Mechanical ventilation	Where mechanical ventilation is not specified in the proposed design: None Where mechanical ventilation is specified in the proposed design, the annual vent fan energy use, in units of kWh/yr, shall equal $(1/e_f) \times [0.0876 \times CFA + 65.7 \times (N_{br} + 1)]$ where: e_f = the minimum fan efficacy, as specified in Table 403.6.2, corresponding to the system type at a flow rate of $0.01 \times CFA + 7.5 \times (N_{br} + 1)$ CFA = conditioned floor area, ft². N_{br} = number of bedrooms.	As proposed
Internal gains	IGain, in units of Btu/day per dwelling unit, shall equal $17,900 + 23.8 \times CFA + 4,104 \times N_{br}$ where: CFA = conditioned floor area, ft². N_{br} = number of bedrooms.	Same as standard reference design.
Internal mass	Internal mass for furniture and contents: 8 pounds per square foot of floor area.	Same as standard reference design, plus any additional mass specifically designed as a thermal storage element[c] but not integral to the building envelope or structure.
Structural mass	For masonry floor slabs: 80 percent of floor area covered by R-2 carpet and pad, and 20 percent of floor directly exposed to room air.	As proposed
	For masonry basement walls: as proposed, but with insulation as specified in Table R402.1.3, located on the interior side of the walls.	As proposed
	For other walls, ceilings, floors, and interior walls: wood frame construction.	As proposed
Heating systems[d, e]	For other than electric heating without a heat pump: as proposed. Where the proposed design utilizes electric heating without a heat pump, the standard reference design shall be an air source heat pump meeting the requirements of Section C403 of the IECC—Commercial Provisions. Capacity: sized in accordance with Section R403.7.	As proposed
Cooling systems[d, f]	As proposed. Capacity: sized in accordance with Section R403.7.	As proposed

(continued)

TABLE R405.4.2(1)—continued
SPECIFICATIONS FOR THE STANDARD REFERENCE AND PROPOSED DESIGNS

<table>
<tr><th>BUILDING COMPONENT</th><th>STANDARD REFERENCE DESIGN</th><th>PROPOSED DESIGN</th></tr>
<tr><td>Service water heating[d, g]</td><td>As proposed.
Use, in units of gal/day = $25.5 + (8.5 \times N_{br})$
where:
N_{br} = number of bedrooms.</td><td>As proposed
Use, in units of gal/day = $25.5 + (8.5 \times N_{br}) \times (1 - HWDS)$
where:
N_{br} = number of bedrooms.
HWDS = factor for the compactness of the hot water distribution system.
<table>
<tr><th colspan="2">Compactness ratio[i] factor</th><th>HWDS</th></tr>
<tr><td>1 story</td><td>2 or more stories</td><td></td></tr>
<tr><td>> 60%</td><td>> 30%</td><td>0</td></tr>
<tr><td>> 30% to ≤ 60%</td><td>> 15% to ≤ 30%</td><td>0.05</td></tr>
<tr><td>> 15% to ≤ 30%</td><td>> 7.5% to ≤ 15%</td><td>0.10</td></tr>
<tr><td>< 15%</td><td>< 7.5%</td><td>0.15</td></tr>
</table></td></tr>
<tr><td>Thermal distribution systems</td><td>Duct insulation: in accordance with Section R403.3.1.
A thermal distribution system efficiency (DSE) of 0.88 shall be applied to both the heating and cooling system efficiencies for all systems other than tested duct systems.
Duct location: same as proposed design.
Exception: For nonducted heating and cooling systems that do not have a fan, the standard reference design thermal distribution system efficiency (DSE) shall be 1. For tested duct systems, the leakage rate shall be 4 cfm (113.3 L/min) per 100 ft² (9.29 m²) of conditioned floor area at a pressure of differential of 0.1 inch w.g. (25 Pa).</td><td>Duct location: as proposed.
Duct insulation: as proposed.
As tested or, where not tested, as specified in Table R405.4.2(2).</td></tr>
<tr><td>Thermostat</td><td>Type: Manual, cooling temperature setpoint = 75°F;
Heating temperature setpoint = 72°F.</td><td>Same as standard reference design.</td></tr>
<tr><td>Dehumidistat</td><td>Where a mechanical ventilation system with latent heat recovery is not specified in the proposed design:
None.
Where the proposed design utilizes a mechanical ventilation system with latent heat recovery:
Dehumidistat type: manual, setpoint = 60% relative humidity.
Dehumidifier: whole-dwelling with integrated energy factor = 1.77 liters/kWh.</td><td>Same as standard reference design.</td></tr>
</table>

For SI: 1 square foot = 0.93 m², 1 British thermal unit = 1055 J, 1 pound per square foot = 4.88 kg/m², 1 gallon (US) = 3.785 L, °C = (°F-32)/1.8, 1 degree = 0.79 rad.

a. Where required by the code official, testing shall be conducted by an approved party. Hourly calculations as specified in the ASHRAE *Handbook of Fundamentals*, or the equivalent, shall be used to determine the energy loads resulting from infiltration.

b. The combined air exchange rate for infiltration and mechanical ventilation shall be determined in accordance with Equation 43 of 2001 ASHRAE *Handbook of Fundamentals*, page 26.24 and the "Whole-house Ventilation" provisions of 2001 ASHRAE *Handbook of Fundamentals*, page 26.19 for intermittent mechanical ventilation.

c. Thermal storage element shall mean a component that is not part of the floors, walls or ceilings that is part of a passive solar system, and that provides thermal storage such as enclosed water columns, rock beds, or phase-change containers. A thermal storage element shall be in the same room as fenestration that faces within 15 degrees (0.26 rad) of true south, or shall be connected to such a room with pipes or ducts that allow the element to be actively charged.

d. For a proposed design with multiple heating, cooling or water heating systems using different fuel types, the applicable standard reference design system capacities and fuel types shall be weighted in accordance with their respective loads as calculated by accepted engineering practice for each equipment and fuel type present.

e. For a proposed design without a proposed heating system, a heating system having the prevailing federal minimum efficiency shall be assumed for both the standard reference design and proposed design.

f. For a proposed design home without a proposed cooling system, an electric air conditioner having the prevailing federal minimum efficiency shall be assumed for both the standard reference design and the proposed design.

g. For a proposed design with a nonstorage-type water heater, a 40-gallon storage-type water heater having the prevailing federal minimum energy factor for the same fuel as the predominant heating fuel type shall be assumed. For a proposed design without a proposed water heater, a 40-gallon storage-type water heater having the prevailing federal minimum efficiency for the same fuel as the predominant heating fuel type shall be assumed for both the proposed design and standard reference design.

(continued)

TABLE R405.4.2(1)—continued
SPECIFICATIONS FOR THE STANDARD REFERENCE AND PROPOSED DESIGNS

h. For residences with conditioned basements, R-2 and R-4 residences, and for townhouses, the following formula shall be used to determine glazing area:
$AF = A_s \times FA \times F$
where:
AF = Total glazing area.
A_s = Standard reference design total glazing area.
FA = (Above-grade thermal boundary gross wall area)/(above-grade boundary wall area + 0.5 × below-grade boundary wall area).
F = (above-grade thermal boundary wall area)/(above-grade thermal boundary wall area + common wall area) or 0.56, whichever is greater.
and where:
Thermal boundary wall is any wall that separates conditioned space from unconditioned space or ambient conditions.
Above-grade thermal boundary wall is any thermal boundary wall component not in contact with soil.
Below-grade boundary wall is any thermal boundary wall in soil contact.
Common wall area is the area of walls shared with an adjoining dwelling unit.

i. The factor for the compactness of the hot water distribution system is the ratio of the area of the rectangle that bounds the source of hot water and the fixtures that it serves (the "hot water rectangle") divided by the floor area of the dwelling.
   1. Sources of hot water include water heaters, or in multiple-family buildings with central water heating systems, circulation loops or electric heat traced pipes.
   2. The hot water rectangle shall include the source of hot water and the points of termination of all hot water fixture supply piping.
   3. The hot water rectangle shall be shown on the floor plans and the area shall be computed to the nearest square foot.
   4. Where there is more than one water heater and each water heater serves different plumbing fixtures and appliances, it is permissible to establish a separate hot water rectangle for each hot water distribution system and add the area of these rectangles together to determine the compactness ratio.
   5. The basement or attic shall be counted as a story when it contains the water heater.
   6. Compliance shall be demonstrated by providing a drawing on the plans that shows the hot water distribution system rectangle(s), comparing the area of the rectangle(s) to the area of the dwelling and identifying the appropriate compactness ratio and *HWDS* factor.

TABLE R405.4.2(2)
DEFAULT DISTRIBUTION SYSTEM EFFICIENCIES FOR PROPOSED DESIGNS[a]

DISTRIBUTION SYSTEM CONFIGURATION AND CONDITION	FORCED AIR SYSTEMS	HYDRONIC SYSTEMS[b]
Distribution system components located in unconditioned space	—	0.95
Untested distribution systems entirely located in conditioned space[c]	0.88	1
"Ductless" systems[d]	1	—

a. Default values in this table are for untested distribution systems, which must still meet minimum requirements for duct system insulation.
b. Hydronic systems shall mean those systems that distribute heating and cooling energy directly to individual spaces using liquids pumped through closed-loop piping and that do not depend on ducted, forced airflow to maintain space temperatures.
c. Entire system in conditioned space shall mean that no component of the distribution system, including the air-handler unit, is located outside of the conditioned space.
d. Ductless systems shall be allowed to have forced airflow across a coil but shall not have any ducted airflow external to the manufacturer's air-handler enclosure.

R405.5 Calculation software tools. Calculation software, where used, shall be in accordance with Sections R405.5.1 through R405.5.3.

R405.5.1 Minimum capabilities. Calculation procedures used to comply with this section shall be software tools capable of calculating the annual energy consumption of all building elements that differ between the *standard reference design* and the *proposed design* and shall include the following capabilities:

1. Computer generation of the *standard reference design* using only the input for the *proposed design*. The calculation procedure shall not allow the user to directly modify the building component characteristics of the *standard reference design*.

2. Calculation of whole-building (as a single *zone*) sizing for the heating and cooling equipment in the *standard reference design* residence in accordance with Section R403.7.

3. Calculations that account for the effects of indoor and outdoor temperatures and part-load ratios on the performance of heating, ventilating and air-conditioning equipment based on climate and equipment sizing.

4. Printed *code official* inspection checklist listing each of the *proposed design* component characteristics from Table R405.4.2(1) determined by the analysis to provide compliance, along with their respective performance ratings such as *R*-value, *U*-factor, SHGC, HSPF, AFUE, SEER and EF.

R405.5.2 Specific approval. Performance analysis tools meeting the applicable provisions of Section R405 shall be permitted to be *approved*. Tools are permitted to be *approved* based on meeting a specified threshold for a jurisdiction. The *code official* shall be permitted to approve such tools for a specified application or limited scope.

R405.5.3 Input values. When calculations require input values not specified by Sections R402, R403, R404 and R405, those input values shall be taken from an *approved* source.

SECTION R406
ENERGY RATING INDEX COMPLIANCE ALTERNATIVE

R406.1 Scope. This section establishes criteria for compliance using an Energy Rating Index (ERI) analysis.

R406.2 ERI compliance. Compliance based on the ERI requires that the rated design meets all of the following:

1. The requirements of the sections indicated within Table R406.2.
2. Maximum ERI of Table R406.5.

TABLE R406.2
REQUIREMENTS FOR ENERGY RATING INDEX

SECTION[a]	TITLE
General	
R401.2.5	Additional efficiency packages
R401.3	Certificate
Building Thermal Envelope	
R402.1.1	Vapor retarder
R402.2.3	Eave baffle
R402.2.4.1	Access hatches and doors
R402.2.10.1	Crawl space wall insulation installation
R402.4.1.1	Installation
R402.4.1.2	Testing
Mechanical	
R403.1	Controls
R403.3 except Sections R403.3.2, R403.3.3 and R403.3.6	Ducts
R403.4	Mechanical system piping insulation
R403.5.1	Heated water calculation and temperature maintenance systems
R403.5.3	Drain water heat recovery units
R403.6	Mechanical ventilation
R403.7	Equipment sizing and efficiency rating
R403.8	Systems serving multiple dwelling units
R403.9	Snow melt and ice systems
R403.10	Energy consumption of pools and spas
R403.11	Portable spas
R403.12	Residential pools and permanent residential spas
Electrical Power and Lighting Systems	
R404.1	Lighting equipment
R404.2	Interior lighting controls
R406.3	Building thermal envelope

a. Reference to a code section includes all of the relative subsections except as indicated in the table.

R406.3 Building thermal envelope. Building and portions thereof shall comply with Section R406.3.1 or R406.3.2.

R406.3.1 On-site renewables are not included. Where on-site renewable energy is not included for compliance using the ERI analysis of Section R406.4, the proposed total building thermal envelope UA, which is sum of U-factor times assembly area, shall be less than or equal to the building thermal envelope UA using the prescriptive U-factors from Table R402.1.2 multiplied by 1.15 in accordance with Equation 4-1. The area-weighted maximum fenestration SHGC permitted in Climate Zones 0 through 3 shall be 0.30.

$$UA_{\text{Proposed design}} = 1.15 \times UA_{\text{Prescriptive reference design}}$$

(Equation 4-1)

R406.3.2 On-site renewables are included. Where on-site renewable energy is included for compliance using the ERI analysis of Section R406.4, the *building thermal envelope* shall be greater than or equal to the levels of efficiency and SHGC in Table R402.1.2 or Table R402.1.4 of the 2018 *International Energy Conservation Code.*

R406.4 Energy Rating Index. The Energy Rating Index (ERI) shall be determined in accordance with RESNET/ICC 301 except for buildings covered by the *International Residential Code*, the ERI reference design ventilation rate shall be in accordance with Equation 4-2.

Ventilation rate, CFM = (0.01 × total square foot area of house) + [7.5 × (number of bedrooms + 1)]

(Equation 4-2)

Energy used to recharge or refuel a vehicle used for transportation on roads that are not on the building site shall not be included in the *ERI reference design* or the *rated design*. For compliance purposes, any reduction in energy use of the rated design associated with on-site renewable energy shall not exceed 5 percent of the total energy use.

R406.5 ERI-based compliance. Compliance based on an ERI analysis requires that the *rated proposed design* and confirmed built dwelling be shown to have an ERI less than or equal to the appropriate value indicated in Table R406.5 when compared to the *ERI reference design*.

TABLE R406.5
MAXIMUM ENERGY RATING INDEX

CLIMATE ZONE	ENERGY RATING INDEX
0-1	52
2	52
3	51
4	54
5	55
6	54
7	53
8	53

R406.6 Verification by approved agency. Verification of compliance with Section R406 as outlined in Sections R406.4 and R406.6 shall be completed by an *approved* third party. Verification of compliance with Section R406.2 shall be completed by the authority having jurisdiction or an *approved* third-party inspection agency in accordance with Section R105.4.

R406.7 Documentation. Documentation of the software used to determine the ERI and the parameters for the *residential building* shall be in accordance with Sections R406.7.1 through R406.7.4.

R406.7.1 Compliance software tools. Software tools used for determining ERI shall be *Approved* Software Rating Tools in accordance with RESNET/ICC 301.

R406.7.2 Compliance report. Compliance software tools shall generate a report that documents that the home and the ERI score of the *rated design* complies with Sections R406.2, R406.3 and R406.4. Compliance documentation shall be created for the proposed design and shall be submitted with the application for the building permit. Confirmed compliance documents of the built *dwelling unit* shall be created and submitted to the code official for review before a certificate of occupancy is issued. Compliance reports shall include information in accordance with Sections R406.7.2.1 and R406.7.2.2.

R406.7.2.1 Proposed compliance report for permit application. Compliance reports submitted with the application for a building permit shall include the following:

1. Building street address, or other *building site* identification.
2. Declare ERI on title page and building plans.
3. The name of the individual performing the analysis and generating the compliance report.
4. The name and version of the compliance software tool.
5. Documentation of all inputs entered into the software used to produce the results for the reference design and/or the rated home.
6. A certificate indicating that the proposed design has an ERI less than or equal to the appropriate score indicated in Table R406.5 when compared to the ERI reference design. The certificate shall document the building component energy specifications that are included in the calculation, including: component level insulation R-values or U-factors; assumed duct system and building envelope air leakage testing results; and the type and rated efficiencies of proposed heating, cooling, mechanical ventilation, and service water-heating equipment to be installed. If on-site renewable energy systems will be installed, the certificate shall report the type and production size of the proposed system.
7. When a site-specific report is not generated, the proposed design shall be based on the worst-case orientation and configuration of the rated home.

R406.7.2.2 Confirmed compliance report for a certificate of occupancy. A confirmed compliance report submitted for obtaining the certificate of occupancy shall be made site and address specific and include the following:

1. Building street address or other *building site* identification.
2. Declaration of ERI on title page and on building plans.
3. The name of the individual performing the analysis and generating the report.
4. The name and version of the compliance software tool.
5. Documentation of all inputs entered into the software used to produce the results for the reference design and/or the rated home.
6. A final confirmed certificate indicating that the confirmed rated design of the built home complies with Sections R406.2 and R406.4. The certificate shall report the energy features that were confirmed to be in the home, including: component-level insulation R-values or U-factors; results from any required duct system and building envelope air leakage testing; and the type and rated efficiencies of the heating, cooling, mechanical ventilation, and service water-heating equipment installed. Where on-site renewable energy systems have been installed on or in the home, the certificate shall report the type and production size of the installed system.

R406.7.3 Renewable energy certificate (REC) documentation. Where on-site renewable energy is included in the calculation of an ERI, one of the following forms of documentation shall be provided to the code official:

1. Substantiation that the RECs associated with the on-site renewable energy are owned by, or retired on behalf of, the homeowner.
2. A contract that conveys to the homeowner the RECs associated with the on-site renewable energy, or conveys to the homeowner an equivalent quantity of RECs associated with other renewable energy.

R406.7.4 Additional documentation. The *code official* shall be permitted to require the following documents:

1. Documentation of the building component characteristics of the *ERI reference design*.
2. A certification signed by the builder providing the building component characteristics of the *rated design*.
3. Documentation of the actual values used in the software calculations for the *rated design*.

R406.7.5 Specific approval. Performance analysis tools meeting the applicable subsections of Section R406 shall be *approved*. Documentation demonstrating the approval of performance analysis tools in accordance with Section R406.7.1 shall be provided.

R406.7.6 Input values. Where calculations require input values not specified by Sections R402, R403, R404 and R405, those input values shall be taken from RESNET/ICC 301.

SECTION R407
TROPICAL CLIMATE REGION COMPLIANCE PATH

R407.1 Scope. This section establishes alternative criteria for residential buildings in the tropical region at elevations less than 2,400 feet (731.5 m) above sea level.

R407.2 Tropical climate region. Compliance with this section requires the following:

1. Not more than one-half of the *occupied* space is air conditioned.
2. The *occupied* space is not heated.
3. Solar, wind or other renewable energy source supplies not less than 80 percent of the energy for service water heating.
4. Glazing in *conditioned spaces* has a *solar heat gain coefficient* (SHGC) of less than or equal to 0.40, or has an overhang with a projection factor equal to or greater than 0.30.
5. Permanently installed lighting is in accordance with Section R404.
6. The exterior roof surface complies with one of the options in Table C402.3 of the *International Energy Conservation Code*–Commercial Provisions or the roof or ceiling has insulation with an *R*-value of R-15 or greater. Where attics are present, attics above the insulation are vented and attics below the insulation are unvented.
7. Roof surfaces have a slope of not less than $^1/_4$ unit vertical in 12 units horizontal (21-percent slope). The finished roof does not have water accumulation areas.
8. Operable fenestration provides a ventilation area of not less than 14 percent of the floor area in each room. Alternatively, equivalent ventilation is provided by a ventilation fan.
9. Bedrooms with *exterior walls* facing two different directions have operable fenestration on exterior walls facing two directions.
10. Interior doors to bedrooms are capable of being secured in the open position.
11. A ceiling fan or ceiling fan rough-in is provided for bedrooms and the largest space that is not used as a bedroom.

SECTION R408
ADDITIONAL EFFICIENCY PACKAGE OPTIONS

R408.1 Scope. This section establishes additional efficiency package options to achieve additional energy efficiency in accordance with Section R401.2.5.

R408.2 Additional efficiency package options. Additional efficiency package options for compliance with Section R401.2.1 are set forth in Sections R408.2.1 through R408.2.5.

R408.2.1 Enhanced envelope performance option. The total *building thermal envelope* UA, the sum of *U*-factor times assembly area, shall be less than or equal to 95 percent of the total UA resulting from multiplying the *U*-factors in Table R402.1.2 by the same assembly area as in the proposed building. The UA calculation shall be performed in accordance with Section R402.1.5. The area-weighted average SHGC of all glazed fenestration shall be less than or equal to 95 percent of the maximum glazed fenestration SHGC in Table R402.1.2.

R408.2.2 More efficient HVAC equipment performance option. Heating and cooling *equipment* shall meet one of the following efficiencies:

1. Greater than or equal to 95 AFUE natural gas furnace and 16 SEER air conditioner.
2. Greater than or equal to 10 HSPF/16 SEER air source heat pump.
3. Greater than or equal to 3.5 COP ground source heat pump.

For multiple cooling systems, all systems shall meet or exceed the minimum efficiency requirements in this section and shall be sized to serve 100 percent of the cooling design load. For multiple heating systems, all systems shall meet or exceed the minimum efficiency requirements in this section and shall be sized to serve 100 percent of the heating design load.

R408.2.3 Reduced energy use in service water-heating option. The hot water system shall meet one of the following efficiencies:

1. Greater than or equal to 0.82 EF fossil fuel service water-heating system.
2. Greater than or equal to 2.0 EF electric service water-heating system.
3. Greater than or equal to 0.4 solar fraction solar water-heating system.

R408.2.4 More efficient duct thermal distribution system option. The thermal distribution system shall meet one of the following efficiencies:

1. 100 percent of ducts and air handlers located entirely within the *building thermal envelope*.
2. 100 percent of ductless thermal distribution system or hydronic thermal distribution system located completely inside the *building thermal envelope*.

3. 100 percent of duct thermal distribution system located in *conditioned space* as defined by Section R403.3.2.

R408.2.5 Improved air sealing and efficient ventilation system option. The measured air leakage rate shall be less than or equal to 3.0 ACH50, with either an Energy Recovery Ventilator (ERV) or Heat Recovery Ventilator (HRV) installed. Minimum HRV and ERV requirements, measured at the lowest tested net supply airflow, shall be greater than or equal to 75 percent Sensible Recovery Efficiency (SRE), less than or equal to 1.1 cubic feet per minute per watt (0.03 m^3/min/watt) and shall not use recirculation as a defrost strategy. In addition, the ERV shall be greater than or equal to 50 percent Latent Recovery/Moisture Transfer (LRMT).

CHAPTER 5 [RE]
EXISTING BUILDINGS

User note:

About this chapter: *Many buildings are renovated or altered in numerous ways that could affect the energy use of the building as a whole. Chapter 5 requires the application of certain parts of Chapter 4 in order to maintain, if not improve, the conservation of energy by the renovated or altered building.*

SECTION R501
GENERAL

R501.1 Scope. The provisions of this chapter shall control the *alteration, repair, addition* and change of occupancy of existing *buildings* and structures.

R501.1.1 General. Except as specified in this chapter, this code shall not be used to require the removal, *alteration* or abandonment of, nor prevent the continued use and maintenance of, an existing *building* or *building* system lawfully in existence at the time of adoption of this code. Unaltered portions of the existing *building* or *building* supply system shall not be required to comply with this code.

R501.2 Compliance. *Additions, alterations, repairs* or changes of occupancy to, or relocation of, an existing *building, building* system or portion thereof shall comply with Section R502, R503, R504 or R505, respectively, in this code. Changes where unconditioned space is changed to *conditioned space* shall comply with Section R502.

R501.3 Maintenance. *Buildings* and structures, and parts thereof, shall be maintained in a safe and sanitary condition. Devices and systems that are required by this code shall be maintained in conformance to the code edition under which installed. The owner or the owner's authorized agent shall be responsible for the maintenance of *buildings* and structures. The requirements of this chapter shall not provide the basis for removal or abrogation of energy conservation, fire protection and safety systems and devices in existing structures.

R501.4 Compliance. *Alterations, repairs, additions* and changes of occupancy to, or relocation of, existing buildings and structures shall comply with the provisions for *alterations, repairs, additions* and changes of occupancy or relocation, respectively, in this code and the *International Residential Code, International Building Code, International Existing Building Code, International Fire Code, International Fuel Gas Code, International Mechanical Code, International Plumbing Code, International Property Maintenance Code, International Private Sewage Disposal Code* and NFPA 70.

R501.5 New and replacement materials. Except as otherwise required or permitted by this code, materials permitted by the applicable code for new construction shall be used. Like materials shall be permitted for *repairs*, provided that hazards to life, health or property are not created. Hazardous materials shall not be used where the code for new construction would not allow their use in *buildings* of similar occupancy, purpose and location.

R501.6 Historic buildings. Provisions of this code relating to the construction, *repair, alteration*, restoration and movement of structures, and *change of occupancy* shall not be mandatory for *historic buildings* provided that a report has been submitted to the code official and signed by the owner, a *registered design professional*, or a representative of the State Historic Preservation Office or the historic preservation authority having jurisdiction, demonstrating that compliance with that provision would threaten, degrade or destroy the historic form, fabric or function of the *building*.

SECTION R502
ADDITIONS

R502.1 General. *Additions* to an existing *building, building* system or portion thereof shall conform to the provisions of this code as those provisions relate to new construction without requiring the unaltered portion of the existing *building* or *building* system to comply with this code. *Additions* shall not create an unsafe or hazardous condition or overload existing *building* systems. An *addition* shall be deemed to comply with this code where the *addition* alone complies, where the existing *building* and *addition* comply with this code as a single building, or where the *building* with the *addition* does not use more energy than the existing *building*. *Additions* shall be in accordance with Section R502.2 or R502.3.

R502.2 Change in space conditioning. Any unconditioned or low-energy space that is altered to become *conditioned space* shall be required to be brought into full compliance with this code.

Exceptions:

1. Where the simulated performance option in Section R405 is used to comply with this section, the annual energy cost of the *proposed design* is permitted to be 110 percent of the annual energy cost otherwise allowed by Section R405.2.

2. Where the Total UA, as determined in Section R402.1.5, of the existing *building* and the *addition*, and any *alterations* that are part of the project, is less than or equal to the Total UA generated for the existing *building*.

3. Where complying in accordance with Section R405 and the annual energy cost or energy use of the *addition* and the existing *building*, and any

alterations that are part of the project, is less than or equal to the annual energy cost of the existing *building*. The *addition* and any *alterations* that are part of the project shall comply with Section R405 in its entirety.

R502.3 Prescriptive compliance. *Additions* shall comply with Sections R502.3.1 through R502.3.4.

R502.3.1 Building envelope. New *building* envelope assemblies that are part of the *addition* shall comply with Sections R402.1, R402.2, R402.3.1 through R402.3.5, and R402.4.

Exception: New envelope assemblies are exempt from the requirements of Section R402.4.1.2.

R502.3.2 Heating and cooling systems. HVAC ducts newly installed as part of an *addition* shall comply with Section R403.

Exception: Where ducts from an existing heating and cooling system are extended to an *addition*.

R502.3.3 Service hot water systems. New service hot water systems that are part of the *addition* shall comply with Section R403.5.

R502.3.4 Lighting. New lighting systems that are part of the *addition* shall comply with Section R404.1.

SECTION R503
ALTERATIONS

R503.1 General. *Alterations* to any building or structure shall comply with the requirements of the code for new construction, without requiring the unaltered portions of the existing building or building system to comply with this code. *Alterations* shall be such that the existing building or structure is not less conforming to the provisions of this code than the existing *building* or structure was prior to the *alteration*.

Alterations shall not create an unsafe or hazardous condition or overload *existing* building systems. *Alterations* shall be such that the existing *building* or structure does not use more energy than the existing building or structure prior to the *alteration*. *Alterations* to existing *buildings* shall comply with Sections R503.1.1 through R503.1.4.

R503.1.1 Building envelope. Building envelope assemblies that are part of the *alteration* shall comply with Section R402.1.2 or R402.1.4, Sections R402.2.1 through R402.2.12, R402.3.1, R402.3.2, R402.4.3 and R402.4.5.

Exception: The following alterations shall not be required to comply with the requirements for new construction provided that the energy use of the building is not increased:

1. Storm windows installed over existing fenestration.
2. Existing ceiling, wall or floor cavities exposed during construction provided that these cavities are filled with insulation.
3. Construction where the existing roof, wall or floor cavity is not exposed.
4. Roof recover.
5. Roofs without insulation in the cavity and where the sheathing or insulation is exposed during reroofing shall be insulated either above or below the sheathing.
6. Surface-applied window film installed on existing single pane fenestration assemblies to reduce solar heat gain provided that the code does not require the glazing or fenestration assembly to be replaced.

R503.1.1.1 Replacement fenestration. Where some or all of an existing fenestration unit is replaced with a new fenestration product, including sash and glazing, the replacement fenestration unit shall meet the applicable requirements for *U*-factor and SHGC as specified in Table R402.1.3. Where more than one replacement fenestration unit is to be installed, an area-weighted average of the *U*-factor, SHGC or both of all replacement fenestration units shall be an alternative that can be used to show compliance.

R503.1.2 Heating and cooling systems. HVAC ducts newly installed as part of an *alteration* shall comply with Section R403.

Exception: Where ducts from an existing heating and cooling system are extended to an *addition*.

R503.1.3 Service hot water systems. New service hot water systems that are part of the *alteration* shall comply with Section R403.5.

R503.1.4 Lighting. New lighting systems that are part of the *alteration* shall comply with Section R404.1.

Exception: *Alterations* that replace less than 10 percent of the luminaires in a space, provided that such alterations do not increase the installed interior lighting power.

SECTION R504
REPAIRS

R504.1 General. *Buildings*, structures and parts thereof shall be repaired in compliance with Section R501.3 and this section. Work on nondamaged components necessary for the required *repair* of damaged components shall be considered to be part of the *repair* and shall not be subject to the requirements for *alterations* in this chapter. Routine maintenance required by Section R501.3, ordinary repairs exempt from *permit*, and abatement of wear due to normal service conditions shall not be subject to the requirements for *repairs* in this section.

R504.2 Application. For the purposes of this code, the following shall be considered to be *repairs*:

1. Glass-only replacements in an existing sash and frame.
2. Roof *repairs*.

3. *Repairs* where only the bulb, ballast or both within the existing luminaires in a space are replaced provided that the replacement does not increase the installed interior lighting power.

SECTION R505
CHANGE OF OCCUPANCY OR USE

R505.1 General. Any space that is converted to a dwelling unit or portion thereof from another use or occupancy shall comply with this code.

Exception: Where the simulated performance option in Section R405 is used to comply with this section, the annual energy cost of the *proposed design* is permitted to be 110 percent of the annual energy cost allowed by Section R405.2.

R505.1.1 Unconditioned space. Any unconditioned or low-energy space that is altered to become a *conditioned space* shall comply with Section R502.

CHAPTER 6 [RE]
REFERENCED STANDARDS

User note:

About this chapter: This code contains numerous references to standards promulgated by other organizations that are used to provide requirements for materials and methods of construction. Chapter 6 contains a comprehensive list of all standards that are referenced in this code. These standards, in essence, are part of this code to the extent of the reference to the standard.

This chapter lists the standards that are referenced in various sections of this document. The standards are listed herein by the promulgating agency of the standard, the standard identification, the effective date and title, and the section or sections of this document that reference the standard. The application of the referenced standards shall be as specified in Section R108.

AAMA

American Architectural Manufacturers Association
1827 Walden Office Square
Suite 550
Schaumburg, IL 60173-4268

AAMA/WDMA/CSA 101/I.S.2/A440—17: North American Fenestration Standard/Specification for Windows, Doors, and Skylights
  R402.4.3

ACCA

Air Conditioning Contractors of America
1330 Braddock Place, Suite 350
Alexandria, VA 22314

ANSI/ACCA 2 Manual J—2016: Residential Load Calculation
  R403.7

ANSI/ACCA 3 Manual S—2014: Residential Equipment Selection
  R403.7

APSP

Pool & Tub Alliance (formerly the APSP)
2111 Eisenhower Avenue, Suite 500
Alexandria, VA 22314

ANSI/APSP/ICC 14—2019: American National Standard for Portable Electric Spa Energy Efficiency
  R403.11

ANSI/APSP/ICC 15a—2011: American National Standard for Residential Swimming Pool and Spas—Includes Addenda A
   Approved January 9, 2013
  R403.12

ASHRAE

ASHRAE
180 Technology Parkway NW
Peachtree Corners, GA 30092

ASHRAE 193—2010(RA 2014): Method of Test for Determining the Airtightness of HVAC Equipment
  R403.3.4.1

ASHRAE—2001: 2001 ASHRAE Handbook of Fundamentals
  Table R405.5.2(1)

ASHRAE—2021: ASHRAE Handbook of Fundamentals
  R402.1.5

REFERENCED STANDARDS

ASTM

ASTM International
100 Barr Harbor Drive, P.O. Box C700
West Conshohocken, PA 19428-2959

C1363—11: Standard Test Method for Thermal Performance of Building Materials and Envelope Assemblies by Means of a Hot Box Apparatus
    R303.1.4.1

E283—2004(2012): Test Method for Determining the Rate of Air Leakage Through Exterior Windows, Curtain Walls and Doors Under Specified Pressure Differences Across the Specimen
    R402.4.5

E779—2010(2018): Standard Test Method for Determining Air Leakage Rate by Fan Pressurization
    R402.4.1.2

E1554/E1554M—E2013: Standard Test Methods for Determining Air Leakage of Air Distribution Systems by Fan Pressurization
    R403.3.5

E1827—: 2011(2017): Standard Test Methods for Determining Airtightness of Building Using an Orifice Blower Door
    R402.4.1.2

E2178—2013: Standard Test Method for Air Permanence of Building Materials
    R303.1.5

CSA

CSA Group
8501 East Pleasant Valley Road
Cleveland, OH 44131-5516

AAMA/WDMA/CSA 101/I.S.2/A440—17: North American Fenestration Standard/Specification for Windows, Doors and Unit Skylights
    R402.4.3

CSA B55.1—2015: Test Method for Measuring Efficiency and Pressure Loss of Drain Water Heat Recovery Units
    R403.5.3

CSA B55.2—2015: Drain Water Heat Recovery Units
    R403.5.3

DASMA

Door & Access Systems Manufacturers Association
1300 Sumner Avenue
Cleveland, OH 44115-2851

105—2017: Test Method for Thermal Transmittance and Air Infiltration of Garage Doors and Rolling Doors
    R303.1.3

HVI

Home Ventilating Institute
1740 Dell Range Blvd, Ste H, PMB 450
Cheyenne, WY 82009

916—18: Airflow Test Procedure
    Table R403.6.2

ICC

International Code Council, Inc.
500 New Jersey Avenue NW 6th Floor
Washington, DC 20001

ANSI/APSP/ICC 14—2019: American National Standard for Portable Electric Spa Energy Efficiency
    R403.11

ANSI/APSP/ICC 15a—2020: American National Standard for Residential Swimming Pool and Spa Energy Efficiency
    R403.12

ANSI/RESNET/ICC 301—2019: Standard for the Calculation and Labeling of the Energy Performance of Dwelling and Sleeping Units using an Energy Rating Index
    R406.4

ICC—continued

ANSI/RESNET/ICC 380—2019: Standard for Testing Airtightness of Building, Dwelling Unit and Sleeping Unit Enclosures; Airtightness of Heating and Cooling Air Distribution Systems, and Airflow of Mechanical Ventilation Systems
  R402.4.1.2

IBC—21: International Building Code®
  R201.3, R303.1.1, R303.2, R402.1.1, R501.4

ICC 400—17: Standard on the Design and Construction of Log Structures
  R402.1

ICC 500—2020: ICC/NSSA Standard for the Design and Construction of Storm Shelters
  R402.5

IEBC—21: International Existing Building Code®
  R501.4

IECC—06: 2006 International Energy Conservation Code®
  R202

IECC—09: 2009 International Energy Conservation Code®
  R406.2

IECC—15: 2015 International Energy Conservation Code®
  Table R406.5

IFC—21: International Fire Code®
  R201.3, R501.4

IFGC—21: International Fuel Gas Code®
  R201.3, R501.4

IMC—21: International Mechanical Code®
  R201.3, R403.3.3, R403.3.4, R403.6, R501.4

IPC—21: International Plumbing Code®
  R201.3, R501.4

IPMC—21: International Property Maintenance Code®
  R501.4

IPSDC—21: International Private Sewage Disposal Code®
  R501.4

IRC—21: International Residential Code®
  R201.3, R303.1.1, R303.2, R402.1.1, R402.2.10.1, R403.3.3, R403.3.4, R403.6, R501.4

IEEE

Institute of Electrical and Electronics Engineers, Inc.
3 Park Avenue, 17th Floor
New York, NY 10016-5997

515.1—2012: IEEE Standard for the Testing, Design, Installation and Maintenance of Electrical Resistance Trace Heating for Commercial Applications
  R403.5.1.2

NFPA

National Fire Protection Association
1 Batterymarch Park
Quincy, MA 02169-7471

70—20: National Electrical Code
  R501.4

REFERENCED STANDARDS

NFRC

National Fenestration Rating Council, Inc.
6305 Ivy Lane, Suite 140
Greenbelt, MD 20770

100—2020: Procedure for Determining Fenestration Products *U*-factors
    R303.1.3

200—2020: Procedure for Determining Fenestration Product Solar Heat Gain Coefficient and Visible Transmittance at Normal Incidence
    R303.1.3

400—2020: Procedure for Determining Fenestration Product Air Leakage
    R402.4.3

RESNET

Residential Energy Services Network, Inc.
P.O. Box 4561
Oceanside, CA 92052-4561

ANSI/RESNET/ICC 301—2019: Standard for the Calculation and Labeling of the Energy Performance of Dwelling and Sleeping Units using an Energy Rating Index
    R406.4, R406.7.1, R406.7.6

ANSI/RESNET/ICC 380—2019: Standard for Testing Airtightness of Building, Dwelling Unit and Sleeping Unit Enclosures; Airtightness of Heating and Cooling Air Distribution Systems, and Airflow of Mechanical Ventilation Systems
    R402.4.1.2, R403.3.5

UL

UL LLC
333 Pfingsten Road
Northbrook, IL 60062

127—2011: Standard for Factory-Built Fireplaces—with Revisions through July 2016
    R402.4.2

515—2015: Standard for Electrical Resistance Trace Heating for Commercial Applications
    R403.5.1.2

US-FTC

United States-Federal Trade Commission
600 Pennsylvania Avenue NW
Washington, DC 20580

CFR Title 16 (2015): R-value Rule
    R303.1.4

WDMA

Window and Door Manufacturers Association
2025 M Street NW, Suite 800
Washington, DC 20036-3309

AAMA/WDMA/CSA 101/I.S.2/A440—17: North American Fenestration Standard/Specification for Windows, Doors and Skylights
    R402.4.3

APPENDIX RA

BOARD OF APPEALS—RESIDENTIAL

The provisions contained in this appendix are not mandatory unless specifically referenced in the adopting ordinance.

User note:

About this appendix: *Appendix RA provides criteria for board of appeals members. Also provided are procedures by which the board of appeals should conduct its business.*

SECTION RA101
GENERAL

RA101.1 Scope. A board of appeals shall be established within the jurisdiction for the purpose of hearing applications for modification of the requirements of this code pursuant to the provisions of Section R110. The board shall be established and operated in accordance with this section, and shall be authorized to hear evidence from appellants and the code official pertaining to the application and intent of this code for the purpose of issuing orders pursuant to these provisions.

RA101.2 Application for appeal. Any person shall have the right to appeal a decision of the code official to the board. An application for appeal shall be based on a claim that the intent of this code or the rules legally adopted hereunder have been incorrectly interpreted, the provisions of this code do not fully apply or an equally good or better form of construction is proposed. The application shall be filed on a form obtained from the code official within 20 days after the notice was served.

RA101.2.1 Limitation of authority. The board shall not have authority to waive requirements of this code or interpret the administration of this code.

RA101.2.2 Stays of enforcement. Appeals of notice and orders, other than Imminent Danger notices, shall stay the enforcement of the notice and order until the appeal is heard by the board.

RA101.3 Membership of board. The board shall consist of five voting members appointed by the chief appointing authority of the jurisdiction. Each member shall serve for **[INSERT NUMBER OF YEARS]** years or until a successor has been appointed. The board members' terms shall be staggered at intervals, so as to provide continuity. The code official shall be an ex officio member of said board but shall not vote on any matter before the board.

RA101.3.1 Qualifications. The board shall consist of five individuals, who are qualified by experience and training to pass on matters pertaining to building construction and are not employees of the jurisdiction.

RA101.3.2 Alternate members. The chief appointing authority is authorized to appoint two alternate members who shall be called by the board chairperson to hear appeals during the absence or disqualification of a member. Alternate members shall possess the qualifications required for board membership, and shall be appointed for the same term or until a successor has been appointed.

RA101.3.3 Vacancies. Vacancies shall be filled for an unexpired term in the same manner in which original appointments are required to be made.

RA101.3.4 Chairperson. The board shall annually select one of its members to serve as chairperson.

RA101.3.5 Secretary. The chief appointing authority shall designate a qualified clerk to serve as secretary to the board. The secretary shall file a detailed record of all proceedings, which shall set forth the reasons for the board's decision, the vote of each member, the absence of a member and any failure of a member to vote.

RA101.3.6 Conflict of interest. A member with any personal, professional or financial interest in a matter before the board shall declare such interest and refrain from participating in discussions, deliberations and voting on such matters.

RA101.3.7 Compensation of members. Compensation of members shall be determined by law.

RA101.3.8 Removal from the board. A member shall be removed from the board prior to the end of their term only for cause. Any member with continued absence from regular meeting of the board may be removed at the discretion of the chief appointing authority.

RA101.4 Rules and procedures. The board shall establish policies and procedures necessary to carry out its duties consistent with the provisions of this code and applicable state law. The procedures shall not require compliance with strict rules of evidence, but shall mandate that only relevant information be presented.

RA101.5 Notice of meeting. The board shall meet upon notice from the chairperson, within 10 days of the filing of an appeal or at stated periodic intervals.

RA101.5.1 Open hearing. All hearings before the board shall be open to the public. The appellant, the appellant's representative, the code official and any person whose interests are affected shall be given an opportunity to be heard.

RA101.5.2 Quorum. Three members of the board shall constitute a quorum.

RA101.5.3 Postponed hearing. When five members are not present to hear an appeal, either the appellant or the appellant's representative shall have the right to request a postponement of the hearing.

RA101.6 Legal counsel. The jurisdiction shall furnish legal counsel to the board to provide members with general legal advice concerning matters before them for consideration. Members shall be represented by legal counsel at the jurisdiction's expense in all matters arising from service within the scope of their duties.

RA101.7 Board decision. The board shall only modify or reverse the decision of the code official by a concurring vote of three or more members.

RA101.7.1 Resolution. The decision of the board shall be by resolution. Every decision shall be promptly filed in writing in the office of the code official within 3 days and shall be open to the public for inspection. A certified copy shall be furnished to the appellant or the appellant's representative and to the code official.

RA101.7.2 Administration. The code official shall take immediate action in accordance with the decision of the board.

RA101.8 Court review. Any person, whether or not a previous party of the appeal, shall have the right to apply to the appropriate court for a writ of certiorari to correct errors of law. Application for review shall be made in the manner and time required by law following the filing of the decision in the office of the chief administrative officer.

APPENDIX RB

SOLAR-READY PROVISIONS—DETACHED ONE- AND TWO-FAMILY DWELLINGS AND TOWNHOUSES

The provisions contained in this appendix are not mandatory unless specifically referenced in the adopting ordinance.

User note:
About this appendix: Harnessing the heat or radiation from the sun's rays is a method to reduce the energy consumption of a building. Although Appendix RB does not require solar systems to be installed for a building, it does require the space(s) for installing such systems, providing pathways for connections and requiring adequate structural capacity of roof systems to support the systems.

SECTION RB101
SCOPE

RB101.1 General. These provisions shall be applicable for new construction where solar-ready provisions are required.

SECTION RB102
GENERAL DEFINITION

SOLAR-READY ZONE. A section or sections of the roof or building overhang designated and reserved for the future installation of a solar photovoltaic or solar thermal system.

SECTION RB103
SOLAR-READY ZONE

RB103.1 General. New detached one- and two-family dwellings, and townhouses with not less than 600 square feet (55.74 m^2) of roof area oriented between 110 degrees and 270 degrees of true north shall comply with Sections RB103.2 through RB103.8.

Exceptions:
1. New residential buildings with a permanently installed on-site renewable energy system.
2. A building where all areas of the roof that would otherwise meet the requirements of Section RB103 are in full or partial shade for more than 70 percent of daylight hours annually.

RB103.2 Construction document requirements for solar-ready zone. Construction documents shall indicate the solar-ready zone.

RB103.3 Solar-ready zone area. The total solar-ready zone area shall be not less than 300 square feet (27.87 m^2) exclusive of mandatory access or setback areas as required by the *International Fire Code*. New townhouses three stories or less in height above grade plane and with a total floor area less than or equal to 2,000 square feet (185.8 m^2) per dwelling shall have a solar-ready zone area of not less than 150 square feet (13.94 m^2). The solar-ready zone shall be composed of areas not less than 5 feet (1524 mm) in width and not less than 80 square feet (7.44 m^2) exclusive of access or setback areas as required by the *International Fire Code*.

RB103.4 Obstructions. Solar-ready zones shall be free from obstructions, including but not limited to vents, chimneys, and roof-mounted equipment.

RB103.5 Shading. The solar-ready zone shall be set back from any existing or new permanently affixed object on the building or site that is located south, east or west of the solar zone a distance not less than two times the object's height above the nearest point on the roof surface. Such objects include, but are not limited to, taller portions of the building itself, parapets, chimneys, antennas, signage, rooftop equipment, trees and roof plantings.

RB103.6 Capped roof penetration sleeve. A capped roof penetration sleeve shall be provided adjacent to a solar-ready zone located on a roof slope of not greater than 1 unit vertical in 12 units horizontal (8-percent slope). The capped roof penetration sleeve shall be sized to accommodate the future photovoltaic system conduit, but shall have an inside diameter of not less than $1^1/_4$ inches (32 mm).

RB103.7 Roof load documentation. The structural design loads for roof dead load and roof live load shall be clearly indicated on the construction documents.

RB103.8 Interconnection pathway. Construction documents shall indicate pathways for routing of conduit or plumbing from the solar-ready zone to the electrical service panel or service hot water system.

RB103.9 Electrical service reserved space. The main electrical service panel shall have a reserved space to allow installation of a dual pole circuit breaker for future solar electric installation and shall be labeled "For Future Solar Electric." The reserved space shall be positioned at the opposite (load) end from the input feeder location or main circuit location.

RB103.10 Construction documentation certificate. A permanent certificate, indicating the solar-ready zone and other requirements of this section, shall be posted near the electrical distribution panel, water heater or other conspicuous location by the builder or registered design professional.

APPENDIX RC

ZERO ENERGY RESIDENTIAL BUILDING PROVISIONS

The provisions contained in this appendix are not mandatory unless specifically referenced in the adopting ordinance.

User Note:

About this appendix: This appendix provides requirements for residential buildings intended to result in net zero energy consumption over the course of a year. Where adopted by ordinance as a requirement, Section RC101 language is intended to replace Section R401.2.

SECTION RC101 COMPLIANCE

RC101.1 Compliance. Existing residential buildings shall comply with Chapter 5. New residential buildings shall comply with Section RC102.

SECTION RC102 ZERO ENERGY RESIDENTIAL BUILDINGS

RC102.1 General. New *residential buildings* shall comply with Section RC102.2.

RC102.2 Energy Rating Index zero energy score. Compliance with this section requires that the rated design be shown to have a score less than or equal to the values in Table RC102.2 when compared to the Energy Rating Index (ERI) reference design determined in accordance with RESNET/ICC 301 for both of the following:

1. ERI value not including on-site power production (OPP) calculated in accordance with RESNET/ICC 301.
2. ERI value including on-site power production calculated in accordance with RESNET/ICC 301 with the OPP in Equation 4.1.2 of RESNET/ICC 301 adjusted in accordance with Equation RC-1.

Adjusted OPP = OPP + CREF + REPC   (Equation RC-1)

where:

CREF = Community Renewable Energy Facility power production—the yearly energy, in kilowatt hour equivalent (kWheq), contracted from a community renewable energy facility that is qualified under applicable state and local utility statutes and rules, and that allocates bill credits to the rated home.

REPC = Renewable Energy Purchase Contract power production—the yearly energy, in kilowatt hour equivalent (kWheq), contracted from an energy facility that generates energy with photovoltaic, solar thermal, geothermal energy or wind systems, and that is demonstrated by an energy purchase contract or lease with a duration of not less than 15 years.

TABLE RC102.2
MAXIMUM ENERGY RATING INDEX[a]

CLIMATE ZONE	ENERGY RATING INDEX NOT INCLUDING OPP	ENERGY RATING INDEX INCLUDING ADJUSTED OPP (as proposed)
1	43	0
2	45	0
3	47	0
4	47	0
5	47	0
6	46	0
7	46	0
8	46	0

a. The building shall meet the requirements of Table R406.2, and the building thermal envelope shall be greater than or equal to the levels of efficiency and SHGC in Table R402.1.2 or R402.1.3 of the 2015 *International Energy Conservation Code*.

INDEX

A

ABOVE-GRADE WALL R202
ACCESS (TO) R202
ACCESS HATCHES R402.2.4
ACCESSIBLE R202
ADDITION
   Defined........................ R202
   Requirements................... R502
ADMINISTRATION .... Chapter 1
AIR BARRIER
   Installation .... R402.4.1.1, Table R402.4.1.1
   Testing R402.4.1.2
AIR INFILTRATION
   Requirements................... R402.4.1.2
AIR LEAKAGE R402.4, R403.3.5, R403.3.6
AIR-IMPERMEABLE INSULATION R202
ALTERATION
   Defined........................ R202
   Requirements................... R503
ALTERNATE MATERIALS................. R102
APPROVED R202
APPROVED AGENCY R202
AUTOMATIC......................... R202

B

BASEMENT WALL
   Defined........................ R202
   Requirements......... R303.2.1, Table R402.1.3, R402.2.8.1, Table R405.5.2(1)
BIOGAS R202
BIOMASS R202
BOARD OF APPEALS R110
   Qualifications of members R110.3
BUILDING.......................... R202
BUILDING SITE R202
BUILDING THERMAL ENVELOPE
   Air tightness R402.4.1
   Compliance documentation R103.2, R401.3
   Defined........................ R202
   Insulation R303.1.1
   Insulation and fenestration criteria R402.1.2
   Performance method .... Table R405.5.2(1)
   Requirements.............. R102.1.1, R402

C

CAVITY INSULATION R202
CEILINGS R402.2.1, R402.2.2
   Specification for standard reference design........... Table R405.5.2(1)
CERTIFICATE...................... R401.3
CHANGE OF OCCUPANCY............... R505
CIRCULATING HOT WATER SYSTEM....... R202
CIRCULATON SYSTEMS................ R403.5.1
CLIMATE TYPES
   Defined.................... Table R301.3(1)
CLIMATE ZONE R202
CLIMATE ZONES .... R301, Figure R301.1, Table R301.1
   By state or territory... Figure R301.1, Table R301.1
   International climate zones............. R301.3, Table R301.3(1), Table R301.3(2)
   Tropical........................ R301.4
   Warm humid .... Table R301.1, R301.2, Table R301.3(1)
CODE OFFICIAL..................... R202
COMMERCIAL BUILDING
   Compliance R101.5
   Defined........................ R202
COMPLIANCE AND ENFORCEMENT....... R101.5
   Compliance report R405.3.2.2
CONDITIONED FLOOR AREA............. R202
CONDITIONED SPACE R202
CONSTRUCTION DOCUMENTS R103
   Amended R103.4
   Approval....................... R103.3.1
   Examination R103.3
   Information required R103.2
   Phased........................ R103.3.3
   Previous R103.3.2
   Retention R103.5
   Thermal envelope depiction.......... R103.2.1
CONTINUOUS AIR BARRIER R202
CONTINUOUS INSULATION (ci)........... R202
CONTROLS
   Heat pump R403.1.2
   Heating and cooling R403.1
   Service water heating............. R403.5
CRAWL SPACE WALL
   Defined........................ R202

INDEX

Requirements . . . . R303.2.1, Table R402.1.2,
Table R402.1.3, R402.1.4,
R402.2.10.1, Table R405.5.2(1)
CURTAIN WALL R202

D

DEFAULT DISTRIBUTION
   SYSTEM EFFICIENCIES . . . . Table R405.5.2(2)
DEFINITIONS. . . . . Chapter 2
DEGREE DAY COOLING . . . . Table R301.3(2)
DEGREE DAY HEATING. . . . . Table R301.3(2)
DEMAND RECIRCULATION WATER SYSTEM
   Defined R202
   Requirements R403.5.1.1.1
DESIGN CONDITIONS . . . . Chapter 3, R302
DIMMER. R202
DOORS
   Attics and crawl spaces. R402.2.4
   Default U-factors . . . . Table R303.1.3(2)
   Opaque R402.3.4
   Performance requirements . . . . . . Table R405.5.2(1)
   SHGC values. . . . . Table R402.1.3
   U-factors R402.1.4
DUCT
   Defined R202
   Insulation. . . . R103.2, R401.3, R403.3.1, R403.3.2,
R403.3.3
   Sealing R103.2, R403.3.4
   Tightness verification
      Postconstruction test R403.3.5
      Rough-in test R403.3.5
   Within conditioned space R403.3.2
DUCT SYSTEM R202
DWELLING UNIT
   Defined R202
   Multiple units R403.8
DWELLING UNIT ENCLOSURE AREA R202
DYNAMIC GLAZING R402.3.2

E

EAVE BAFFLE
   Installation. R402.2.3
ELECTRICAL POWER AND LIGHTING R404
ENERGY ANALYSIS R202
ENERGY ANALYSIS, ANNUAL
   Defined R202
   Documentation R405.3
   Requirements R405.3
ENERGY COST
   Defined R202

Energy Rating Index R202, R406.4
Energy Rating Index compliance alternative . . . R406
ERI-based compliance R406.5
ENERGY RECOVERY VENTILATION SYSTEMS
   Requirements . . . . Table R405.5.2(1)
ENERGY SIMULATION TOOL R202
ENVELOPE, BUILDING THERMAL R202
ENVELOPE DESIGN PROCEDURES. R402
EQUIPMENT EFFICIENCIES. R103.2, R401.3
EQUIPMENT ROOM
   For fuel burning appliance. R402.4.4
ERI REFERENCE DESIGN R202
EXISTING BUILDINGS . . . . Chapter 5
EXTERIOR WALL
   Defined R202
   Thermal performance R402, R402.1.2,
Table R405.5.2(1)

F

FEES R104
   Refunds. R104.5
   Related fees R104.4
   Schedule of permit fees. R104.2
FENESTRATION R303.1.3, R402.3, R402.3.2,
R402.4.3
   Default U-factors . . . . Table R303.1.3(1)
   Defined R202
   Rating and labeling R303.1.3, R402.1.2
   Replacement. R402.3.5
   Requirements . . . . Table R402.1.3
FENESTRATION PRODUCT, SITE-BUILT R202
FIREPLACES. R402.4.2
FLOORS
   Above-grade . . . . Table R405.5.2(1)
   Insulation. R402.2.6
   Slab-on-grade insulation requirements R402.2.9
FOUNDATIONS
   Requirements . . . . Table R402.4.1.1,
Table R405.5.2(1)
FURNACE EFFICIENCY . . . . Table R405.5.2(1)

G

GLAZED FENESTRATION R402.3.2, R402.3.3

H

HEAT PUMP. R403.1.2
HEATED SLAB R202
HEATING AND COOLING LOADS R302.1,
R403.1.2

INDEX

HIGH-EFFICACY LAMPS................. R202
HISTORIC BUILDING............... R202, R501.6
HOT WATER
   Piping insulation..................... R403.5.2
HOT WATER BOILER
   Outdoor temperature setback........... R403.2
HVAC SYSTEMS
   Tests
      Postconstruction.................. R403.3.6
      Rough-in-test..................... R403.3.6

I

**IDENTIFICATION (MATERIALS,
EQUIPMENT AND SYSTEM)**........... R303.1
INDIRECTLY CONDITIONED SPACE........ R202
INFILTRATION......................... R202
INFILTRATION, AIR LEAKAGE........... R402.4,
                              Table R405.5.2(1)
   Defined............................. R202
INSPECTIONS......................... R105
INSULATED SIDING..................... R202
INSULATION
   Air-impermeable......... R202, Table R402.4.1.1
   Basement walls..................... R402.2.8.1
   Ceilings with attic spaces............. R402.2.1
   Ceilings without attic spaces.......... R402.2.2
   Crawl space walls................... R402.2.10.1
   Duct............................. R403.3.1
   Eave baffle........................ R402.2.3
   Floors..................... R402.2.6, R402.2.7
   Hot water piping..................... R403.5.2
   Identification................. R303.1, R303.1.2
   Installation............. R303.1.1, R303.1.1.1,
      R303.1.2, R303.2, Table R402.4.1.1
   Masonry veneer..................... R402.2.11
   Mass walls......................... R402.2.5
   Mechanical system piping............. R403.4
   Product rating...................... R303.1.4
   Protection of exposed foundation....... R303.2.1
   Protection of piping insulation......... R403.4.1
   Requirements......... Table R402.1.3, R402.2
   Slab-on-grade floors................. R402.2.9
   Steel-frame ceilings, walls
      and floors.......... R402.2.6, Table R402.2.6
   Sunroom........................... R402.2.12

L

LABELED
   Defined............................. R202
   Requirements............. R303.1.3, R402.4.3
LIGHTING SYSTEMS.................... R404
   Recessed..................... R402.4.5, R404
LISTED............................. R202
LOG HOMES............ R402.1, Table R402.4.1.1
LOW-ENERGY BUILDINGS............... R402.1
LOW-VOLTAGE LIGHTING............... R202
LUMINAIRE
   Sealed........................... R402.4.5

M

MAINTENANCE INFORMATION........... R303.3
MANUAL............................ R202
MANUALS.................... R101.5.1, R303.3
MASONRY VENEER
   Insulation......................... R402.2.11
MASS
   Wall.............................. R402.2.5
MATERIALS AND EQUIPMENT........... R303
**MECHANICAL SYSTEMS
AND EQUIPMENT**................. R403, R405.1
MECHANICAL VENTILATION............ R403.6,
      Table R403.6.1, Table R405.5.2(1)
MULTIPLE DWELLING UNITS............ R403.8

O

OCCUPANCY
   Requirements.................. R101.4, R101.5
OCCUPANT SENSOR CONTROL.....R202, R404.2
ON-SITE RENEWABLE ENERGY...R202, R403.5.1,
      R403.10.3, R406.3.2, R406.4
OPAQUE DOOR................. R202, R402.3.4

P

PERFORMANCE ANALYSIS.............. R405
PERMIT............................ R202
PIPE INSULATION............. R403.4, R403.5.2
PLANS AND SPECIFICATIONS.......... R103
POOLS............................ R403.10
   Covers........................... R403.10.3
   Heaters........................... R403.10.1
   Time switches..................... R403.10.2
PROPOSED DESIGN
   Defined............................ R202
   Requirements......... R405, Table R405.5.2(1)

INDEX

PUMPS
   Time switches R403.5.1.1, R403.10.2

R

RATED DESIGN R202
READILY ACCESSIBLE R202
READY ACCESS (TO) R202
REFERENCED STANDARDS . . . . . . R108, Chapter 6
RENEWABLE ENERGY
   CERTIFICATE (REC) R202
RENEWABLE ENERGY RESOURCES R202
REPAIR
   Defined R202
   Requirements R504
REROOFING R202
RESIDENTIAL BUILDING
   Compliance R101.5
   Defined R202
   Energy Rating Index alternative R406
   Simulated performance alternative R405
ROOF ASSEMBLY
   Defined R202
   Requirements R303.1.1.1, R402.2.2,
                                    Table R405.5.2(1)
ROOF RECOVER R202
ROOF REPAIR R202
ROOF REPLACEMENT R202
R-VALUE (THERMAL RESISTANCE) R202

S

SCOPE R101.2
SERVICE HOT WATER
   Requirements R403.5
SERVICE WATER HEATING R202
SHEATHING, INSULATING (see INSULATING SHEATHING)
SHGC (see SOLAR HEAT GAIN COEFFICIENT)
SHUTOFF DAMPERS R403.6
SIMULATED PERFORMANCE
   ALTERNATIVE R405
   Documentation R405.3
   Mandatory requirements R405.2
   Performance-based compliance R405.3
   Report R405.3.2
   Software tools R405.3.1
SIZING
   Equipment and system R405.5.1
SKYLIGHTS R303.1.3, R402.1.2, R402.3,
                                    Table R405.5.2(1)
SNOW MELT SYSTEM CONTROLS R403.9

SOLAR HEAT GAIN COEFFICIENT
   (SHGC) . . . . R103.2, Table R303.1.3(3),
                   R401.3, Table R402.1.3, R402.1.4,
                 R402.3.2, R402.3.3, R402.3.5, R402.5
   Defined R202
STANDARD REFERENCE DESIGN
   Defined R202
   Requirements . . . . R405, Table R405.5.2(1)
STANDARDS, REFERENCED . . . . . Chapter 6, R108
STEEL FRAMING R402.2.6
STOP WORK ORDER R109
   Authority R109.1
   Emergencies R109.3
   Failure to comply R109.4
   Issuance R109.2
SUNROOM R402.2.12, R402.3.5,
                                    Table R405.5.2(1)
   Defined R202
   Insulation R402.2.12
SWIMMING POOLS R403.10

T

THERMAL DISTRIBUTION
   EFFICIENCY (TDE) R202
THERMAL ISOLATION R402.2.12, R402.3.5,
                                    Table R405.5.2(1)
   Defined R202
THERMAL RESISTANCE [see R-VALUE (THERMAL RESISTANCE)]
THERMAL TRANSMITTANCE [see U-FACTOR (THERMAL TRANSMITTANCE)]
THERMOSTAT
   Controls R403.1
   Defined R202
   Programmable R403.1.1
TIME SWITCHES R403.10.2
TOTAL BUILDING PERFORMANCE
   Residential R405
TOWNHOUSE R202
TROPICAL CLIMATE REGION R301.4, R407.2

U

U-FACTOR (THERMAL TRANSMITTANCE) . . . R202

V

VALIDITY R107
VAPOR RETARDER R402.1.1
VENTILATION . . . . R403.6, Table R403.6.1,
                                Table R405.5.2(1), R407.2

Defined R202
VENTILATION AIR R202
VISIBLE TRANSMITTANCE (VT)
   Default glazed fenestration . . . . . Table R303.1.3(3)
   Defined R202

W

WALL
   Above-grade, defined R202
      Standard reference design . . . . Table R405.4.2(1)
   Basement, defined R202
      Installation R402.2.8.1
      Standard reference design . . . . Table R405.4.2(1)
   Crawl space, defined R202
      Installation R402.2.10.1
      Standard reference design . . . . Table R405.4.2(1)
   Exterior, defined R202
   Mass R402.2.5
   Steel-frame . . . . R402.2.6, Table R402.2.6
   With partial structural sheathing R402.2.7
WALLS ADJACENT TO
UNCONDITIONED SPACE R202
WATER HEATING R401.3, R403.5, R405.1,
                                    Table R405.5.2(1)
WHOLE HOUSE MECHANICAL VENTILATION SYSTEM
   Defined R202
   System fan efficacy R403.6.2

Z

ZONE R202

EDITORIAL CHANGES – SECOND PRINTING

Page C3-20, **TABLE C301.1—continued:** column 1, row 41 now reads . . . 4B Catron

Page C3-36, **Section C303.1.3:** paragraph 3, lines 10 through 12 now read . . . from Table C303.1.3(3). For Tubular Daylighting Devices, VT_{annual} shall be measured and rated in accordance with NFRC 203.

Page C4-3, **TABLE C402.1.3:** column 13, row 18, line 2 now reads . . . 48

Page C4-3, **TABLE C402.1.3:** column 14, row 18, line 2 now reads . . . 24

Page C4-4, **TABLE C402.1.4:** column 1, row 7 now reads . . . Mass[f]

Page C4-4, **TABLE C402.1.4:** column 1, row 18 now reads . . . Heated slabs

Page C4-4, **TABLE C402.1.4:** column 1, row 21 now reads . . . Swinging door[g]

Page C4-4, **TABLE C402.1.4:** column 1, row 22 now reads . . . Garage door < 14% glazing[h]

Page C4-4, **TABLE C402.1.4:** Note f has been deleted

Page C4-4, **TABLE C402.1.4:** Note f now reads . . . f. " Mass walls" shall be in accordance with Section C402.2.2.

Page C4-4, **TABLE C402.1.4:** Note g now reads . . . g. Swinging door *U*-factors shall be determined in accordance with NFRC-100.

Page C4-4, **TABLE C402.1.4:** Note h now reads . . . h. Garage doors having a single row of fenestration shall have an assembly *U*-factor less than or equal to 0.44 in Climate Zones 0 through 6 and less than or equal to 0.36 in Climate Zones 7 and 8, provided that the fenestration area is not less than 14 percent and not more than 25 percent of the total door area.

Page C4-7, **TABLE C402.3:** row 1 now reads . . . Three-year-aged solar reflectance[b] of 0.55 and 3-year aged thermal emittance[c] of 0.75

Page C4-10, **Section C402.4.4:** line 5 now reads . . . sidelit *daylight* zones.

Page C4-13, **Section C402.5.11.1:** line 2 now reads . . . comply with Section C403.14.

Page C4-37. **Section C403.4.2.3:** line 2 now reads . . . and stop controls shall be provided for each HVAC

Page C4-44. **Section C403.7.4.2:** Exception 6, line 3 now reads . . . 0, 3C, 4C, 5B, 5C, 6B, 7 and 8.

Page C4-50. new section added and now reads . . .

C403.10.6 Heat recovery for space conditioning in healthcare facilities. Where heating water is used for space heating, a condenser heat recovery system shall be installed provided that all of the following are true:

1. The building is a Group I-2, Condition 2 occupancy.
2. The total design chilled water capacity for the Group I-2, Condition 2 occupancy, either air cooled or water cooled, required at cooling design conditions exceeds 3,600,000 Btu/h (1100 kw) of cooling.
3. Simultaneous heating and cooling occurs above 60°F (16°C) outdoor air temperature.

The required heat recovery system shall have a cooling capacity that is not less than 7 percent of the total design chilled water capacity of the Group I-2, Condition 2 occupancy at peak design conditions.

Exceptions:

1. Buildings that provide 60 percent or more of their reheat energy from on-site renewable energy or site-recovered energy.
2. Buildings in Climate Zones 5C, 6B, 7 and 8.

Page C4-58, **TABLE C404.5.2.1:** column 8, row 2 now reads . . . PE-RT SDR 9

Page C4-60, **Section C405.2.1.2:** Item 2, lines 3 and 4 now read . . . unoccupied setpoint of not more than 50 percent of full power within 20 minutes after all

Page C4-60, **Section C405.2.1.4:** lines 3 and 4 now read . . . uniformly reduce lighting power to an occupied setpoint not more than 50 percent of full power within

Page C4-62, **Section C405.2.4.2:** Item 3, lines 7 through 9 now read . . . tration to the nearest full height wall, or up to 0.5 times the height from the floor to the top of the fenestration, whichever is less, as indicated

Page C4-63, **FIGURE C405.2.4.2(1)** now reads as shown

Page C4-72, **Section C405.9.2.1** now reads . . . **C405.9.2.1 Energy recovery.** Escalators shall be designed to recover electrical energy when resisting overspeed in the down direction.

Page C4-78, **Section C406.3:** section title now reads . . . **C406.3 Reduced lighting power.**

Page C4-80, **TABLE C406.1(4):** column 11, row 6 now reads . . . 1

Page C-4-83, **TABLE C406.12(2)**: column 4 now reads . . .

IDLE ENERGY RATE
400 watts
530 watts
670 watts
800 watts
6,250 Btu/h
8,350 Btu/h
10,400 Btu/h
12,500 Btu/h

Page C4-84, **TABLE C406.12(3)**: column 2, row 3 now reads . . . $\leq .50$ kW

Page C4-84, **TABLE C406.12(3)**: column 2, row 4 now reads . . . $\leq .70$ kW

Page C4-84, **Section C407.2**: Item 2, line 1 now reads . . . 2. An annual energy cost that is less than or equal to 80

Page R2-2, definition **HIGH-EFFICACY LIGHT SOURCES** now reads . . .

HIGH-EFFICACY LIGHT SOURCES. Any lamp with an efficacy of not less than 65 lumens per watt, or luminaires with an efficacy of not less than 45 lumens per watt.

Page R3-20, **TABLE R301.1—continued**: column 1, row 41 now reads . . . 4B Catron

Page R4-2, **TABLE R402.1.2**: column 4, row 7 now reads . . . 0.40

Page R4-3, **TABLE R402.1.3**: column 6 now reads . . .

WOOD FRAME WALL R-VALUE[g]
13 or 0&10ci
13 or 0&10ci
13 or 0&10ci
20 or 13&5ci[h] or 0&15ci[h]
30 or 20&5ci[h] or 13&10ci[h] or 0&20ci[h]
30 or 20&5ci[h] or 13&10ci[h] or 0&20ci[h]
30 or 20&5ci[h] or 13&10ci[h] or 0&20ci[h]
30 or 20&5ci[h] or 13&10ci[h] or 0&20ci[h]

Page R4-3, **TABLE R402.1.3**: column 9, row 7 now reads . . . 15ci or 19 or 13&5ci

Page R4-3, **TABLE R402.1.3**: column 11, row 7 now reads . . . 15ci or 19 or 13&5ci

Page R4-3, **TABLE R402.1.3**: column 9, row 8 now reads . . . 15ci or 19 or 13&5ci

Page R4-3, **TABLE R402.1.3**: column 11, row 8 now reads . . . 15ci or 19 or 13&5ci

Page R4-3, **TABLE R402.1.3**: column 9, row 9 now reads . . . 15ci or 19 or 13&5ci

Page R4-3, **TABLE R402.1.3**: column 11, row 9 now reads . . . 15ci or 19 or 13&5ci

Page R4-3, **TABLE R402.1.3**: Note c, line 2 now reads . . . 19 or 13&5ci" means R-15 continuous insulation (ci) on the interior or exterior surface of the wall; or R-19 cavity insulation on the interior side of the wall;

Page R4-3, **TABLE R402.1.3**: Note g, line 1 now reads . . . g. The first value is cavity insulation; the second value is continuous insulation. Therefore, as an example, "13&5" means R-13 cavity insulation plus R-5

Page R4-4, **Section R402.2.1** title now reads . . . **R402.2.1 Ceilings with attics.**

Page R4-4, **Section R402.2.1**: lines 8 and 9 now read . . . ceiling or attic, installing R-49 over 100 percent of the ceiling or attic area requiring insulation shall satisfy the

Page R4-4, **Section R402.2.4.1**: lines 8 through 10 now read . . . a wood-framed or equivalent baffle, retainer, or dam shall be installed to prevent loose-fill insulation from spilling into living space from higher to lower sections

Page R4-7, **TABLE R402.4.1.1**: column 2, row 9, lines 1 and 2 now read . . . Duct and flue shafts to exterior or unconditioned space shall be sealed.

Page R4-9, **Section R402.4.1.2**: Exception 2, lines 8 and 9 now read . . . 0.2 inch w.g. (50 Pa), shall be permitted in all climate zones for:

Page R4-11, **Section R403.3.5**: new paragraph added after Item 2 now reads . . . A written report of the results of the test shall be signed by the party conducting the test and provided to the *code official*.

Page R4-11, **Section R403.3.5**: Exception line 2 now reads . . . required for ducts serving ventilation systems that are

Page R4-12, **Section R403.6** now reads . . . **R403.6 Mechanical ventilation.** The *buildings* complying with Section R402.4.1 shall be provided with ventilation that complies with the requirements of Section M1505 of the *International Residential Code* or *International Mechanical Code*, as applicable, or with other *approved* means of *ventilation*. Outdoor air intakes and exhausts shall have automatic or gravity dampers that close when the *ventilation* system is not operating.

Page R4-13, **Section R404.1.1**: line 3 now reads . . . C405.5.

Page R4-13, **Section R405.2**: Item 2, line 1 now reads . . . The building thermal envelope shall be greater than

Page R4-13, **Section R405.3.2**: paragraph 2, line 2 now reads . . . dance with Sections R405.3.2.1 and R405.3.2.2.

Page R4-14, **TABLE R405.2**: column 1, row 28 now reads . . . R404.2

Page R4-16, **TABLE R405.4.2(1)—continued**: column 2, row 6, line 7 now reads . . . e_f = the minimum fan efficacy, as specified in Table 403.6.2,

Page R4-17, **TABLE R405.4.2(1)—continued**: column 2, row 2, line 2 now reads . . . Use, in units of gal/day = 25.5 + (8.5 × N_{br})

Page R4-17, **TABLE R405.4.2(1)—continued**: column 3, row 9, line 2 now reads . . .

> Duct location: as proposed.
>
> Duct insulation: as proposed.
>
> As tested or, where not tested, as specified in Table R405.4.2(2).

Page R4-18, **Section R405.5.1**: Item 2 line 4 now reads . . . dance with Section R403.7.

Page R4-19, **TABLE R406.2**: column 1, row 27 now reads . . . R404.2

Page R4-19, **Section R406.3.2**: line 6 now reads . . . R402.1.4 of the 2018 *International Energy Conservation*

Page R4-21, **Section R408.2.3**: Item 1, line 1 now reads . . . 1. Greater than or equal to 0.82 EF fossil fuel service

Page Appendix RC-1, **TABLE RC102.2**: Note a, lines 3 and 4 now read . . . efficiency and SHGC in Table R402.1.2 or R402.1.3 of the 2015 *International Energy Conservation Code.*

For the complete errata history of this code, please visit: https://www.iccsafe.org/errata-central/

Building Confidence, Building Community

Additional Code Resources from ICC
Tools to help you learn, interpret and apply the I-Codes®

Significant Changes to the 2021 International Codes®
Practical resources that offer a comprehensive analysis of the critical changes since the previous edition

Authored by code experts, these useful tools are "must-have" guides to the many important changes in the 2021 International Codes. Key changes are identified then followed by in-depth, expert discussion of how the change affects real world application. A full-color photo, table or illustration is included for each change to further clarify application.

Significant Changes to the International Building Code, 2021 Edition	Search #7024S21
Significant Changes to the International Residential Code, 2021 Edition	Search #7101S21
Significant Changes to the International Fire Code, 2021 Edition	Search #7404S21
Significant Changes to the International Energy Conservation Code, 2021 Edition	Search #7808S21
Significant Changes to the International Plumbing Code/International Mechanical Code/International Fuel Gas Code, 2021 Edition	Search #7202S21

2021 Code Essentials
A straightforward, focused approach to code fundamentals using non-code language

A user-friendly, concise approach that facilitates understanding of the essential code provisions. These invaluable companion guides contain detailed full-color illustrations to enhance comprehension, references to corresponding code sections, and a glossary of code and construction terms that clarify their meaning in the context of the code.

Building Code Essentials: Based on the 2021 IBC	Search #4031S21
Residential Code Essentials: Based on the 2021 IRC	Search #4131S21
Fire Code Essentials: Based on the 2021 IFC	Search #4431S21
Energy Code Essentials: Based on the 2021 IECC	Search #4831S21
Existing Building Code Essentials: Based on the 2021 IEBC	Search #4552S21
Plumbing Code Essentials: Based on the 2021 IPC	Search #4231S21
Mechanical Code Essentials: Based on the 2021 IMC	Search #4031S21

Study Companions
The ideal learning tool for exam prep or everyday application

These comprehensive study guides provide practical learning assignments helpful for independent study or instructor-led programs in the workplace, college courses, or vocational training programs. Each book is organized into study sessions with clear learning objectives, key points for review, code text and commentary applicable to the specific topic, and hundreds of illustrations. A helpful practice quiz at the end of each session allows you to measure your progress along the way. The answer key lists the code section referenced in each question for further information.

2021 International Building Code Study Companion	Search #4017S21
2021 International Residential Code Study Companion	Search #4117S21
2021 International Fire Code Study Companion	Search #4407S21
2021 International Plumbing Code Study Companion	Search #4217S21
2021 International Mechanical Code Study Companion	Search #4317S21
2021 International Fuel Gas Code Study Companion	Search #4607S21
2021 International Energy Conservation Code Study Companion	Search #4807S21
2021 Permit Technician Study Companion	Search #4027S21
2021 Special Inspection Study Companion	Search #4032S21
2021 Accessibility Study Companion	Search #4123S21

Browse the latest code tools at shop.iccsafe.org

ICC EVALUATION SERVICE

☑ **Specify** and
☑ **Approve** *with* **Confidence**

When facing new or unfamiliar materials, look for an ICC-ES Evaluation Report or Listing before approving for installation.

ICC-ES® **Evaluation Reports** are the most widely accepted and trusted technical reports for code compliance.

ICC-ES **Building Product Listings** and **PMG Listings** show product compliance with applicable standard(s) referenced in the building and plumbing codes as well as other applicable codes.

When you specify or approve products or materials with an ICC-ES report, building product listing or PMG listing, you avoid delays on projects and improve your bottom line.

ICC-ES is a subsidiary of ICC®, the publisher of the codes used throughout the U.S. and many global markets, so you can be confident in their code expertise.

www.icc-es.org | 800-423-6587

Look for the Trusted Marks of Conformity

A subsidiary of

Proctored Remote Online Testing Option (PRONTO™)

Convenient, Reliable, and Secure Certification Exams

Take the Test at Your Location

Take advantage of ICC PRONTO, an industry leading, secure online exam delivery service. PRONTO allows you to take ICC Certification exams at your convenience in the privacy of your own home, office or other secure location. Plus, you'll know your pass/fail status immediately upon completion.

 With PRONTO, ICC's Proctored Remote Online Testing Option, take your ICC Certification exam from any location with high-speed internet access.

 With online proctoring and exam security features you can be confident in the integrity of the testing process and exam results.

 Plan your exam for the day and time most convenient for you. PRONTO is available 24/7.

 Eliminate the waiting period and get your results in private immediately upon exam completion.

 ICC was the first model code organization to offer secured online proctored exams—part of our commitment to offering the latest technology-based solutions to help building and code professionals succeed and advance. We continue to expand our catalog of PRONTO exam offerings.

Discover ICC PRONTO and the wealth of certification opportunities available to advance your career: www.iccsafe.org/MeetPRONTO

Lead the Way to Energy Efficiency

Subscribe to the New ICC Digital Codes Energy Collection

The International Code Council's New Energy Collection is brought to you in collaboration with ASHRAE and other energy leaders, such as RESNET and AMCA International. This new collection provides the most complete (and growing) collection of Energy Titles available with a **Premium Collections** subscription.

ENERGY COLLECTION

Unlock All Codes with ICC Digital Codes Premium Complete

ICC's Digital Codes is the most trusted and authentic source of model codes and standards, which conveniently provides access to the latest code text from across the United States. With a Premium subscription, users can further enhance their code experience with powerful features such as team collaboration and access sharing, bookmarks, notes, errata and revision history, and more.

Start Your Free 14-Day Premium trial at codes.iccsafe.org/trial

Learn More at codes.iccsafe.org/codes/energy